Best Practices in Global Water Policy

AF736308

This volume challenges the assumption that dominant water-management solutions are universally effective, neutral and scientifically validated.

Global water policy tends to frame certain concepts and tools – such as river basin organisations, water pricing, drip irrigation, nature-based solutions and wastewater reuse – as 'best practices'. These ready-made solutions, endorsed by policymakers, development agencies and experts, are widely disseminated across countries with limited critical scrutiny. This book examines how these dominant solutions, while often appearing rational and evidence-based, can obscure trade-offs, negative externalities and unintended consequences. From irrigation modernisation projects that inadvertently increase water consumption due to the rebound effect to water user associations that fail to address deeper governance and equity issues, this book engages with 20 'best practices' to illustrate how universal solutions often ignore the complexity of water politics, institutional settings, cultural specificities and ecological realities. While not dismissing the value of best practices outright, this book argues for a more sceptical, politically aware and context-sensitive approach to water management. It calls for a shift in the burden of proof: rather than accepting a best practice as effective unless proven otherwise, water governance should start by scrutinising the potential externalities, risks and limitations of any proposed intervention.

This book is essential reading for students, academics and professionals in water policy, water governance and environmental management more broadly.

François Molle is Emeritus Director of Research at the French National Research Institute for Sustainable Development (IRD), France. He has 40 years of experience in development research in topics such as the analysis of irrigation systems, the governance of river basins and groundwater, water policy and the political ecology of the interaction between society, technology and the environment.

Sylvain Barone is Director of Research at the French National Research Institute for Agriculture, Food and Environment (INRAE), France. For the past 15 years, his research has focused on water policy and governance, environmental policy and the politicisation of ecological issues.

Earthscan Water Text series

Environmental Health Engineering in the Tropics
Water, Sanitation and Disease Control
Sandy Cairncross and Sir Richard Feachem

Flood Risk Management
Global Case Studies of Governance, Policy and Communities
Edited by Edmund Penning-Rowsell and Matilda Becker

Water Ethics
A Values Approach to Solving the Water Crisis
David Groenfeldt

The Water Footprint of Modern Consumer Society
Arjen Y. Hoekstra

Water Governance, Justice and the Right to Water
Edited by Farhana Sultana and Alex Loftus

Digital Water
Enabling a More Resilient, Secure and Equitable Water Future
William Sarni

Legal Geographies of Water
The Spaces, Places and Narratives of Human-Water Relations
Cristy Clark

Best Practices in Global Water Policy
Revisiting Water Mantras
Edited by François Molle and Sylvain Barone

For more information about this series, please visit: www.routledge.com/Earthscan-Water-Text/book-series/ECEWT

Best Practices in Global Water Policy

Revisiting Water Mantras

Edited by François Molle and Sylvain Barone

Routledge
Taylor & Francis Group
LONDON AND NEW YORK

earthscan
from Routledge

Designed cover image: Getty Images

First published 2026
by Routledge
4 Park Square, Milton Park, Abingdon, Oxon OX14 4RN

and by Routledge
605 Third Avenue, New York, NY 10158

Routledge is an imprint of the Taylor & Francis Group, an informa business

© 2026 selection and editorial matter, François Molle and Sylvain Barone; individual chapters, the contributors

The right of François Molle and Sylvain Barone to be identified as the authors of the editorial material, and of the authors for their individual chapters, has been asserted in accordance with sections 77 and 78 of the Copyright, Designs and Patents Act 1988.

All rights reserved. No part of this book may be reprinted or reproduced or utilised in any form or by any electronic, mechanical, or other means, now known or hereafter invented, including photocopying and recording, or in any information storage or retrieval system, without permission in writing from the publishers.

For Product Safety Concerns and Information please contact our EU representative GPSR@taylorandfrancis.com. Taylor & Francis Verlag GmbH, Kaufingerstraße 24, 80331 München, Germany.

Trademark notice: Product or corporate names may be trademarks or registered trademarks, and are used only for identification and explanation without intent to infringe.

British Library Cataloguing-in-Publication Data
A catalogue record for this book is available from the British Library

ISBN: 978-1-041-25192-7 (hbk)
ISBN: 978-1-041-25190-3 (pbk)
ISBN: 978-1-003-74553-2 (ebk)

DOI: 10.4324/9781003745532

Typeset in Optima
by Newgen Publishing UK

Contents

Contributors

Jason Alexandra is a senior research fellow at the Institute for Climate, Energy & Disaster Solutions (ICEDS) of the Australian National University, who researches adaptative water governance, climate risk assessment and water policy reform. Jason has extensive experience in water and natural resources governance in particular as a senior executive responsible for ecosystem monitoring, climate science and risk assessment in Australia's Murray Darling Basin.

Vazken Andréassian, a hydrologist at the National Research Institute for Agriculture, Food and Environment (INRAE), France, specializes in the mathematical modelling of river catchment behaviour. His work has applications in flood forecasting, low-flow forecasting, forest hydrology and climate change impact assessment. His interests also extend to the international management of water resources and science communication for both adults and children.

Catherine Baron is a professor of development studies at the Political Science Institute of Toulouse/LEREPS, France. Her research focuses on water governance and water as a common in the Global South (West Africa, Indonesia and Thailand).

Sylvain Barone is a political scientist and research director at the French National Research Institute for Agriculture, Food and Environment (INRAE). His research focuses on environmental and water politics and policy, and on what they reveal about the structuring of power, the state and democracy, mainly in France and the US.

Bernard Barraqué is CNRS emeritus research director at the International Research Centre on Environment and Development (CIRED). He is a specialist of water policies in Europe: WSS services' sustainability, water resources allocation and conservation, and the link between both in terms of impure public goods (club goods/common goods); and looking for solutions to close the financing gap of water policy.

Régis Barraud is a full professor of geography at the University of Poitiers, France. He contributes to the interdisciplinary field of political ecology. More specifically, he carries out his research on environmental controversies (dam dismantling, rewilding).

Kate Bayliss is Research Associate at SOAS University of London and works in political economy. Her main research interest is in the provisioning of essential services such as water, energy, health and housing. For the past two decades she has worked extensively on privatisation, financialisation and social equity in the UK and the Global South.

Gabrielle Bouleau is a French political scientist at the National Research Institute for Agriculture, Food and Environment (INRAE), France. Senior researcher in the LISIS research unit, she investigates water issues. Her research revolves around environmental policies, water politics and political ecology with a specific focus on expertise.

Anne-Laure Collard is a researcher in sociology at the National Research Institute for Agriculture, Food and Environment (INRAE), France. She works on the relationship between water and society through the mediation of technical objects, in contexts of water scarcity. She also explores the conditions for interdisciplinary dialogue.

Marine Colon is a public-sector engineer and lecturer at AgroParisTech, France. She is a scientist of public management, specialising in the management of public water and sanitation services and working on utility transformation in Europe, Africa and Asia.

Alexandre Gaudry is a PhD candidate in spatial planning and urbanism at LEREPS (Sciences Po Toulouse) and AgroParisTech, France, in partnership with the French Development Agency (AFD). His thesis examines access to drinking water beyond centralised networks in cities of the Global South, with Dakar and Bandung as case studies.

Marie-Anne Germaine is a professor of geography at the Université Paris Nanterre, France. Her research focuses on humans and nature relations in everyday spaces. She has coordinated several research programs examining the social and political aspects of ecological restoration of rivers and the removal of dams and small weirs.

Florence Habets is a hydroclimatologist at the Centre national de la recherche scientifique (CNRS), and Professor at the Laboratory of Geology of l'Ecole Normale Supérieure (ENS), France. She is interested in the evolution of water resources in France, in particular seasonal forecasts of groundwater resources and the cumulative impacts of small reservoirs on hydrology.

Audrey Joslin's is an associate professor in the Department of Geography and Geospatial Sciences at Kansas State University, USA. Her research explores environmental governance, land management and human–environment relationships. Her work on water and fire governance bridges political ecology and applied geography to explore how landscapes, livelihoods and knowledge are shaped through conservation practice.

Doug Kenney is Director of the Western Water Policy Program at the University of Colorado, USA. His research primarily focuses on transboundary water management in the American West, with a particular emphasis on the Colorado River Basin.

M. Dinesh Kumar is Executive Director of Institute for Resource Analysis and Policy, India. He has three and a half decades of experience working on land and water issues within India and internationally. He has authored 12 books and 10 edited volumes on water, agriculture, energy and climate related topics.

Bruce Lankford is Emeritus Professor of Water and Irrigation Policy at the University of East Anglia, UK, with more than 40 years' of experience in irrigation and water resources management. His interests cover irrigation policy in Sub-Saharan Africa; design for irrigation manageability; irrigation efficiency, water savings and the paracommons; irrigated catchment resilience and serious games in irrigation and water resource management.

Rebecca Lave is a Professor of Geography at Indiana University Bloomington, USA, and a past president of the American Association of Geographers. Her research takes a Critical Physical Geography approach, combining political economy, science and technology studies and fluvial geomorphology to analyse stream restoration, the politics of environmental expertise and non-structural approaches to flood attenuation.

Rens de Man, over the last 25 years, has built up broad professional experience as strategic advisor, researcher and mutual gains practitioner in varying international contexts. De Man has published and presented various papers at international conferences. He has a background in hydrology, social sciences and water diplomacy.

Sylvain Massuel is a quantitative hydrogeologist at the French National Research Institute for Sustainable Development (IRD), France. He works on the evolution of groundwater reserves in relation to societies, in order to shed light on water governance. He develops approaches based on interdisciplinary work with the human and social sciences, rooted in the field.

Matthew McCartney is a principal researcher at the International Water Management Institute (IWMI), specialising in water resources and wetland and hydro-ecological studies. He is currently IWMIs Research Group leader on Sustainable Water Infrastructure and Ecosystems. A hydrologist by training, his personal research focuses on nature-based solutions and ecosystem services across catchments and landscapes and the need to rapidly transform agriculture and food systems to ensure resilience and sustainability.

David McDonald is Professor of Global Development Studies at Queen's University, Canada, and Director of the Municipal Services Project.

François Molle has 40 years of experience in the management and governance of hydro-systems (irrigated schemes, aquifers, river basins, ...) and water policy, especially in Brazil, Southeast Asia and the Middle-East and North Africa. He serves as co-editor in chief of the *Water Alternatives* journal and is an emeritus researcher at the French National Research Institute for Sustainable Development (IRD).

Erin O'Donnell, a non-Indigenous person living on unceded Wurundjeri Woi-wurrung Country, has over 20 years' experience in the public and private sectors. As a senior lecturer and ARC Research Fellow at the University of Melbourne Law School, Australia, Erin explores the opportunities and potential pitfalls of legal personhood for rivers through investigation of implementation of river personhood in Australia, Chile, Colombia, Ecuador, India, Bangladesh, Canada, the USA and Aotearoa New Zealand.

Chris Perry worked for the World Bank for 20 years, mostly in South Asia and the Middle East. He then worked for the International Water Management Institute (IWMI), as head of research. After retirement he was an editor of the *Agricultural Water Management* journal.

Edwin Rap is a freelance trainer and researcher in water governance and the anthropology of infrastructure. He has 30 years of experience in Latin America, Africa, Asia and Europe and has published widely in high quality journals.

Jean Carlo Rodríguez de Francisco is an environmental and social scientist at the German Institute of Development and Sustainability, Germany. He works at the intersection of water, biodiversity and climate. His research explores environmental change, justice and governance, with a current focus on examining the environmental justice dimensions of nature-based solutions projects in Latin America and Europe.

Dik Roth is a social anthropologist and associate professor (emeritus) at Wageningen University, the Netherlands. He has done extensive research in Indonesia and has been involved in several research programs in South Asia. His scientific interests include legal anthropology, policy processes, land and water governance, and flood risk governance.

Julián Suárez is a PhD candidate in law at the University College Cork, School of Law, Ireland. With a background in judicial review litigation, he is currently seeking solicitor qualification before the Law Society of Ireland. His doctoral research focuses on the substantive and procedural implications of rights of nature norms, and challenges and opportunities for rights-based environmental protection.

Jean-Philippe Venot is a human geographer at the French National Research Institute for Sustainable Development (IRD) looking at irrigation and water governance in the Global South. His most recent research interrogates how water researchers make sense and practice the notion of engagement in arenas that ambition to co-produce knowledge for just transformation.

Jeroen Warner is an associate professor at the Social Sciences Group, Wageningen University, the Netherlands. His key research interests are in the field of hydropolitics, water disasters and water governance. He was a special visiting professor at the University of Sao Paulo and has led major EU and Dutch Science Foundation projects.

Joe Williams is a human geographer at the School of Geographical Sciences, University of Bristol, UK, and honorary researcher at the School of Geography and Planning, Cardiff University, UK. With interests in political ecology, infrastructure and development, he currently leads a project on the growing use of desalination in the Global South, particularly Kenya, South Africa and Morocco. Joe Williams is supported by an ERC-assessed, UKRI-funded grant (EP/Y036530/1).

Foreword

Margreet Zwarteveen

"We have the solutions. Now we just need to find the problems that match them". is a joke that my colleagues at IHE Delft – an institute that gathers water experts from many disciplines and countries under one roof – sometimes tell each other. Or perhaps, it is a bit more than a joke: it is a recognition that much thinking, or indeed much science and engineering, in water is characterised by a distinct form of solutionism: rather than diagnosing water problems to understand why they occur, many water experts instead ask: what to do, or what can we do?

This book is a powerful antidote to this solutionism of water experts. It mobilises scholarly research to critically discuss 20 'water mantras' to show that none of them are as universally applicable, generically successful and politically neutral as policy discourses have it.

The chapters provide succinct empirically informed critical overviews of those technologies and technical interventions (drip irrigation, desalinisation, wastewater treatment and reuse, water harvesting, large dams), policy concepts (safe yield, re-forestation, room-for-the river, free-flowing rivers or river restoration, green growth, rivers as legal persons, off-grid) and institutional arrangements (water pricing, river basin organisations, water user organisations, privatisation) that figure most prominently in global policy discourses as remedies to present-day water challenges. At the risk of giving it more importance than it perhaps deserves, the recent Report of the Global Commission on the Economics of Water is a telling illustration of the continued popularity of many of these remedies. Even though the report claims to offer a radically new framing of water – as a global common good – and even though it much more explicitly foregrounds justice as compared to most other water policy documents (most of which emphasise productivity and efficiency), when it comes to operationalising its recommendations, the report repeats many of the same familiar technical and institutional recipes that have already been circulating for a long time in the water world (even when some of them have changed names). These include many of the very recipes that this book discusses.

Beyond providing critical discussions of the effectiveness of these recipes, the chapters in this book together are also useful in revealing the particular logic of solutionist reasoning and analysis – the science that justifies and informs them. By making sense of water realities through the lens of a proposed solution or best practice, it is a logic that tends to use ideal-typical futures as the yard-stick against which the present is measured and assessed, resulting in analyses that note a difference (a gap) between what is and what should or could be, or between the potential (often determined on the basis of results obtained in laboratories, experiments and pilot plots) and actual performance of water systems. The proposed technological or institutional 'best practices' then become the self-evident remedy to closing the identified gaps. This solutionist gap-logic is compelling in its simplicity, while its promise of an achievable better future is very effective in mobilising financial and political support for whatever best practice or solution proposed.

This book clearly shows that science based on solutionist gap logics is poorly suited for assessing the effectiveness and impacts of interventions. For one, it treats complex realities as if they are ordered and knowable, as if consisting of repeatable, perceivable and predictable cause-and-effect relations. While recognising the comfort this may provide to researchers and professionals alike, most chapters of this book show that this comfort is false. The socio-hydrological dynamics that popular technical or institutional solutions interfere with are typically only coherent in retrospect and seldom repeat. Pretending order or artificially creating it for the sake of measurement and prediction is therefore bound to yield false promises. Secondly, following solutionist gap logics tends to direct the attention away from the water problems themselves, to make the performance of the proposed 'best practices' – the proposed technical or institutional arrangements – the main concern. The question of how to make these arrangements work as-designed or planned takes centre-stage, pushing concern about whether or not they actually solve the problems at stake to the background. A third and related concern is that solutionist gap logics tend to vest far too much transformative power in the solution. This fuels the erroneous belief that one solution can achieve the same or very similar results – whether it is improving the efficiency of water use, offering better protection against floods or restoring the health of rivers or aquifers or something else – irrespective of when and where it is applied or implemented. This is a belief that, of course, suits international expert-consultants as it makes it easier for them to sell their advice.

If solutionist science is poor, what kind of science or engineering is needed to develop and test effective solutions to water problems? The last chapter of this book provides some useful suggestions, starting with the reminder that there never is one objectively best solution, as assessment of which solution 'works' (best) or what effectiveness means (and which descriptors, metrics, scales, indicators, etc. to use to measure it) depends on the problem diagnosis or framing used, choice of which is always partly social and political. Solutions do often not become popular because of their effectiveness, but because of how well they align with vested political and economic interests and cultural preferences. In pointing this out, this book implicitly cautions against too much closeness – in either financial, cultural or political terms – between those who promote a particular remedy and those who test and assess its effectiveness.

In all, this book provides a tremendously useful go-to reference about whether and how favoured water solutions work in practice, and with what possible 'side-effects'. Submerging myself in its 22 engaging, meticulously researched and eloquently written chapters has been a delight. This book equips anyone who reads it with useful ammunition to effectively engage in contemporary water policy discussions. It should sit on the actual or virtual shelves of anyone involved in water advice or policymaking, serving as a much-needed shield against the dangerous simplifications and depoliticisations that most 'best practices' entail.

1 INTRODUCTION: Engaging with Water Policy Best Practices

François Molle and Sylvain Barone

Water traditionally fell within the remit of engineering, hydraulics and hydrology (Linton, 2010) – disciplines that focus on the mastery of water resources for human use, i.e., domestic and industrial supply, flood control, irrigation, shipping and hydropower. When water was seen as abundant, its extraction and pollution had minimal third-party consequences, allowing largely unregulated exploitation. However, as the pressure on the resource intensified, and the heyday of the 'hydraulic mission' began to fade, the need for more diverse and sophisticated responses – technical, economic and institutional – became apparent. Water managers and policymakers were asked to produce new 'solutions'.

1.1 The Nature and Hegemony of 'Best Practices'

The world of water, like all policy domains and social spheres, is permeated by buzzwords, hegemonic concepts, models and received ideas about what constitute the 'right' solutions or responses to specific problems. These ready-made responses, tools or 'best practices'[1] – mainstream policy solutions that are promoted as being the most effective and replicable based on expert consensus – circulate widely and are often taken up uncritically due to their perceived global validation. Their strength lies in having all the trappings of evidence, neutrality, and universality, while – paradoxically – they purport to be tested and rooted in particular success stories.[2] They form the sanctioned discourse.

Asked whether adopting drip irrigation or reusing treated wastewater would be desirable, most people will instinctively say 'yes'. Few would question a move towards increased 'efficiency', the establishment of a river basin organisation, or involving stakeholders in management. Greater efficiency, reduced waste, or an augmented resource managed more rationally are all accepted as obviously 'good things' and as such seem to evade controversy.

These prevailing solutions, and the narratives that accompany them, often form the frontline in the implementation of key water management principles or doctrines, such as Integrated Water Resources Management (IWRM), the

DOI: 10.4324/9781003745532-1

water-energy-food nexus and water security. The dominance, even hegemony, of these concepts and approaches has reinforced the widespread dissemination of the received wisdom and the standard solutions attached to them. This is particularly evident in the case of IWRM, which, in response to a perceived lack of actionability, has given birth to 'toolboxes' of best practices that range from river basin management to water pricing (Molle, 2008; Allouche, 2016; Kramer & Pahl-Wostl, 2014).

Like 'sustainable development', IWRM in practice is more concerned with efficiency than with highlighting the contradictions between ecology and economics. While seemingly technical and seen as rooted in unequivocal rationality, IWRM-related and other policy recommendations and 'solutions' often have ideological underpinnings. Following the 3rd World Water Forum in Kyoto in 2003, Merrett (2003) identified what he called the 'Kyoto consensus' ('best understood as the waterish extension of the Washington consensus') with these key features:

> Big dams are bad whereas small dams are good; [...] water resources management should be carried out on the catchment scale; [...] irrigation projects should be turned over to water user associations; [...] public water utilities are ripe for privatization; [...] water resources management should focus on demand management; [...] water use should everywhere be priced.

More recent additions include, among others, so-called non-conventional resources (such as desalination and the reuse of treated wastewater), payments for environmental services, and nature-based solutions.

Over time, some of these ideas come to be institutionalised, and, in a process of 'obliteration by incorporation' (Merton, 1968), it becomes unclear who exactly first named them and promoted them to the implicit status of 'best practice'. Their dissemination reveals forms of compliance and standardisation among organisations responsible for water management at various scales – either through adherence to dominant sectoral or professional norms ('normative isomorphism'), through mimicry aimed at reducing learning costs in contexts of uncertainty ('mimetic isomorphism'), or through the formal and informal pressure exerted by other organisations, foremost among them, in the water sector, intergovernmental organisations ('coercive isomorphism') (DiMaggio & Powell, 1983). Some ideas tend to fade away after a few years, while others endure for decades (IWRM); still others come and go every decade or so (water pricing). Once they are embedded in policy frameworks, institutionalised best practices resist scrutiny. This is not because they are unquestionably true, but rather because they have become woven into the institutional environment that shapes the beliefs of the very people who then go on to propagate them. Their persistence is reinforced by a degree of validation in specific contexts, making them far harder to challenge than an outright falsehood.

1.2 Why Best Practices Are So In Demand

Belief in best practices is akin to 'solutionism' – that is, the tendency to frame complex societal problems as issues that can be addressed with straightforward, often technological fixes (Morozov, 2013); as such, they share some of the same appeal.

Best practices offer common-sense causalities and ready-made solutions, and thus they seem to answer complex questions in domains characterised by high uncertainty and thick politics (Ostrom, 2007). They reduce the discomfort of navigating complex, ambiguous problems. Associated with instrumental governance, best practices often convey the idea that there is a managerial or technical optimum that can be packaged as a rational, scientific, neutral, and replicable solution. As Evans (2004) aptly noted, 'The intellectual difficulties of translating the complexities of institutional analysis into concrete suggestions for development policy and practice are sufficiently daunting to make retreat to simpler models a constant temptation'.

Solutionism and its associated sociotechnical imaginaries (Hommes et al., 2022), whether consciously or not, are instrumental in depoliticising the issue at hand by rendering it technical and avoiding controversy. 'Solutions', and the narratives behind them, tend to obscure negative externalities, trade-offs, and the existence of power struggles and hierarchies. Indeed, the solutions most often put forward in the water sector are those that make it possible to get around the most difficult political trade-offs. A solution will be all the more valued if it does not threaten the balance of social interests, especially where these are powerful and well organised. Typical best practices therefore include supply augmentation options such as desalination, storage, inter-basin transfers, or wastewater reuse, rather than, for example, rationing or reallocation, which actually require the politically challenging reality of reducing usage by existing beneficiaries. Depoliticised and instrumental best practices are thus attractive to politicians and decision-makers.

Blueprints, models, or packaged standard solutions also perfectly fit the interests of donor agencies, providing them with a powerful, coherent, legible, and consistent agenda for action at both the macro and micro levels (Mosse, 2005; Venot & Veldwisch, 2017). 'Nothing fits the internal organization needs of an "assistance" agency better than an objectively quantifiable "gap"' (Pritchett & Woolcock, 2004) and the sanctioned solutions into which resources can be poured. Lengthy and always uncertain studies into local complexity are antithetical to how development or infrastructure projects work. Aid bureaucracies, donors, or politicians under pressure need project or policy designs that are replicable, easily justified, and associated with well-defined time-related deliverables and cash outflow schedules. This demands a pre-packaged solution with predefined targets that can be uniformly applied, and where the 'solution' is preferably capital-intensive and quick rather than reform-intensive and slow.

The centrality and usefulness of such panaceas and approved best practices are manifested and magnified in times of crisis, such as drought or flooding, when leaders call on administrations to come up with 'appropriate responses' quickly. Officials must then hastily assemble lists of measures that effectively convey the political message that 'something is being done'. These will tend to rest on received ideas and seemingly uncontroversial ways of improving the situation, since predominant concepts, best practices, and policy models are, by definition, 'low risk'. If the measure being taken is seen as sound and conforming to best practices, then any failure will more likely be ascribed to the context, implementation problems, or 'politics' and thus will be less likely to be blamed on those in charge or the solution itself. When President Macron of France called on the relevant ministries to generate an urgent 'Water Plan' following the historic drought of 2022, it was unsurprising that they came up with 53 measures taken straight from the best practices canon.

1.3 Why They Are Problematic

Best-practice measures do not come out of the blue. There are, of course, cases where they have delivered positive results. It would be foolish to dismiss any of them on principle, whether storage, micro-irrigation, nature-based approaches, economic tools, or the reuse of treated wastewater. Every solution revisited in this book has proven beneficial somewhere and is certainly part of 'the solution'. So why are they problematic?

A key problem with standardised solutions is that they are often promoted on the basis of a very narrow view of what they achieve. In other words, their benefits are overemphasised while negative externalities are downplayed or ignored altogether. Storage, drip irrigation, weir removal, or the reuse of treated wastewater often have clear advantages that obfuscate what is lost – the latter only becoming visible when we change the social and spatial focal length.

Best practices, often pushed through 'project pipelines' and straightjacketing timeframes, do not do well with complexity. Feek (2007) wonders

> [how] you decide what is "best" when all practice – all development action, including communication interventions, addressing priority development issues – takes place in different contexts, with different purposes, different population groups, and significantly different opportunities, involving challenges within widely varying cultural, political, and resource environments.

Indeed, projects or best practices often look promising in their conceptual stages because they simplify real-world complexities and are based on idealised assumptions; their (often top-down) implementation, however, generally brings out a host of unforeseen local, political, social, cultural, ontological, environmental, and technical challenges. This is what anthropologist Jean-Pierre Olivier de Sardan (2021) calls the 'revenge of context'.[3]

It has been recognised, particularly in the social sciences, that researchers are *uncertainty creators* (as they reveal complexity and ambiguous causalities, and question conventional wisdom), while policymakers need to act as *uncertainty reducers* (Cleaver & Franks, 2008). Faced with humans' 'cognitive miser' (Fiske & Taylor, 1991), they seek cognitive shortcuts and simplifications. But there is an unavoidable trade-off here: 'Debunk the language, simplify the ideas. No nuances, no problems, just solutions', said a senior international water policy maker cited by Cleaver and Franks. The need to simplify the world can become all the more defective when fuelled by an urge to rush, an aversion to nuances, or hardened ideologies or worldviews that only champion specific solutions.

Bronk and Jacoby have emphasised the dangers of analytical monocultures, harmonised regulations, and global definitions of 'best practice' in a world riven with uncertainty and instability; these analytical monocultures 'tend to become particularly entrenched (and impervious to criticism) when embedded in widely shared technocratic practices, models, and data collection methods shaped by these same ideational frameworks' (2016). Policy ideas can become 'sticky' or reach a lock-in situation when certain circumstances prevail; these can include a difficulty in assessing outcomes, low levels of news coverage or scrutiny, a persuasive and persistent causal story, people becoming cognitively and emotionally invested in the means, interest groups supporting particular policy ideas, or risk aversion (Jones et al., 2014).

A problem germane to analytical monocultures is that they are very much part of the dominant, top-down, infrastructure-based expert culture that underpins conventional development and continues to display limited sharing of decision-making power. Because the solutions are already defined, they tend to crowd out not only alternative options but also other worldviews or ontologies and any diversity of rationalities and values (Arce & Long, 2000; Mosse, 2005; Escobar, 2011). In other words, the problem is not so much that simplification happens, but that particular ways of simplifying become dominant, foregrounding certain values, metrics, and ideologies while reducing others to the status of 'context'.

The establishment of a doxa with its accepted 'solutions' serves a number of social, economic, organisational, and political interests. Although the process by which best practices emerge is not clear-cut, they are often actively promoted by certain actors. They open up markets for consulting and water companies; they strengthen the political and professional positions of decision-makers and managers in the sector; and they generate rents for scientific and research organisations, water experts, or territories that are considered to be 'model students' of water management (Barone, 2018). Coalitions form and are strengthened by the funding they are able to attract (see Drapier, 2019, for the case of weir removal in the United States). Once this web of interests solidifies, the best practice is institutionalised and becomes very hard to challenge. It tends to remain unquestioned even when contradictions emerge and/or stakeholders become disillusioned with its application. Instead, any failure is likely to fuel more projects and initiatives that aim to make it work – or work better.

When interests take precedence, disclaimers or caveats stressing the need to 'adapt solutions to context' sound hollow. Generalising Oya's (2009) observation on the 'win-win scenarios' identified by the *World Development Report 2008*, promising tools or policies are often accompanied by 'a caveat in the form of reality check usually starting with a "but", which emphasises the challenges in achieving these desired [objectives]'. The potential for and benefits of solar irrigation, for example, will be promoted as a 'game changer for climate resilience in agriculture' (see for example Dirwal et al., 2024), but with the proviso that an 'appropriate regulatory framework' needs to be in place to avoid groundwater overexploitation. Likewise, billions of dollars have been sunk into irrigation modernisation programmes with the (late) warning that appropriate measures should be taken to avoid the concomitant expansion of the irrigated area (the rebound effect). These "wishful disclaimers" allow best practices to be promoted, while the key associated problems are treated as secondary and manageable side effects, and are shifted outside or beyond the responsibility of the provider of the 'solution'.

The centrality of best practices is also in line with the principles of New Public Management (Pal & Clark, 2013) and its emphasis on the alleged appropriateness of measuring the performance of public organisations using quantitative indicators. The putative validity and universality of water policy narratives and their solutions legitimise comparison, particularly between countries. 'Audit culture' (Power, 1997), with its indicators and benchmarking, validates comparisons in terms of lag or lack and stymies reflection. Israel, for example, recycles 90% of its wastewater and Spain 20%; France, with its 1%, is thus seen as clearly 'lagging behind', regardless of the differences between the countries.[4] If per capita water storage in this country is so low compared with Spain, the US or Australia, should we not build more dams to achieve water security (Grey & Sadoff, 2007)? Figures, graphs, and maps compare rates of cost recovery in water services, the percentage of leaks in urban networks or drip irrigation systems, and even the degree of participation or the excellence of water governance. Countries with no river basin organisations are seen as yet to be converted to the virtues of integrated management. When attention focuses on benchmarking[5] and on who is 'lagging behind', the choice and validity of the means is no longer an issue.

1.4 Embodying Success

Policy models or best practices are always closely accompanied by exemplars or 'success stories' (Molle, 2008), which are very much in demand in the business sector (Strang & Macy, 2001; Pal & Clark, 2013), as well as by governments and development or aid actors (Chambers, 1997). Success stories suggest that there are models or templates that are worth being replicated. They have the power to 'simplify and control complex environments' (Cooper & Packard, 1997) by structuring debate, defining what data is relevant, and delineating acceptable alternatives.

Success stories are also marketable and compelling to audiences. They fit into narratives of human ingenuity and progress, legitimise the actors concerned, and can help secure future funding. Robert Chambers (1997) emphasised the centrality of what he called 'islands of salvation', that is, particular cases of implementation that are decreed as successful and lend themselves well to field visits by outsiders.

Arguably, even discounting the possible self-interest of the entity who claims to have achieved it, success is not easily defined. Assessments depend on the timeframe and which sustainability criteria are taken into account. Cernea (1987), for example, re-evaluated 25 World Bank projects that had been deemed successful at completion and found that only 12 of them had actually achieved long-term sustainability. Village wells, water kiosks, or handpumps are considered indicators of a project's success when introduced but not after a few years, when many have been abandoned. Dams may be counted as a success when the focus is on energy production and not on their negative externalities, such as population displacement and ecological impact.

Success, in other words, depends largely on the indicators used to measure it and, more fundamentally, on who imposes the dominant narrative. When a World Bank official was asked by Rap (2006) why he considered the Mexican irrigation management transfer case to have been a success, he pondered before answering, 'because nobody denies that it is a success'. If the storyline is strong enough, even those who admit they are facing implementation difficulties will often rush to proclaim that they know it has been 'successful elsewhere' (Chambers, 1997).

As Bruno Latour (1988) has shown, the success of a particular idea or project is not inherent in the initial design; rather, it arises from the ability of a network of actors to recruit support and impose their logic and coherence on opponents. As Mosse (2005) also stated, 'the more interests that are tied up with their particular interpretations, the more stable and dominant development's policy models become'.

These observations do not dismiss, of course, the possibility of learning from what is generally presented as a success. There are certainly lessons to be learned from specific 'models', whether these be water markets (Australia), irrigation management transfer (Mexico), river basin organisations (France's *Agences de l'Eau*), payment for environmental services (New York City), nature-based solutions for flood control (Netherlands), urban water supply (Singapore), drip irrigation (Israel), or wastewater reuse (Namibia). Nonetheless, we contend that learning from past mistakes and failures remains very much underdeveloped.

1.5 Learning from Failure

As we have seen above, what does and does not work often involves a degree of subjectivity. However, carrying out thorough stock-taking studies on topics such as, say, groundwater overexploitation, water user groups, river basin organisations, or irrigation water pricing leads to the conclusion that there are few, rather than many, 'success stories'. Such investigations also reveal that descriptions of cases branded successful are often quite superficial or partial and rarely expose the complexities of the context and the actors' interests and strategies. If 'what works' is more elusive than what does not, then there is a need for increased scrutiny on the flipside of these policy tools.

But understanding failure is as complex as explaining success. In the Amhara region of Ethiopia, for example, 4,000 rainwater harvesting ponds were created between 2003 and 2008, of which the majority were non-functional by 2009. Eguavoen (2009) identified various factors responsible for this outcome, including poor site selection, problems with design, technical issues, and inadequate institutional and incentive structures. More recently, Dickson et al. (2023) investigated the causes of failure of conservation projects and came up with a taxonomy that included no fewer than 57 items.

Some analysts have shown that development projects are particularly prone to failure when 'best practices' from high-income countries are imposed in low-income settings without adapting to the local context. Pritchett and Woolcock (2004) argue that rigid project design and overly ambitious goals often result in limited success or outright failure. A standardised and inflexible framework is bound to conflict with the environmental and social complexities onto which it will be overlaid. Development agencies and policymakers often shift priorities faster than a project can be implemented, tested, or evaluated. This makes it difficult to assess the original objectives, and valuable lessons from its success or failure are overlooked.

Robert Chambers has very eloquently reflected upon dealings with success and failure in development practice. He points out that

> [e]rrors which are recognised and embraced can lead quickly to better understanding and performance. [...] Embedded errors go deeper, last longer, and do more damage. Often they reflect widely held views and are generalised. Often they fit what powerful people want to believe.
>
> (Chambers, 1997:15)

In other words, errors creep in when technical, physical, or social complexities are overlooked. But they can also stem from what we choose to ignore and what we cannot see because of self-interest, deep-seated biases, or narrow rationalities – when special interest groups, bureaucracies, and/or political decision-makers do not directly produce ignorance (Proctor, 2012).

Indeed, the odd mechanism through which failure is actually taken as an indication that more of the solution is needed to finally achieve success has been cogently unpacked by Robert Fletcher (2023) regarding the rise and fall of neoliberal nature conservation (see also Long & van der Ploeg, 1989).[6] Other instances have been analysed in that field, for example, regarding payment for environmental services (Büscher, 2014), community-based natural resources management (Blaikie, 2006), the REDD programme (Reducing Emissions from Deforestation and Forest Degradation; Svarstad & Benjaminsen, 2017) and drip irrigation (Venot, 2016).

As Strang and Macy (2001) warn, the biases and self-interest involved when success (rather than failure) is attended to can generate unhelpful, faddish cycles. In contrast, nurturing a healthy sceptical/critical approach and applying multiple evaluative principles can 'foster a generative friction that disrupts received categories of business as usual and makes possible an ongoing recombination of resources [...] [and] new ways of recognizing opportunities' (Stark, 2009:16). Acknowledging the value of scrutinising failure, we may follow 'Montaigne's heuristic' as illustrated by the following short story:

> The Atheist, when he was in Samothrace [...] was shown many vows and votive portraits from those who had survived ship-wrecks and was then asked, 'You, there, who think that the gods are indifferent to human affairs, what have you to say about so many men saved by their grace?' – 'It is like this,' he replied, 'there are no portraits here of those who stayed and drowned – and they are more numerous!'
>
> (Montaigne, quoted by Strang & Macy, 2001)

In other words, look not only at the celebrated success stories of the saved but also at the conveniently disregarded, arguably more numerous, cases of failure – of the drowned.

1.6 Why Revisit Best Practices?

This book emerges from many years of engagement in the field of water policy and governance, shaped by countless interactions with officials, decision-makers, politicians, journalists, and everyday water users. Across these encounters, we observed a striking uniformity: the widespread and often uncritical promotion of received ideas and seemingly obvious solutions.

We were particularly struck by the frequency of implementation vagaries and perceived 'misfits' – mismatches between policy tools and real-world conditions – which we believe deserve closer scrutiny. While best practices are routinely accompanied by disclaimers noting that they are not one-size-fits-all and should be adapted to local contexts, such caveats are often superficial and self-serving. In practice, the bureaucratic and technocratic machinery surrounding policy implementation rarely allows for meaningful adaptation or fine-tuning.

We contend that 'the devil is in the details' and that it is precisely both structural and local contextual specificities – often overlooked or flattened in planning – that derail the anticipated success of best practices. This volume, thus, aims to unpack the complexity obscured by the toolkits and standardised solutions so prevalent in the water sector and to expose the limitations and unintended consequences of 20 ubiquitous 'best practices' and 'solutions'. However, beyond 'improved solutionism', our broader aim is to reopen, in each case, the 'universe of possibilities': to question dominant values, challenge taken-for-granted rationalities, and encourage deeper reflection on what plural and context-sensitive water governance might look like.

Rather than start with a best practice and end with recommendations to mitigate its collateral impact, we advocate *inverting the burden of proof*. In other words, we propose to begin by scrutinising the *potential* social, economic, and environmental externalities of a planned measure or intervention based on the most common shortcomings already documented and analysed in this book. If the context and risks are deemed favourable, then a more detailed examination of the specifics of the intervention can proceed. The goal, then, is to displace the usual best practice statement (i.e., 'Although it has to be adapted to the context and may face some limitations, Policy A is promising and a recommended way forward') with a more sceptical and reflexive stance (i.e., 'Although it has proven to be of value in particular contexts, Policy A commonly faces difficulties and produces unanticipated effects that must be foregrounded and assessed first').

While we hope to contribute to an enhanced understanding of the usual policy responses, we are aware that the homogenisation of policies and the attractiveness and stickiness of best practices have deeper roots that speak particularly to their ideological underpinnings and ontological biases. Those are not the subject of this book, nor do we address any related questions long grappled with by sociology and political sciences, however relevant they may be. Thus, we do not focus on the mechanisms by which best practices or hegemonic concepts emerge or are consolidated, echoing neither scholarly interest in epistemic communities (Haas, 1992) nor policy or discourse coalitions (Hajer, 1995), global knowledge networks (Stone, 2004), policy bubbles (Jones et al., 2014), or social and cognitive processes (Douglas, 1986). We do not address the question of why some of these concepts seem to last forever while others come and go in faddish cycles often unrelated to their actual 'success'. And we do not address policy transfer or diffusion – how particular policy tools or measures travel across contexts and countries (Stone, 2012; Mukhtarov, 2022).

This said, we certainly acknowledge the pervasive and pivotal importance of the political economy logic behind development practice and state rule in how 'solutions' and 'best practices' are adopted, implemented, and sustained. These aspects are often deeply contextual and not easily captured in chapters that are generic by design. Yet, the chapters also point to the typical bundle of interests and ideologies associated with the best practices examined. We will return to these aspects in the conclusion.

The best practices discussed here are of different kinds. Some are primarily affected by weak internal consistency or causalities (such as water pricing, sustainable groundwater use, and relationships between forests and water) or wishful thinking (private finance for water supply and sanitation, or ecological modernisation), while others are legitimised by general and largely unquestioned principles (e.g., river basin organisations, water user associations, and payment for ecosystem services) or by a focus on benefits that overlooks or obscures negative aspects (e.g., desalination, wastewater reuse, storage, drip irrigation, and river weir removal). Overall, most best practices are bedevilled by implementation and contextual challenges. The shortcomings and problems associated with each 'solution' vary widely, however, depending on local contexts, governance patterns, technical capacity, the availability of funding, or the soundness of the causalities considered.

The entries in this book come in the form of short statements that are recognisably common in the water sector. Each is preceded by the word *'revisiting'*, which signals the central objective of bringing experience and critical analysis to bear, *deliberately focusing on the 'flipside of the coin'*. The chapters include a first section that recalls the context in which each particular best practice emerged, gained traction, and was disseminated. A second section then exposes the practice's respective flaws, limitations, pitfalls, and pervasive unintended or counterintuitive outcomes. These brief chapters, akin to encyclopaedia entries, are *not* intended as comprehensive or exhaustive reviews but, rather, as warnings and entry points to a sceptical examination. Bibliographies are thus intentionally *selective*, pointing readers to key references for further exploration.

Received ideas, we hope to explain, offer no easy shortcut or substitute for careful examination and analysis. We hope the commentaries provided by the contributors to this volume will fuel and nurture discussions and productive arguments. Indeed, our aim is to promote informed, plural, and holistic decision-making and reopen debate. If the obvious solutions are, after all, not so obvious, collective discussion on what is and is not desirable becomes possible again. This is a question of real democracy, in that it hinges on the constant and tireless examination of diversity and complexity but also of how power shapes the formation of knowledge.

Acknowledgements

The editors would like to thank Frances Cleaver, Chris Perry, Peter Mollinga, Margreet Zwarteveen, Marie-Anne Germaine, and Luke Whaley for their comments on earlier drafts of this introduction. This book was published with the financial support of AgroParisTech.

Notes

1 Although critiques of this expression have led to the use of softer and less presumptuous formulations such as "best fit", "smart practices" or "good practices", arguably the same spirit prevails (Pal & Clark, 2013).
2 As Panastasiou (2024) underlines the contradiction inherent in best practices, which 'profess to be both situated in local practice and to hold universal relevance'.
3 Meinzen-Dick (2007) also observed,

> Over the last fifty years, a series of institutional arrangements has been presented as panaceas to water management. ... Each of these approaches has failed to live up to expectations, largely because of the variability of local situations and the difficulty of transplanting institutions from one place to another.

4 This reasoning is commonly heard in France. An anecdote, told by an eyewitness, has it that President Macron in a field visit before launching his National Water Plan (*'Plan Eau'*), was told that Spain was reusing 20% of its TWW while France sorely 'lagged behind' with only 1%. This prompted him to declare that he wanted a 'target at 10%'. Benchmarking is performative and does not need to ask whether there are good reasons for this situation.
5 A graphic example of this is provided by a progress report on water sector reforms in Asia (Mosley, 2004) that established a list of 41 "policy elements" and estimated the degree to which different countries had achieved them. They included "legislation adopted", "private sector investment", "cost recovery", "river basin/aquifer organisations" and "rights and responsibilities of stakeholders" (cited in Molle, 2008).
6 Long and van der Ploeg (1989) also believe that it is even possible to 'argue that a certain degree of failure is strategic in the reproduction of intervention itself'.

References

Allouche, J. (2016). The birth and spread of IWRM: A case study of global policy diffusion and translation. *Water Alternatives, 9*(3), 412–433.

Arce, A., & Long, N. (2000). Reconfiguring modernity and development from an anthropological perspective. In A. Arce & N. Long (Eds.), *Anthropology, development and modernity* (pp. 1–32). Routledge.

Barone, S. (2018). Building a narrative on environmental policy success: Reflections from a watershed management experience. *Critical Policy Studies, 12*(2), 135–148.

Blaikie, P. (2006). Is small really beautiful? Community-based natural resource management in Malawi and Botswana. *World Development, 34*(11), 1942–1957.

Bronk, R., & Jacoby, W. (2016). *Uncertainty and the dangers of monocultures in regulation, analysis, and practice* (MPIfG Discussion Paper No. 16/6). MPIfG.

Büscher, B. (2014). Selling success: Constructing value in conservation and development. *World Development, 57*, 79–90.

Cernea, M. (1987). Farmer organization and institution building for sustainable development. *Regional Development Dialog, 8*(2), 1–24.

Chambers, J. M., Massarella, K., & Fletcher, R. (2022). The right to fail? Problematizing failure discourse in international conservation. *World Development, 150*, 105723.

Chambers, R. (1997). *Whose reality counts? Putting the first last.* Intermediate Technology Publications.

Cleaver, F., & Franks, T. (2008). Distilling or diluting? Negotiating the water research-policy interface. *Water Alternatives, 1*(1), 157–176.
Conn, S. (2019). *Nothing succeeds like failure: The sad history of American business schools*. Cornell University Press.
Cooper, F., & Packard, R. M. (1997). Introduction. In F. Cooper & R. M. Packard (Eds.), *International development and the social sciences: Essays on the history and politics of knowledge* (pp. 1–41). University of California Press.
Cornwall, A. (2007). Buzzwords and fuzzwords: Deconstructing development discourse. *Development in Practice, 17*(4–5), 471–484.
Dickson, I., Butchart, S. H., Catalano, A., Gibbons, D., Jones, J. P., Lee-Brooks, K., ... Sutherland, W. J. (2023). Introducing a common taxonomy to support learning from failure in conservation. *Conservation Biology, 37*(1), e13967.
DiMaggio, P. J., & Powell, W. W. (1983). The iron cage revisited: Institutional isomorphism and collective rationality in organizational fields. *American Sociological Review, 48*(2), 147–160.
Dirwal, T. L., Nkosi, M., & Choruma, D. Jr. (2024). Solar irrigation: A game changer for climate resilience in agriculture. IWMI. www.cgiar.org/news-events/news/solar-irrigation-a-game-changer-for-climate-resilience-in-agriculture
Douglas, M. (1986). *How institutions think*. Syracuse University Press.
Drapier, L. (2019). *Approche géographique comparée du démantèlement des seuils et des barrages sur les deux rives de l'Atlantique: projet écologique, politiques publiques et riverains (Sélune, Orne, Musconetcong, Wood-Pawcatuck, Mousam)* [Unpublished doctoral dissertation]. Université de Paris-Nanterre.
Eguavoen, I. (2009). *The acquisition of water storage facilities in the Abay River Basin*. Ethiopia.
Escobar, A. (2011). *Encountering development: The making and unmaking of the Third World*. Princeton University Press.
Evans, P. (2004). Development as institutional change: The pitfalls of monocropping and the potentials of deliberation. *Studies in Comparative International Development, 38*(4), 30–52.
Feek, W. (2007). Best of practices? *Development in Practice, 17*(4–5), 653–655.
Fiske, S. T., & Taylor, S. E. (1991). *Social cognition* (2nd ed.). McGraw-Hill.
Fletcher, R. (2023). *Failing forward: The rise and fall of neoliberal conservation*. University of California Press.
Grey, D., & Sadoff, C. W. (2007). Sink or swim? Water security for growth and development. *Water Policy, 9*(6), 545–571.
Haas, P. M. (1992). Introduction: Epistemic communities and international policy coordination. *International Organization, 46*(1), 1–35.
Hajer, M. A. (1995). *The politics of environmental discourse: Ecological modernization and the policy process*. Clarendon Press.
Hommes, L., Hoogesteger, J., & Boelens, R. (2022). (Re)making hydrosocial territories: Materializing and contesting imaginaries and subjectivities through hydraulic infrastructure. *Political Geography, 97*, 102698.
Jones, B. D., Thomas III, H. F., & Wolfe, M. (2014). Policy bubbles. *Policy Studies Journal, 42*(1), 146–171.
Kramer, A., & Pahl-Wostl, C. (2014). The global policy network behind integrated water resources management: Is it an effective norm diffusor? *Ecology and Society, 19*(4). www.jstor.org/stable/26269658

Latour, B. (1988). *Science in action: How to follow scientists and engineers through society*. Harvard University Press.

Linton, J. (2010). *What is water? The history of a modern abstraction*. UBC Press.

Long, N., & van der Ploeg, J. D. (1989). Demythologizing planned intervention: An actor perspective. *Sociologia Ruralis, 29*(3/4), 226–249.

Meinzen-Dick, R. (2007). Beyond panaceas in water institutions. *Proceedings of the National Academy of Sciences, 104*(39), 15200–15205.

Merrett, S. (2003). Virtual water and the Kyoto consensus: A water forum contribution. *Water International, 28*(4), 540–542.

Merton, R. K. (1968). *Social theory and social structure*. Free Press.

Molle, F. (2008). Nirvana concepts, narratives and policy models: Insights from the water sector. *Water Alternatives, 1*(1), 131–156.

Morozov, E. (2013). *To save everything, click here: The folly of technological solutionism*. Public Affairs.

Mosley, P. (2004). *Water sector reform: A progress report*. Asian Development Bank.

Mosse, D. (2005). *Cultivating development: An ethnography of aid policy and practice*. Pluto Press.

Mukhtarov, F. (2022). A review of water policies on the move: Diffusion, transfer, translation or branding? *Water Alternatives, 15*(2), 290–306.

Olivier de Sardan, J.-P. (2021). *La revanche des contextes: des mésaventures de l'ingénierie sociale en Afrique et au-delà*. Karthala.

Ostrom, E. (2007). A diagnostic approach for going beyond panaceas. *Proceedings of the National Academy of Sciences, 104*(39), 15181–15187.

Oya, C. (2009). The World Development Report 2008: Inconsistencies, silences, and the myth of 'win-win' scenarios. *The Journal of Peasant Studies, 36*(3), 593–601.

Pal, L. A., & Clark, I. D. (2013). *Best practices in public management: A critical assessment*. Working Paper for the Best Practices in Public Management Project.

Papanastasiou, N. (2024). The politics of generating best practice knowledge: Epistemic practice and rendering space technical in a European Commission working group on education policy. *Environment and Planning C: Politics and Space, 42*(2), 179–197.

Power, M. (1997). *The audit society: Rituals of verification*. Oxford University Press.

Pritchett, L., & Woolcock, M. (2004). Solutions when the solution is the problem: Arraying the disarray in development. *World Development, 32*(2), 191–212.

Proctor, R. N. (2012). *Golden holocaust: Origins of the cigarette catastrophe and the case for abolition*. University of California Press.

Rap, E. (2006). The success of a policy model: Irrigation management transfer in Mexico. *The Journal of Development Studies, 42*(8), 1301–1324.

Redford, K. H., Padoch, C., & Sunderland, T. (2013). Fads, funding, and forgetting in three decades of conservation. *Conservation Biology, 27*(3), 437–438.

Stark, D. (2009). *The sense of dissonance: Accounts of worth in economic life*. Princeton University Press.

Steiner-Khamsi, G. (2012). Understanding policy borrowing and lending: Building comparative policy studies. In G. Steiner-Khamsi & F. Waldow (Eds.), *World yearbook of education* 2012 (pp. 3–17). Routledge.

Stone, D. (2004). Knowledge networks and global policy. In S. Maxwell & D. Stone (Eds.), *Global knowledge networks and international development* (pp. 109–125). Routledge.

Stone, D. (2012). Transfer and translation of policy. *Policy Studies, 33*(6), 483–499.

Strang, D., & Macy, M. W. (2001). In search of excellence: Fads, success stories, and adaptive emulation. *American Journal of Sociology*, *107*(1), 147–182.
Svarstad, H., & Benjaminsen, T. A. (2017). Nothing succeeds like success narratives: A case of conservation and development in the time of REDD. *Journal of Eastern African Studies*, *11*(3), 482–505.
Venot, J.-P. (2016). A success of some sort: Social enterprises and drip irrigation in the developing world. *World Development*, *79*, 69–81.
Venot, J.-P., & Veldwisch, G. J. (2017). Mythes sociotechniques et développement. *Anthropologie & Développement*, *46–47*, 27–48. https://doi.org/10.4000/anthropodev.585

2 *REVISITING*: Drip Irrigation Will Save Water

François Molle and Jean-Philippe Venot

2.1 Drip Irrigation: The Advent of an Uncontroversial Solution

The benefit of the 'little and often' application of water to the root zone of crops has been recognised since the advent of agriculture. Modern-day drip irrigation (applying water through a system of pipes perforated with drippers), however, can be traced back to improvements in plastic manufacturing in the aftermath of the Second World War. In the decades that followed, with the advent of the Green Revolution, drip irrigation came to be seen not only as a means to increase crop yields but also as a way to extend irrigation to areas previously deemed unfit for agriculture. This was particularly true in semi-arid and arid regions with sandy soils, such as the Middle East – notably Israel, where small-scale production lines evolved into global industry leaders – and the United States.

The 'coming of age' of drip irrigation came later – in the 1980s and 1990s – at the intersection of two concomitant narratives: one stressing the low performance of (mostly state-) managed, large irrigation systems in need of 'modernisation', and another warning of a looming 'global water crisis' and calling for wiser use of water resources, in line with the global discourse of sustainable development. In that context, high nominal irrigation-efficiency rates, mostly obtained in highly controlled environments, such as laboratories or experimental plots managed by universities and industrial firms, **led to drip irrigation being seen as the embodiment** of the 'more crop per drop' promise popularised in the early 2000s by the former secretary general of the United Nations, Kofi Annan. Many international organisations such as the FAO and the World Bank, and national ministries alike rushed to promise quixotic water savings simplistically estimated based on the ratio between initial and improved static farm-level efficiencies – whereby, typically, an increase from 40% to 80% in irrigation efficiency would reduce overall water requirements by half.

Over time, the issue of drip irrigation shifted away from the technical domain of irrigation engineering – where public engineers and researchers focused primarily on maximising plot-level efficiency – and entered the

DOI: 10.4324/9781003745532-2

interconnected worlds of policy and industry. Many countries (including China, India, Morocco and Spain) implemented large-scale subsidy programmes to support farmers' conversion to drip irrigation. In Morocco, the *Plan Maroc Vert* offered subsidies of 80% to 100% of the cost of installing drip irrigation systems, depending on the applicant; Spain's *Plano de Choque* had subsidies of 60%, and many Indian states began subsidising drip irrigation installation at 80%. The predicted water savings (or demand-reduction) were enshrined in policy documents (e.g. UNEP's Blue Plan [Benoit and Comeau, 2012]). In parallel, the 2000s saw large drip irrigation equipment manufacturers such as Netafim, Jain Irrigation Systems Ltd., and Chinadrip Irrigation Equipment take a prominent place in a multi-billion business. These companies now spearhead R&D and have activities in the global North and the global South alike, alongside a multitude of smaller-scale equipment manufacturers targeting a growing base of farmers. Drip irrigation's global appeal was further cemented in the 2000s, when international NGOs, development banks and agencies began promoting low-cost drip irrigation kits specifically designed to alleviate poverty among smallholders in the global South (Venot, 2016). Over time, a network of academic, development, and economic actors coalesced – all committed to, and interested in, promoting the alleged benefits of drip irrigation (Pérez-Blanco et al., 2021).

The aura that drip irrigation acquired lies in its malleability and the fact that it can be framed as an uncontroversial solution to a range of ever-changing international policy issues, such as food security, poverty alleviation, environmental conservation and agricultural and, maybe most obviously, irrigation modernisation (Venot et al., 2017). But also central to its appeal is specific knowledge-making research, namely measurements in dedicated laboratories or experimental fields under very controlled conditions. These conditions hardly reflect the diverse realities of farmers' practices and behaviours and fail to address their effect at the river-basin level. This will guide this chapter's attempt to revisit the ubiquitous narrative that drip irrigation is a straightforward and safe path to saving water.

2.2 What Drip Irrigation Does and Does Not Do

The rhetoric of water-saving associated with drip irrigation has been challenged by a growing body of evidence. Observations and analyses carried out at the plot, farm and system levels have qualified or undermined received wisdom. Before reviewing these insights, it is crucial to define the terms of the water balance under consideration (Perry et al., 2009; Lankford, 2013), whereby:

> Withdrawals **W** (from surface or groundwater) = evaporation **E** (of soil or water bodies) + transpiration **T** (of plants) + transpiration **TA** (of ancillary vegetation) + return flows **RF** (to aquifers or streams)

A key water-management issue is the distinction between *gross* water withdrawals/diversions (W) and net consumption/depletion (ET). When considering the plot or system level, the former is the water volume applied, while the latter combines crop transpiration (T) and soil evaporation (E) (ET is the water really consumed and returned to the atmosphere). The difference is the return flow (RF) to the drainage system or aquifer, less what is taken up by ancillary vegetation (TA) (taken at zero in what follows, for the sake of simplicity).[1] While T is beneficial and should be sustained or enhanced, E and TA are seen as losses to be avoided. Efficiency is defined as the fraction ET/W.

After looking at the potential benefits of drip irrigation, this chapter highlights several unexpected or underestimated consequences of modernising irrigation by moving to drip systems – from the plot to the river basin or aquifer level.

2.2.1 The Visible Benefits of Drip Irrigation

The benefits of drip irrigation are most evident in capital- and labour-intensive, high-value agriculture enjoying secure access to water. Individual groundwater use, in particular, comes with a secure, flexible and pressurised supply that is suitable for the use of drip irrigation for intensive agriculture. Under such controlled conditions, using drip irrigation allows the farmer (or investor) to reduce labour costs, control fertilisation ('fertigation' is the technique of adding liquid fertiliser to water), enhance product quality and finally increase yield and revenue. Thus, it corresponds to very specific, plot-level practices, water access, capital endowment and type of agriculture. Such conditions are not necessarily easy to reproduce in collective pressurised networks, as the example of the Jordan Valley attests. A further benefit of drip irrigation is that it can be used to irrigate sandy soil.

When transitioning from gravity to drip irrigation on a large scale – the focus of this chapter – the interdependencies between drip-based intensive agriculture and the broader hydrological cycle and value chains are revealed.

2.2.2 Reduced Water Application?

Quite counterintuitively, the transition to drip irrigation – assuming the cropping pattern remains unchanged – does not necessarily reduce the volume W of water *applied*. While this tends to be the case for large-scale investors, smallholder practices are reportedly different. A closer look at how farmers use drip systems in Morocco showed that overirrigation is common, with plot-level efficiency varying between 25% and 90% (Benouniche et al., 2014). Overirrigation is mostly due to non-uniform pressure and drip discharge (caused by poor system design or clogged drippers), leading farmers to apply excess water to ensure that the full cultivated area receives enough moisture. In arid regions, farmers accustomed to gravity irrigation often let water drip through emitters until they see surface pounding (visible assurance that the crop's water needs are met). Another common case in sub-Saharan

Africa is to see farmers reverting to using hoses to irrigate their fields in the morning because their drip irrigation kit is designed with labour requirements that are incompatible with the other job they have during the day. The practice of removing drippers to turn them into small water jets has also been observed.

Problems with filtration and inappropriate fertilisation (overapplication) are also common. A World Bank project evaluation (2009) in Morocco highlighted that drip irrigation is often misused and comes with technical requirements, such as cleaning filters and drippers, replacing dripper lines, mastering fertigation, etc. However, in practice, these technical requirements are frequently neglected, leading to uneven water distribution and inefficiency. Neglect may also reflect the fact that saving water is not a prime concern. Even in advanced irrigation systems, such as in the Capitanata in Italy, California or Spain, these problems mount up and result in performance levels far below expectation. While extension services and technical support could help improve these practices – such as sensors to inform farmers when there is enough moisture in the soil – implementing them is challenging, particularly in smallholder farming. The result is highly varied plot-level efficiency in drip irrigation, often falling far below the theoretical maximum of 90%.

2.2.3 Net Consumption Is Unlikely To Be Reduced

In the budget above, ET corresponds to the overall *net consumption* of water. Since productivity is directly linked to T, it is not meant to be decreased and may even increase. Two effects tend to cancel each other out:

a first, drip irrigation does not necessarily reduce (non-beneficial) soil evaporation (E) significantly. This depends on several factors, such as dripper density, soil coverage by crop, soil texture, the use of plastic mulch, etc. Less soil is wet than with surface irrigation, but it is wet for longer periods. Traditional olive groves in soil with good retention capacity in North Africa can be irrigated by gravity every three weeks, as opposed to twice a week or more with drip.

b second, crop transpiration (T) tends to increase due to more frequent and timely irrigation (Burt et al., 2001; Ward and Pulido-Velazquez, 2008; Perry et al., 2009) compared with previous systems (for example those where farmers practised deficit irrigation or gravity irrigation with poor land-levelling).

Working in California, Burt et al. (2001) showed how challenging it was to compare soil evaporation (E) and crop transpiration (T) for different types of crops, crop cover/density, soil and irrigation techniques, and the significant discrepancies involved. In certain cases, such as almond orchards in the Westlands Water District, ET was estimated to be 10%–15% higher with drip irrigation than other methods. Where soil has good retention capacity and/

or wetting is frequent, both E and T are likely to be higher with drip irrigation: 'when one combines possible higher evaporation with almost certain higher transpiration, the overall ET can be expected to be higher under drip/micro than for surface and sprinkler methods in many cases' (Burt et al., 2001).

While the distribution of E and T varies, total ET is often basically unchanged. Importantly, Burt et al. (2001) found that in Californian almond plantations, E represented only 11% of annual ET. Therefore, possible reductions in E are unlikely to exceed increases in T, and the conclusion that shifting to drip irrigation is ET-neutral is conservative. Although this percentage will vary with soil coverage, plant shading and wetting patterns, it reminds us that E is often rapidly reduced as plants grow. Indeed, a recent literature review by Perry and Steduto (2017) flatly found 'no documented examples of water savings for field crops in terms of consumptive use' and confirmed the 'null-hypothesis'. Even with variations in crop and location, significant net savings should not be expected from variations in overall ET (see also, Pérez-Blanco et al., 2020).

2.2.4 Drip Irrigation Is Accompanied by Shifts in Cropping Patterns

The transition to drip irrigation is often associated with changes in cropping patterns. It is generally part of an intensification package that makes particular sense for vegetable or fruit trees, and farmers making the transition will often adopt the following strategies (see Molle and Tanouti, 2017):

- 'densifying' the orchard or olive grove by adding (young) trees between old trees (in North Africa, the density of olive trees will typically be multiplied by four);
- adopting mixed cropping, e.g. adding fruit tree rows between olive trees or vegetables and fodder crops between fruit tree rows (especially when these are young);
- uprooting old trees and planting the plot anew, with a higher density of new tree varieties that are more productive but also more water-intensive;
- a full conversion of production, e.g. from cereals to arboriculture, characterised by far higher water requirements distributed throughout the year. This also 'rigidifies' demand, as fruit trees must be irrigated to avoid substantial capital loss (Scott et al., 2014).

Combinations of these effects have been well documented, particularly in Spain and the US. The impact on water consumption of the shift to higher-efficiency irrigation systems, together with the irrigation of new acreage, even led the state of Montana to sue Wyoming for having reduced the amount of water in the Yellowstone River that should have flowed to Montana in violation of the interstate compact (MacDonnell, 2012).

Furthermore, through the production of high-value crops, farmers have enough capital to deepen their wells whenever necessary (Perry and Steduto, 2017). The shift to cash crops and more profitable agriculture removes any brake on tapping groundwater further, even with higher pumping costs (see Chapter 9 on agricultural water pricing). Cumulatively, these facts mean the total water *consumed* ET at the plot level is likely to increase.

2.2.5 The Rebound Effect

Another phenomenon challenges the idea that drip irrigation reduces water depletion. While drip systems reduce the volume of water applied per unit of surface area, this "saved" water is often used to expand cultivation (Berbel et al., 2015). Since many drip irrigation systems rely on individual wells, the total volume of water available is often determined by the well's capacity and the pumping system – and remains largely unchanged in the short term. As a result, farmers can be tempted to extend irrigation to additional plots – whether they own them or rent them from neighbours – that were previously rain-fed or left fallow due to insufficient water supply.

Known as the 'rebound effect' (or Jevon's paradox), this phenomenon is typical of regions where water is scarce relative to land and has been widely observed, for example in Spain, Morocco, Chile, India, Pakistan, Israel, China and the US (Berbel et al., 2015). It results in higher rates of consumption, notwithstanding combinations with other forms of intensification. Sese-Minguez et al. (2017), for example, found that in the Canyoles watershed, Spain, drip irrigation was associated with intensification, a shift to trees, and expansion into formerly rain-fed areas. A second dimension of the rebound effect is that intensification and expansion often increase the economic return to water, which may in turn lead to increasing demand (this, of course, depends on water being available). Moreover, the fact that drip irrigation can be used in sandy soil and desertic conditions has led to increased water abstraction (e.g. in the coastal areas of Peru, North Africa, and the Middle East).

The expansion of consumptive use is highly problematic, particularly in water-stressed basins where public subsidies for drip irrigation are introduced precisely 'to increase efficiency and save water' (Molle et al., 2019). The perverse effect of (subsidised) drip irrigation due to the rebound effect – at the farm or system level – was identified by EU water experts and researchers in the early 2010s (European Commission, 2012). In line with EU recommendations, France and Spain introduced policies to limit the reallocation of the expected gross water savings from irrigation modernisation to 50%, with the other half to be left to the environment. The standard recommendation is for the reduction in gross abstraction to be accompanied by regulation on the expansion of irrigation (Grafton et al., 2018), but this caveat is very hard to implement in practice. Indeed, an economy-wide assessment for Spain showed that 100% of the savings from efficiency improvements were reallocated to other uses (Freire-González, 2019), a value that reached 89% for a sample of 30 Chinese provinces (Xu et al., 2021), despite those areas needing to see an increase in environmental allocation rather than the opposite.

2.2.6 What Happens to the Water That Is 'Saved'?

As discussed earlier, water consumption (ET) at the plot level can be assumed to remain largely stable (assuming no change in cropping patterns). Hence, any reduction in the amount of water applied (W) actually represents a decrease in return flows (RF), whether to neighbouring plots, the drainage system or the underlying aquifer. To assess the impact of switching to drip irrigation, therefore, it is necessary to zoom out and examine the effect on the entire water system from which the irrigators draw their supply (Simon et al., 2015). If an irrigation system 'saves' water by reducing river or groundwater abstraction and application by ΔV, then two questions are raised: (1) where does the ΔV volume that is no longer abstracted go?; (2) assuming, for the sake of simplicity, that total ET is optimal and unchanged, who is affected by the corresponding reduction in the overall return flow? (Lankford, 2013).

Regarding the first question, the ideal situation in water-short and overallocated basins is that ΔV is left in the river or aquifer to contribute to environmental health. In contrast, the water status of the basin is harmed if ΔV is used to (1) expand the irrigated area (the rebound effect discussed above), (2) supply downstream water-short systems or (3) allow new consumptive use. These lead to the same or more *depletion* at the basin level. While beneficial and justifiable economically, it only increases the pressure on resources and the water deficit in the basin, if any.

The second question is whether and how the reduced return flow affects existing prior-appropriators. While plot- and farm-levels dynamics are heterogeneous and may affect the spatial and temporal distribution of the return flow (Lankford, 2012), at the macro level we may distinguish between surface water return flows (SWRF) to the drainage system, and groundwater return flows (GWRF) to the aquifer (Figure 2.1).

Reduced return flow to drains and the river system will impact downstream SWRF-dependent users and wetlands (Figure 2.1). Reduced aquifer recharge means a *worsening of the net balance of the aquifers,* which will impact GWRF-dependent users and aquatic ecosystems (Figure 2.1). This is chiefly the case for alluvial/sedimentary plains and deltas (often with the best soil) and unconfined aquifers in sedimentary plateaus/basins that are recharged by direct vertical infiltration from surface water (or rainfall). In the case of karst or fractured-rock aquifers, the connection is less obvious, and it is often unclear whether return flows in fact return to the source from which farmers are pumping.

Users depending on the return flow will either suffer economic loss or be forced to compensate for the shortfall. The latter means using water from other sources (typically groundwater for SWRF-dependent users) or improving system efficiency (for example, by also adopting drip irrigation), both of which reduce their own return flows. All cases result in water shortages and/or shift the locus of pressure on the resource spatially and socially.

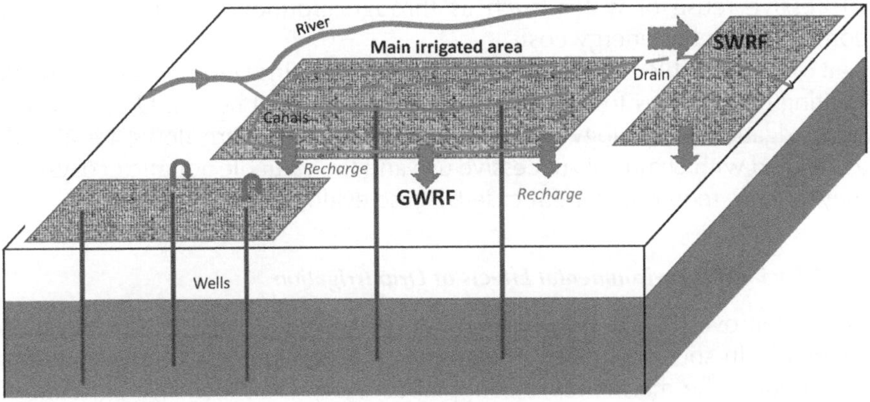

Figure 2.1 Simplified sketch of water resources around a surface irrigation scheme.

Note: This is a very simplified, yet typical, form of hydrologic interdependence. In a river basin, flows are varied, complex and never completely known.

In many situations, Farmers served by surface irrigation schemes where water is short often supplement their canal water with groundwater (where available, but this is generally true in sedimentary plateaus, alluvial plains and deltas). In such conditions, 'losses' from the percolation of irrigation water, along with recharge from rainfall, have long been (more than) fully tapped by numerous wells, typically causing groundwater levels to drop 0.5 to 1 m/year. Hence, any change in irrigation that reduces both the application of surface water and the level of recharge reduces groundwater availability, hence affects anyone using it, often the very farmers shifting to drip.

It is of course possible that there are *no* prior-appropriators of the return flow, or that this flow is abundant, but this is unlikely where water is short, 'water-saving' policies are introduced, and drip irrigation is promoted. It is also possible that the return flow goes to 'sinks' (e.g. saline aquifers or the sea) and cannot be recovered, in which case the shift to drip irrigation is certainly beneficial in terms of resource conservation but still impacts even saline ecosystems.

These broad arguments, while clearly of macro-level significance, must be qualified to allow for the high diversity of cases that can be observed (Gleick et al., 2011; Lankford, 2012). A more elaborate approach would account for, in particular, the fraction of the return flow that is consumed and does not reach the drain or aquifer (TA), and whether or not the ET was optimal before (and/or after) the shift to drip irrigation (Huffaker and Whittlesey, 2000; Lankford, 2012). Unproductive soil evaporation may be high in the case of frequent flood irrigations. Return flows may be delayed in time (e.g. baseflow reaching rivers when demand is low) or partially taken up by natural vegetation. Additionally, the quality of return flow often deteriorates, potentially harming soil, the ecosystem and crop yield. The successive reuse of water (such as through groundwater pumping) often comes at a higher energy cost.

It is preferable, therefore, to implement irrigation systems while limiting return flows to what is needed to ensure soil health. Unfortunately, new irrigation technology is often deployed where return flows are already associated with chains of successive use and the valuable aquatic ecosystems they sustain, requiring careful case-by-case evaluation.

2.2.7 Social and Environmental Effects of Drip Irrigation

One often-overlooked drawback of drip irrigation is the plastic waste it generates. In some countries, inexpensive but low-quality dripper lines that are flooding the market must be replaced every year. Intensive agricultural systems, such as those in Spain, use machines to simultaneously plant seeds, lay dripper lines and spread plastic mulch at the start of each season. Once used, these lines often end up in plastic dumps or are sometimes shredded and left in the soil, contributing to long-term environmental pollution.

Drip irrigation also has a very high energy demand. In southern Spain, the modernisation of irrigation has led to a 20% reduction in water usage per unit of irrigated land since 1950. However, over the same period, related energy consumption surged by 149% (Fernández García et al., 2014). Similar trends have been observed elsewhere, including the US, Syria, India and Morocco.

There is also a social cost of drip irrigation. Smallholder and family farmers tend to be risk-averse and struggle with government policies that promote the transition to capital-intensive, commercial agriculture. When tied to volatile markets and complex value chains, such policies can push these farmers into financial distress or debt. More generally, the introduction of drip irrigation tends to effect unexpected technical and institutional 'bricolage' (Mirhanoğlu et al., 2023; Venot et al., 2017). Here again, careful examination of local dynamics is in order.

2.3 Conclusion

The reflexions presented in this chapter suggest that focusing on plot-level efficiency alone may be deceptive. It is true that, in the large majority of cases, drip irrigation brings benefits in terms of land productivity (yield), total production, and farm income. Whether by adopting cash crops or through intensification or expansion, the net benefit to farmers is increased, as is national production. However, these private benefits accrue only to certain farmers and may come at a substantial environmental and social costs.

Although it is widely presented and promoted as leading to substantial water-saving, transitioning to drip irrigation has almost no effect on the depletion of water at the plot level and can even increase it. Furthermore, the farm-level changes it brings in terms of cropping patterns, crop densification and expansion of irrigation all increase water consumption. Where water is abundant, these changes have a positive economic impact overall. However, because of its alleged 'water savings', drip irrigation is often promoted and subsidised in water-scarce regions. In overexploited river basins or aquifers, water allocation is akin to a zero-sum game (or a negative-sum game, when aquifers are depleted): by increasing net water consumption, the processes of intensification/expansion reduce the return flow to the underlying aquifers, the drainage and river systems and, consequently, negatively impact the users and aquatic ecosystems that depend on them.

The problem is exemplified in Spain and Morocco, both of which have invested billions of euros to achieve water savings while the result has been exactly the opposite. In the name of efficiency, drip irrigation – together with lining and piping – transforms nature and its complex circulation of water into a reticulated network overlaying a desiccating landscape. Despite this conclusion, and while this chapter will look like familiar terrain to those acquainted with this controversy, official news and policy documents from various countries (North and South) continue to announce 'water savings' extrapolated from fallacious cross-multiplications on plot-level efficiencies. The reason why the misconception has endured over more than three decades, beyond the vested interested of the web of interconnected actors that has developed with it, is perhaps that zooming out leads to the unpalatable conclusion that less water is available than claimed. Eye-catching statements that drip irrigation can 'double production with half the water' or that 'water savings from drip can be allocated to other uses' deserve to be interrogated critically every time and in every location.

Further Readings

Lankford, B. A. (Ed.). (2012). Irrigation efficiency and productivity: Scales, systems and science [Special issue]. *Agricultural Water Management, 108*, C. www.sciencedirect.com/journal/agricultural-water-management/vol/108/suppl/C

Pérez-Blanco, C. D., Loch, A., Ward, F., Perry, C., & Adamson, D. (2021). Agricultural water saving through technologies: A zombie idea. *Environment Research Letters, 16*, 114032. https://doi.org/10.1088/1748-9326/ac2fe0

Perry, C., & Steduto, P. (2017). *Does improved irrigation technology save water? A review of the evidence*. FAO.

Venot, J.-P., Kuper, M., & Zwarteveen, M. (Eds.). (2017). *Drip irrigation for agriculture: Untold stories of efficiency, innovation and development*. Taylor & Francis.

Note

1 TA can be quite substantial and also partly beneficial (e.g. uptake by home-gardens near canals or drains). A fraction of RF corresponding to the need for plot salt removal to maintain soil productivity is also beneficial and should be pooled with ET. Word limit does not allow these or other complexities to be factored in.

Selected References

Benoit, G., & Comeau, A. (2012). *A sustainable future for the Mediterranean: The Blue Plan's environment and development outlook*. Routledge.

Benouniche, M., Kuper, M., Hammani, A., & Boesveld, H. (2014). Making the user visible: Analysing irrigation practices and farmers' logic to explain actual drip irrigation performance. *Irrigation Science, 32*(6), 405–420. https://doi.org/10.1007/s00271-014-0438-0

Berbel, J., Gutiérrez-Martín, C., Rodríguez-Díaz, J. A., Camacho, E., & Montesinos, P. (2015). Literature review on rebound effect of water saving measures and analysis of a Spanish case study. *Water Resources Management, 29*(3), 663–678. https://doi.org/10.1007/s11269-014-0839-0

Burt, C. M., Howes, D. J., & Mutziger, A. (2001). *Evaporation estimates for irrigated agriculture in California* (ITRC Paper P 01-002). Irrigation Training and Research Center.

European Commission. (2012). *A blueprint to safeguard Europe's water resources*. European Commission.

Fernández García, I., Rodríguez Díaz, J. A., Camacho Poyato, E., Montesinos, P., & Berbel, J. (2014). Effects of modernization and medium-term perspectives on water and energy use in irrigation districts. *Agricultural Systems, 131*, 56–63. https://doi.org/10.1016/j.agsy.2014.08.002

Freire-González, J. (2019). Does water efficiency reduce water consumption? The economy-wide water rebound effect. *Water Resources Management, 33*(6), 2191–2202.

Gleick, P. H., Christian-Smith, J., & Cooley, H. (2011). Water-use efficiency and productivity: Rethinking the basin approach. *Water International, 36*(7), 784–798. https://doi.org/10.1080/02508060.2011.631873

Grafton, R. Q., Williams, J., Perry, C. J., Molle, F., Ringler, C., Steduto, P., Udall, B., Wheeler, S., Wang, Y., Garrick, D., & Allen, R. (2018). Paradox of irrigation efficiency and the global water crisis. *Science, 361*(6404), 748–750.

Huffaker, R., & Whittlesey, N. (2000). The allocative efficiency and conservation potential of water laws encouraging investments in on-farm irrigation technology. *Agricultural Economics, 24*(1), 47–60.

Lankford, B. (2012). Fictions, fractions, factorials and fractures: On the framing of irrigation efficiency. *Agricultural Water Management, 108*, 27–38.

Lankford, B. (2013). *Resource efficiency complexity and the commons: The paracommons and paradoxes of natural resource losses, wastes and wastages*. Routledge.

MacDonnell, L. J. (2012). Montana v. Wyoming: Sprinklers, irrigation water use efficiency and the doctrine of recapture. *Golden Gate University Environmental Law Journal, 5*(2), 265–295.

Mirhanoğlu, A., Özerol, G., Hoogesteger, J., van den Broeck, P., & Loopmans, M. (2023). Socio-material bricolage—Co-shaping of irrigation institutions and infrastructures. *International Journal of the Commons, 17*(1), 69–86.

Molle, F., Sanchis-Ibor, C., & Avella Reus, L. (Eds.). (2019). *Irrigation in the Mediterranean: Technologies, institutions, policies*. Springer.

Molle, F., & Tanouti, O. (2017). Squaring the circle: Agricultural intensification vs. water conservation in Morocco. *Agricultural Water Management, 192*, 170–179.

Pérez-Blanco, C. D., Hrast-Essenfelder, A., & Perry, C. (2020). Irrigation technology and water conservation: A review of the theory and evidence. *Review of Environmental Economics and Policy, 14*(2), 216–239.

Pérez-Blanco, C. D., Loch, A., Ward, F., Perry, C., & Adamson, D. (2021). Agricultural water saving through technologies: A zombie idea. *Environmental Research Letters, 16*(11), 114032.

Perry, C., Steduto, P., Allen, R. G., & Burt, C. M. (2009). Increasing productivity in irrigated agriculture: Agronomic constraints and hydrological realities. *Agricultural Water Management, 96*, 1517–1524. https://doi.org/10.1016/j.agwat.2009.05.005

Scott, C. A., Vicuña, S., Blanco-Gutiérrez, I., Meza, F., & Varela-Ortega, C. (2014). Irrigation efficiency and water-policy implications for river basin resilience. *Hydrology and Earth System Sciences, 18*(4), 1339–1348. https://doi.org/10.5194/hess-18-1339-2014

Sese-Minguez, S., Boesveld, H., Asins-Velis, S., van der Kooij, S., & Maroulis, J. (2017). Transformations accompanying a shift from surface to drip irrigation in the semi-arid Cànyoles watershed, Valencia, Spain. *Water Alternatives, 10*(1), 81–89.

Simons, G. G., Bastiaanssen, W. W., & Immerzeel, W. W. (2015). Water reuse in river basins with multiple users: A literature review. *Journal of Hydrology, 522*, 558–571.

Venot, J.-P. (2016). A success of some sort: Social enterprises and drip irrigation in the developing world. *World Development, 79*, 69–81.

Ward, F. A., & Pulido-Velazquez, M. (2008). Water conservation in irrigation can increase water use. *PNAS, 105*(47), 18215–18220. https://doi.org/10.1073/pnas.0805554105

World Bank. (2009). *Project performance assessment report*. Morocco Water Resources Management Project. Report No. 48732.

Xu, H., Yang, R., & Song, J. (2021). Agricultural water use efficiency and rebound effect: A study for China. *International Journal of Environmental Research and Public Health, 18*(13), 7151. https://doi.org/10.3390/ijerph18137151

3 *REVISITING*: Reusing Treated Wastewater Reduces Water Scarcity

Anne-Laure Collard

3.1 Principles and Enthusiasm for Reusing Wastewater

With ever-increasing pressure on water resources, finding alternative sources has emerged as a promising option. Reusing treated wastewater (TWW) is a notable example, and the practice ('water reuse' in what follows) is attracting global interest. It consists of treating effluents and diverting them for a variety of uses, including irrigation, industrial processes, potable water supply, recreation, groundwater recharge and environmental enhancement. Water reuse is classified as 'indirect' when TWW is discharged into the environment for later abstraction, and 'direct' when it is captured immediately downstream of the treatment plant (WWTP) and routed to its end use. The objective in both cases is to address water scarcity by reallocating that which is currently discharged into receiving environments (Thompson et al., 1992), a process that generally requires additional treatment technology to ensure that the TWW meets quality standards.

Around the world, the practice of water reuse has grown significantly since the 1970s. Today, it is widely proposed by international institutions as one of the key adaptation strategies to address the impacts of climate change. In 2024, for example, the International Network of Basin Organizations (INBO) incorporated water reuse into its strategy for implementing Integrated Water Resources Management (IWRM) as 'Reconciling the natural and urban water cycles: basin-scale strategy for wastewater reuse and other non-conventional water resources for water security'.

The degree of global enthusiasm for water reuse is often assessed through national rankings based on the percentage of TWW reused, for example, 90% in Israel, a projected 70% by 2030 in Saudi Arabia, 40% in Singapore and less than 1% in France (which is considered to be 'lagging behind'). Furthermore, success stories are often cited to showcase the potential of water reuse (Lazarova et al., 2013). Notable examples include the treatment and reuse of wastewater in Amman, Jordan, for agricultural irrigation in the Jordan Valley; the water reuse for potato cultivation on Noirmoutier Island, France, that has been in place for over four decades, and the supply of

DOI: 10.4324/9781003745532-3

potable water to the residents of Windhoek, Namibia, sourced from TWW reuse since the 1960s.

The growing interest in water reuse is largely driven by its many promises: securing water availability, saving water and providing a source of nutrients and energy, among others (UN, 2017). These benefits are based on a widely shared consensus; discharging wastewater, treated or otherwise, into natural environments (rivers, seas) is increasingly seen as a waste – a loss of valuable resources. This is where the logic of water reuse and recycling comes in: aiming to make better use of these resources. The view also gains its strength from a link to the widely promoted concept of the circular economy (Nika et al., 2020).

Nevertheless, water reuse remains a contested practice and encounters multiple barriers to its effective implementation. A large part of the literature focuses on issues of trust, awareness, perceptions, cultural barriers or understanding of water quality (Nemeroff et al., 2020). Indeed, many reuse projects for drinking water have been rejected by local populations, as in Toowoomba, Australia, and San Diego, California, due to the social and cultural perceptions of risk (Ormerod & Scott, 2012). Another strand of scholarship addresses institutional issues (Riazi et al., 2023; Rodrigues et al., 2024) but often only to emphasise the need for reforms, awareness raising and adequate regulatory frameworks or economic tools. In California, for example, there is significant support from organised groups, industry associations, lobbying groups, regulatory bodies, and academic institutions to engage in extensive institutional work to boost the legitimacy and acceptance of water reuse (Binz et al., 2016).

Moreover, most often, when the main difficulty is convincing farmers or other potential users or regulators that water reuse is a good idea, there is no space for discussion of whether it is indeed the case. Hence, this chapter explores and underscores often concealed or overlooked points of consideration, after a brief historical overview of the institutional emergence of water reuse.

3.2 Short Review of the Emergence of Water Reuse as a Best Practice

The idea of reusing liquid waste from cities is not new. California, often listed in the literature as an early pioneer, adopted its first regulations regarding the use of TWW in agriculture in 1918 (Olivieri et al., 2020). According to Asano and Levine (1996), the 1960s were an 'era of wastewater reclamation, recycling, and reuse', notably inspired by the Windhoek experience. However, the real turning point came in the 1970s when effluents (re)emerged as an environmental and health concern. This was due to the widespread and uncontrolled application of wastewater to land and the direct discharge of industrial effluents into natural environments, leading to environmental pollution and the spread of epidemics. In addition, urban population growth increased the volume of wastewater requiring management. Improvements in wastewater treatment technology and the revision of regulatory standards in several countries then provided the necessary safeguards to ensure the safe use of TWW, thereby enabling the deployment of what is called today 'water reuse'. Beyond improving the quality of discharged water, water reuse also offered local authorities an opportunity to invest in treatment infrastructure to protect the environment and potentially generate revenue by selling the treated water, thus providing a return on investment. The reallocation of TWW is also appealing to governments looking for solutions to meet the growing demand for water in the agricultural sector (Lazarova et al., 2013).

Around the same time, international organisations started to address water reuse by providing a regulatory framework for its implementation (mainly for agricultural purposes). After an early technical report on the issue in 1973, and in collaboration with the World Bank, the World Health Organization (WHO) published its first guidelines in 1989 (WHO, 1989). A panel of experts established quality and usage standards that are still in place today as a global reference (water-quality classes A, B, and C). The WHO's objective was to ensure the practicality of water reuse by finding a balance between risks to human health and economic feasibility. In parallel, an International Organization for Standardization (ISO) process helped to formalise implementation. The UN's Food and Agriculture Organization (FAO) also promoted water reuse and stressed the importance of using higher-quality water in agriculture (FAO, 1992).

The 2010s saw water reuse adopted more broadly worldwide and expanding to a variety of uses. In 2017 the UN endorsed the promise of water reuse in its annual World Water Development Report (UN, 2017). TWW was presented as an 'untapped resource', while the idea of reuse was integrated into the concept of the 'circular economy' and promoted as a paradigm shift. In this transition, water reuse was seen as necessary and subsequently widely adopted by other institutions, such as the World Bank, which sought

> to encourage a paradigm shift in which the value proposition of wastewater in a circular economy is recognized. [...] In this context, wastewater is and should be considered a valuable resource from which energy and nutrients can be extracted, as well as an additional source of water.
>
> (Rodriguez et al., 2020)

The European Commission also backed water reuse, particularly through new regulations on urban wastewater. The Urban Wastewater Treatment Directive framed water reuse as a strategic response to drought:

> the water resources of the Union are increasingly under pressure, resulting in permanent or temporary water scarcity in some areas. The Union's capacity to respond to these growing pressures on water resources could be enhanced through wider reuse of treated urban wastewater.
>
> (Document 52022PC0541, EC, 26/10/2022)

Academics, private and state actors and regulatory bodies are gradually aligning around the issue of water reuse. The International Water Association (IWA) has been a major driving force in making it an established practice and in 2011 published the inaugural issue of its journal, *Water Reuse*. It also hosts a regular 'International Conference on Water Reclamation and Reuse' with attendees that include public institutions, academics, water treatment companies and engineering consultants. The IWA works to show that there is a long history of reusing wastewater in different societies, thereby aiming to bolster the acceptance of water reuse by suggesting it is only a modernised version of a well-established practice (IWA, 2006). In their 2013 publication *Milestones in Water Reuse: The Best Success Stories*, Lazarova & Asano aimed to 'provide policy makers and regulators with a good understanding of water reuse and help them to consider recycled water as safe and how it can be used based on global experience'. Associated epistemic communities are becoming increasingly consolidated while diversifying, reflecting a growing plurality of knowledge and expertise around water reuse (e.g. WateReuse Association, 'REUSE EUROMED 2024' Conference). Finally, water reuse is creating new opportunities for private business – not just established water treatment companies, but also start-ups with innovative technologies. In 2016, the industry association for the European water reuse sector (Water Reuse Europe) was founded 'to bring together commercial and public organisations from across Europe who operate in the water reuse sector and form a single unifying body through which best practice could be shared and develop a single voice on important issues'. Water reuse appears as a new profit-making avenue.

3.3 Empirical Approach to Water Reuse

In this section, a systematic and nuanced examination of water reuse in practice reveals that realising its projected benefits is far from straightforward. Indeed, it involves numerous decisions around water reallocation, economic models and risk management (Cirelli, 2012; Williams et al., 2023; March et al., 2023; Mayaux et al., 2025). As emphasised by Bahers and Rutherford (2024), 'The emergence of circularity on the agenda of urban policies is currently driven by a largely technocentric and management focus, without much reflection on [...] socio-political issues'.

3.3.1 A Safe Alternative Water Source?

Much effort goes into 'cleansing' the image of TWW given its 'dirty' origin. Deploying the positive vocabulary of recycling is one tactic (Wessels, 2023), while regulatory standards and advances in wastewater treatment technology are used to define an acceptable risk of coexisting with these waters.

Yet, norms and standards cannot guarantee against all risks: regulatory frameworks can only cover what we know about, and we only monitor what we are able to measure. Thus, devising a framework is faced with scientific uncertainty as well as limitations in metrology and resources. Choices must be made in terms of risks to be monitored, indicators and threshold values. This inevitably generates grey areas. For example, the WHO guidelines include microbiological risk to human health but not chemical risk. There are various reasons for this: the absence of an immediate, acute or measurable reaction following exposure, unlike with the contamination of the human body by bacteria or parasites; the relatively small contribution TWW is perceived to make to the population's overall chemical exposure, compared to intensive agriculture; a lack of data and research, and therefore of consensus on thresholds; and the difficulty of measuring concentrations that are often very low.

Moreover, the inherent materiality of TWW constantly fluctuates, evolving over time on its journey from WWTP to the point of use. Molecules that were degraded and rendered harmless at one stage can later recombine or transform into compounds that are once again pathogenic to humans or the environment. This 'vitality' (Bennett, 2010) of TWW cannot be captured by existing regulatory frameworks. For instance, European regulation does not require the systematic monitoring of the risk associated with the accumulation or environmental persistence of residual compounds, their potential to recombine or the transfer of resistance genes into the environment (Helmecke et al., 2020). Where water reuse does pose such risks, there is currently no mechanism in place to reverse the process: once TWW is introduced it becomes extremely difficult to prevent its use or contain potential contamination.

The WHO (1989) recommended a new 'multibarrier approach' in light of the perceived stringency and cost of achieving the early quality standards. This approach allowed for the implementation of successive improvement steps, such as UV treatment, temporary storage, drip irrigation systems and vegetable washing, to gradually achieve the standards along the entire reuse chain from WWTP to the point of use. While this risk-management strategy makes it possible to envisage TWW reuse even where the primary treatment is insufficient, it also shifts responsibility to the chain of stakeholders, including those who may lack the technical or institutional capacity to assume such obligations. This technocratic approach also assumes (1) the possibility of a monitoring and enforcement capacity down to the individual plot, and (2) that the effect of each 'barrier' can be estimated (and translated into an 'improvement coefficient'). Drip irrigation, for example, belies this assumption: a barrier in that it avoids direct human and crop contact with water, it may also contaminate water when biofilms develop in the drippers.

Regarding the level of technical guarantee, any treatment process inevitably lets undesirable elements through (micropollutants, heavy metals, pharmaceutical products) that are potentially dangerous depending on their concentration or ability to recompose (Mordechay et al., 2021). However, beyond the limitations inherent in treatment technologies, water quality also, if not primarily, depends on the resources available to system operators: financial resources to cover infrastructure operation and maintenance costs, and human resources for monitoring and regulatory compliance. Such capacities are rarely found in small cities of the Global South, in contrast with cases such as Los Angeles's plan to recycle 100% of its wastewater by 2035 for groundwater recharge and reuse, after 'several layers of treatment, including membrane filtration, reverse osmosis, and ultraviolet advanced oxidation' (Mika et al., 2018).

In other words, despite regulatory standards and technical knowledge, the practice of water reuse often means bringing into circulation water that can never be fully controlled.

3.3.2 Treated Wastewater as an Agronomic Resource – and Salt as an Undesirable Component

A key benefit that is expected from applying the circular-economy principle to water reuse is being able to recover the nutrients contained in wastewater. The use of TWW in agriculture is often said to be a smart and sustainable way to meet crop nutrient demand while reducing the reliance on expensive chemical fertilisers. To support this agronomic interest, reference is often made to the long-standing history of wastewater spreading, such as on 'sewage farms', neglecting historical debates about the public health and environmental risks associated with the practice (as seen near Paris, Tucson, Berlin and in Iraq) (Uekötter, 2016). Moreover, recovering desirable nutrients in adequate concentrations is far from straightforward, since meeting regulatory water-quality standards often requires treatment processes that remove precisely those nutrients. Indeed, experiments to assess the agronomic potential of water reuse have not proven particularly promising, and further research is needed to balance nutrient inputs while promoting water-efficient practices.

Further, it is important to note that not all nutrients are beneficial: while nitrogen and phosphorus are generally favourable for plant growth, salts are not. With regular use in high concentrations, they can degrade soil fertility, and the percolation of saline water can lead to groundwater contamination. The issue has already been acknowledged in several leading water reuse countries, including Saudi Arabia, Israel and Spain, and is now emerging in experimental reuse sites in France (such as the Irrialt'eau site in Aude). In Israel the desalination of TWW has even been considered a mitigation strategy (Brenner, 2011).

Finally, TWW typically retains high levels of suspended solids and organic matter that can eventually clog precision irrigation systems – no small technical challenge for widespread use in agriculture (Moulia et al., 2024).

3.3.3 Using Treated Wastewater to Save Water: The Risk of Driving Up Consumption

Arguments in favour of water reuse as a water-saving strategy are often based on the idea of substituting freshwater withdrawn from the natural ecosystem with treated urban wastewater for uses where potable quality water is not required (Hernández-Chover et al., 2023). This form of 'virtuous substitution' is presented as a way of making more water available to the aquatic environment.

To be effective in terms of water conservation, the approach must involve structural or regulatory measures, such as revising abstraction permits and monitoring withdrawals from natural sources. The need to ensure the economic viability of water reuse conflicts with the idea of substitution because it often requires the identification of additional 'benefits', that is, creating new uses for the reclaimed water in agriculture or recreation. This is the case, for example, with golf courses in Marrakech (Morocco) and fruit tree growing in France, as well as securing intensive water-demanding farming systems vulnerable to drought, such as in Andalusia (Spain) or Israel. In such situations, water reuse resembles an end-of-pipe solution that tends to reinforce or expand existing consumption models rather than representing a paradigm shift.

3.3.4 Who Will Benefit from Treated Wastewater?

Currently, TWW is typically discharged into the natural environment, and return flows from WWTPs are often vital to maintaining minimum river flows and supporting the surrounding ecosystem. This is especially evident in urbanised areas in Southern Europe or California, where treated effluents contribute significantly to streamflow in summer. Although this water already has an environmental function, in many cases it is neglected or contested when deciding on reallocation. Water reuse can therefore damage or even jeopardise aquatic ecosystems. In coastal areas, too, the argument for prioritising water reuse often rests on the idea of capturing water before it is 'lost to the sea'. However, this logic fails to account for the impact on marine biodiversity of intensified reuse practices – something that remains poorly understood and warrants further study (Garcia et al., 2024).

The principle of direct water reuse, where TWW discharge is in part diverted away from the receiving environment, is likely to raise questions around (re) allocation. In regions with scant water resources, such as the Middle East and North Africa, or countries such as Pakistan and India, most untreated or poorly TWW is *already* used by farmers downstream. Indeed, treating and using wastewater *does not add* new water to the system but rather represents a reallocation of the resource, unless the same farmers benefit from it, which is often not the case: in Dar es Salaam (Tanzania), for instance, urban farmers were excluded from a reuse scheme since the authorities sought to remove them from the area. In Rajkot (India) and San Luis Potosí (Mexico), long-standing water rights to urban wastewater held by peri-urban farming communities were not even officially recognised.

The narrative of the circular economy promotes water reuse as a means to restore the metabolic rupture between urban areas where wastewater is produced and rural areas where it can be reused. However, the principle proves difficult to operationalise when demand is geographically distant from the WWTP due to the prohibitive cost of conveyance infrastructure. In practice, the choice of beneficiary is largely driven and limited by the uses/crops available (e.g. golf courses, urban parks, fodder) and/or economic criteria reflecting geographical limitations (proximity to the WWTP, plot configuration vis-à-vis the design and cost of the pressurised network) and/or users' financial capacity or willingness to pay for TWW.

Even when potential users have been identified according to such technical and economic criteria, they can reject TWW due to concerns over water quality (including salinity), cost, the 'yuck factor' (where the water is intended for human consumption) or opposition to the model of intensive agriculture that water reuse may be seen to support, as in Tiznit (Morocco). Farmers located near a WWTP may be given no choice but to use the TWW, particularly where the scheme has strong political backing, such as in the Jordan Valley (Jordan) or the peri-urban area of Sousse (Tunisia).

Moreover, reallocating TWW based on techno-economic criteria often ignores the collective and organisational dynamics of the area, potentially generating inequity, marginalisation, or environmental injustice.

3.3.5 Closing the Loop for Secure Access to Water

Water reuse 'loops' are praised for decoupling supply from natural variability, offering protection from climatic fluctuations and thereby helping to secure access to water, but this principle relies on a fragile assumption: that the volume of water entering the WWTP is stable, or even increasing, due to factors such as tourism or urban growth. Yet, the water supply of major cities around the world has become increasingly uncertain, as evidenced by recent crises in Cape Town, São Paulo, and Mexico City. In addition, the volume of TWW available for reuse is often overestimated (Expósito et al., 2024).

The idea that the production and use of TWW would be stabilised and guaranteed by urban water consumption also misses a key point: water reuse will be secured not only as long as it can be produced but also as long as its quality is compatible with actual uses. In case of (commonplace) technical shortcomings, lack of spare parts, flooding or inflow of toxic water into the plant, water quality can be impaired and reuse discontinued. Agricultural uses, such as fruit tree irrigation, might be at risk when facing an interruption in supply, and areas with limited technical or financial capacity struggle to achieve the promised security.

3.3.6 An Expensive Solution: Who Pays and How?

Water reuse faces a fundamental financial challenge: it is expensive in terms of investment costs and operation and maintenance, and it is seldom introduced or sustained without substantial public subsidies. The WWTP is capital-intensive in itself, but so are the pressurised networks often necessary to connect the plant with the intended users, who are not necessarily located in its vicinity.

Where the financial balance of the sanitation sector is already precarious, water reuse can exacerbate existing vulnerabilities. Many regions are already struggling to maintain an adequate level of sanitation services. This is particularly true where the WWTP's maintenance costs were neglected during project planning, as observed in Mexico (Cáñez-Cota & Rentería, 2023) or where public spending has been curtailed, such as in Tunisia during the 1980s. In Lebanon, half of the country's mostly small-scale WWTPs are non-operational due to a lack of municipal funds and technical expertise. In Israel, the responsibility for maintenance originally fell to local authorities, who neglected infrastructure upkeep, leading to a decline in water quality. This prompted the Ministry of Finance to intervene by establishing 'sewage corporations within local and regional councils, an intermediate step toward the privatization of sewage services' (Teschner & Negev, 2013). The need for current WWTP maintenance is often underestimated, as is the need for continuous water quality monitoring, including regular testing of regulatory parameters. These obligations pose significant technical and financial challenges, especially for small- and medium-sized municipalities, such as those in Tunisia (Mayaux & Ennabih, 2023).

The price of TWW rarely reflects its real cost, leading to budgetary imbalance. This is true for the investment in the WWTPs and distribution infrastructure which thus rely heavily on public subsidies (for example from the EU or regional authorities, in France), but also for Operation & Maintenance costs. The TWW is often sold below its actual cost in order to encourage uptake, to facilitate the quick launch of a politically backed project or to make it attractive compared with other water sources. Wherever groundwater in particular is available, the price of TWW will have to be lower than the cost of its extraction. In some instances, the shortfall between price and cost is absorbed by all consumers of potable water – with limited transparency. In the Barcelona metropolitan region (Spain), reclaimed water is currently provided free of charge, with all associated costs shouldered by the Catalan public water agency.

3.4 Conclusion

While the literature on water reuse focuses mainly on technical issues, questions of acceptability or the need to provide 'robust institutional arrangements and legal measures' (Riazi et al., 2023), it overlooks key problems. Contrary to widespread belief, handling TWW is not straightforward. On the one hand, guaranteeing its quality and safety means controlling a materiality that is in constant flux – in soil, pipes and irrigation material, and over time. Despite advances in treatment technologies and regulatory standards, such safeguards remain inherently limited by the unstable and ambiguous nature of the substance itself. Moreover, the institutions tasked with monitoring water quality are frequently underfunded and understaffed, undermining their ability to carry out long-term oversight reliably. In addition to technical hurdles, water reuse involves politically charged decisions about water reallocation, determining who benefits and who is excluded, often through processes that lack transparency. In many cases, treated water is not a new resource; it is 'simply' reallocated from earlier uses. Despite being considered a secure resource, the supply of TWW remains dependent on the availability of water for urban use and on the technical and financial capacity to monitor and ensure adequate water quality.

Given current pressures on water resources, there is a clear and growing interest in rethinking the role of TWW. However, as with any so-called 'best practice', it is essential to examine how water reuse is implemented: its objectives, the intended beneficiaries and the sociopolitical and material conditions under which it is undertaken. We call for a critical case-by-case examination of the benefits of water reuse, rather than taking them for granted.

Further Readings

Binz, C., Harris-Lovett, S., Kiparsky, M., Sedlak, D. L., & Truffer, B. (2016). The thorny road to technology legitimation—Institutional work for potable water reuse in California. *Technological Forecasting and Social Change, 103*, 249–263.

Mayaux, P.-L., Montginoul, M., & Ait Mouheb, N. (2025). Recycling water, but at what price? The political economy of wastewater reuse. *International Journal of Water Resources Development*, 1–20.

Uekötter, F. (2016). City meets country: Recycling ideas and realities on German sewage farms. *Journal for the History of Environment and Society, 1*(1), 89–107.

Williams, J., Beveridge, R., & Mayaux, P.-L. (2023). Unconventional waters: A critical understanding of desalination and wastewater reuse. *Water Alternatives, 16*(2), 429–443.

Selected References

Asano, T., & Levine, A. D. (1996). Wastewater reclamation, recycling and reuse: Past, present, and future. *Water Science and Technology, 33*(10–11), 1–14.

Bahers, J.-B., & Rutherford, J. (2024). Urban infrastructures, metabolic resource flows and the contradictions of circular economy 'solutions' in Nantes and Gothenburg. *Urban Studies, 62*(9), 1897–1918.

Bennett, J. (2010). *Vibrant matter: A political ecology of things*. Duke University Press.

Brenner, A. (2011). Limitations and challenges of wastewater reuse in Israel. In F. F. Quercia & D. Vidojevic (Eds.), *Clean soil and safe water* (pp. 3–9). Springer Netherlands.

Cáñez-Cota, A., & Rentería, C. (2023). Why are municipal wastewater treatment plants abandoned in Mexico? When a more money policy approach is not enough. *Water Resources and Economics, 43*, 100226.

Cirelli, C. (2012). At the end of the pipe: Change and continuity in sanitation systems in Mexico. *Flux, 87*(1), 39–50.

Expósito, A., Lorenzo Lopez, A. M., & Berbel, J. (2024). How much does reclaimed wastewater cost? A comprehensive analysis for irrigation uses in the European Mediterranean context. *Water Reuse, 14*(3), 434–447.

FAO. (1992). *Wastewater treatment and use in agriculture—FAO irrigation and drainage paper 47*. FAO.

Garcia, T., Carlotti, F., Lepoint, G., Guilloux, L., Tesán-Onrubia, J. A., Grassi, B., Russias, V., & Bănaru, D. (2024). The decadal variability of elemental, isotopic, and biochemical compositions of coastal Mediterranean zooplankton responds to environmental changes. *Marine Biology, 171*(5), 1–18.

Helmecke, M., Fries, E., & Schulte, C. (2020). Regulating water reuse for agricultural irrigation: Risks related to organic micro-contaminants. *Environmental Sciences Europe, 32*(1), 4.

Hernández-Chover, V., Castellet-Viciano, L., Fuentes, R., & Hernández-Sancho, F. (2023). Circular economy and efficiency to ensure the sustainability in the wastewater treatment plants. *Journal of Cleaner Production, 384*, 135563.

IWA. (2006). *1st IWA international symposium on water and wastewater technologies in ancient civilizations*. www.itia.ntua.gr/en/getfile/1224/1/documents/2006wwtac_1.pdf

Lazarova, V., Asano, T., Bahri, A., & Anderson, J. (2013). *Milestones in water reuse*. IWA Publishing.

March, H., Gorostiza, S., & Saurí, D. (2023). Redrawing the hydrosocial cycle through treated wastewater reuse in the Metropolitan Area of Barcelona. *Water Alternatives, 16*(2), 463–479.

Mayaux, P.-L., & Ennabih, A. (2023). Troubled waters: The fraught political economy of wastewater reuse in Morocco and Tunisia. *Water Alternatives, 16*(2), 541–562.

Mika, K., Gallo, E., Porse, E., Hogue, T., Pincetl, S., & Gold, M. (2018). LA sustainable water project: Los Angeles city-wide overview. UCLA: Sustainable LA Grand Challenge. Retrieved from https://escholarship.org/uc/item/4tp3x8g4

Mordechay, E. B., Mordehay, V., Tarchitzky, J., & Chefetz, B. (2021). Pharmaceuticals in edible crops irrigated with reclaimed wastewater: Evidence from a large survey in Israel. *Journal of Hazardous Materials, 416*, 126184.

Moulia, V., Heran, M., Lesage, G., Hamelin, J., Pinta, J., Gazon, A., …, & Ait-Mouheb, N. (2024). Biofilm growth dynamics in a micro-irrigation with reclaimed wastewater in the field scale. *Journal of Environmental Management, 370*, 122976.

Nemeroff, C., Rozin, P., Haddad, B., & Slovic, P. (2020). Psychological barriers to urban recycled water acceptance: A review of relevant principles in decision psychology. *International Journal of Water Resources Development, 36*(6), 956–971.

Nika, C. E., Vasilaki, V., Expósito, A., & Katsou, E. (2020). Water cycle and circular economy: Developing a circularity assessment framework for complex water systems. *Water Research, 187*, 116423.

Olivieri, A. W., Pecson, B., Crook, J., & Hultquist, R. (2020). California water reuse—Past, present and future perspectives. In P. Verlicchi (Ed.), *Advances in chemical pollution, environmental management and protection* (Vol. 5, pp. 65–111). Elsevier.

Ormerod, K. J., & Scott, C. A. (2012). Drinking wastewater: Public trust in potable reuse. *Science, Technology, & Human Values, 38*(3), 351–373.

R3VOLUTION project. (2025). We revolutionize industrial water management! https://r3volution.eu/

Riazi, F., Fidélis, T., Matos, M. V., Sousa, M. C., Teles, F., & Roebeling, P. (2023). Institutional arrangements for water reuse: Assessing challenges for the transition to water circularity. *Water Policy, 25*(3), 218–236.

Rodrigues, P. M., Pinto, F. S., & Marques, R. C. (2024). A framework for enabling conditions for wastewater reuse. *Sustainable Production and Consumption, 46*, 355–366.

Rodriguez, D. J., Serrano, H. A., Delgado, A., Nolasco, D., & Saltiel, G. (2020). *From waste to resource: Shifting paradigms for smarter wastewater interventions in Latin America and the Caribbean*. World Bank.

Teschner, N. A., & Negev, M. (2013). The development of water infrastructures in Israel: Past, present and future. In S. B. Medgal, R. G. Varady, & S. Eden (Eds.), *Shared borders, shared waters: Israeli-Palestinian and Colorado river basin water challenges* (pp. 7–19). UNESCO-IHE.

Thompson, K., Cooper, R. C., Olivieri, A. W., Eisenberg, D., Pettegrew, L. A., & Danielson, R. E. (1992). City of San Diego potable reuse of reclaimed water: Final results. *Desalination, 88*(1–3), 201–214.

UN World Water Development Report. (2017). *Wastewater: The untapped resource*.

Wessels, M. T. (2023). What's in a name? Politicising wastewater reuse in irrigated agriculture. *Water Alternatives, 16*(2), 563–580.

WHO. (1989). *Health guidelines for the use of wastewater in agriculture and aquaculture: Report of a WHO Scientific Group* (Technical Report Series No. 778). WHO.

4 *REVISITING*: Desalination Is the New and Inexhaustible Water Source

Joe Williams

In May 2024, delegates gathered in Bali, Indonesia, for the 10th World Water Forum to discuss the world's ongoing water crisis – and specifically, because this was a World Water Forum event, how private corporations might make money by providing 'solutions' to the water crisis. At the beginning of the conference, the room was addressed in an unscripted and somewhat rambling speech by billionaire and techno-triumphalist Elon Musk (2024), who made a number of bold assertions:

> We are 70% water, 30% earth by surface area. So what that means is that the potential for solving any given water issue is extremely good because there is so much water [...]. Desalination has become very inexpensive, so the availability of freshwater is simply about energy and transport. When I talk to well-read and smart people [...] they will often say that the water crisis is insolvable, but in fact it is very solvable. I think we've got a great water future ahead of us.

Although desalination accounts for a relatively small proportion of potable water treatment globally, the sector has grown rapidly in recent years, and there is every indication this trend will continue (Williams, 2022). Most people agree that in the coming decades, desalination will be an increasingly important source of water in an increasingly diverse range of places and will shape societies' relationships with water in important ways (Borgomeo, 2024).

The excerpt above illustrates a popular narrative: that it is possible to solve problems of water stress and scarcity by converting seawater (and other salty waters, like certain kinds of groundwater) to freshwater through the process of desalination. Musk makes grand claims about the potential of desalination technology to create a virtually unlimited water supply. He went on to make equally grand claims about the potential of solar energy to power the desalination process, such that the availability of plenteous energy from the sun translates to abundant freshwater from the sea. Such ideas are promoted by what we might understand as a *global desalination assemblage*. This includes (sometimes loosely) connected actors in governments that have close ties

DOI: 10.4324/9781003745532-4

with the desalination industry (particularly the United States, Spain, Israel, Singapore, China, Japan, and countries in the Arabian Gulf), global desalination corporations (such as Veolia, Acciona, Metito, and Coxabengoa, to name some of the largest), and industry organisations (including Global Water Intelligence and the International Desalination Association) and scientific and research networks (see the journal *Desalination*, for example). Desalination is thus positioned as *the* route out of the global water crisis.

This is a pervasive and growing narrative, but there is a continuum here. Not all proponents agree that desalination is a global silver bullet (Brannstrom et al., 2022). For others, desalination can be part of a resilient portfolio of water sources or can be used to address water challenges in specific contexts. Nevertheless, desalination is (and has been recurrently over the last 70 years) associated with a great deal of optimism about its capacity to create new water resources and solve water challenges. While the use of desalination is certainly growing very rapidly, in this chapter we push back against claims that it represents a 'solution' to water crisis. There is nothing inherently good or bad in a technology. What matters is the social context in which a technology is used (and, importantly, *not used*): how it is being developed and by whom; whose interests it serves; and how access to, or use of, that technology is mediated. This, of course, is true of water desalting technologies. The chapter begins by sketching out the social, political, and technological history of desalination, before challenging (or at least contextualising) some of the key claims that surround its contemporary use.

4.1 A Brief History of the Desalination 'Solution'

Narratives surrounding desalination as a solution to the world's water problems emerged in the mid-twentieth century. Before this time, desalination had only been used on a small-scale, mostly on ships and in contexts where there were few water source options. Desalination on a municipal scale was extremely rare, due to the expense, energy intensity, and technological complexity of the process. By the 1950s and 1960s, however, distillation technologies[1] had developed enough to make desalination on a larger scale possible. Since then, we have seen, broadly speaking, three waves of development.

4.1.1 Wave 1: 1950s–1970s

The mid-twentieth century saw the first serious wave of techno-triumphalism about desalination. In 1955, the United States government set up the Office of Saline Water. Conceived via the Saline Water Conversion Act of 1952, the Office's remit was to facilitate research and development and promote the use of desalination around the world. It went on to lead the development of the desalting industry until the mid-1970s. During this time, the federal government ran the Saline Water Conversion Program, the purpose of which was to develop desalination capacity in the arid US Western regions, particularly California, although test locations using a variety of technologies were located throughout the country. Throughout the 1960s and '70s, US desalination programmes became increasingly international in focus, particularly bound up with the geopolitics of the Cold War (Low, 2020). There was even some cooperation between the USA and USSR on desalination R&D sharing, such was the strategic interest in this technology. In 1967, the United States hosted a major diplomatic event in Washington, DC: the International Conference on Water for Peace. Desalination featured heavily in the proceedings, and many of the papers delivered expounded on the potential of desalination to bring about an age of abundance and prosperity across the First, Second and Third Worlds.

'Few technical developments in history', argued one delegate in the opening remarks, 'have attracted the attention and raised the hopes of so many as the promise of desalting the sea and brackish water' (Ramey, 1968, p. i).

Throughout this time, the utopianism associated with desalination was closely tied to notions of modernity, technological progress, and the domination of nature (O'Neill & Boyer, 2023). The promise of desalination, in particular, was bound up with the promise of cheap energy from nuclear power. However, actual installed desalination remained low, reaching only 2 million m^3/day by 1975 (IDRA, 2024).

4.1.2 Wave 2: 1970s–2000s

This second wave saw the decline of the promise of desalination in North America and southern Europe – mostly due to energy and cost – and its rise in the Middle East, particularly the Arabian Peninsula. Much of this, again, was bound up with the politics of the Cold War, as the United States transferred technology and research expertise to its allies in exchange for political influence in the region (Koch, 2023). During the 1970s and 80s, the first generation of large-scale desalting plants were developed in the Middle East, which has remained the centre of the industry to this day. These facilities typically used distillation (i.e., heat) methods, taking advantage of cheap fossil fuels, and were often twinned with thermoelectric power plants, which co-produced water and electricity. As well as enabling the urban development of the region, desalination also became a central technology for the exercise of political power and state-building in the region (Al-Aghbari, 2021). Desalination, fossil fuels, urban development, and the exercise of authoritarian political power became inseparably entangled (McDonnell, 2014).

In the 1990s and early 2000s, Israel also embarked on an ambitious desalination programme through the state-owned company Israel Desalination Enterprises (IDE). This was partly an attempt by Israel to overcome water constraints in the region and circumvent the contested politics of water resources shared with Palestine and Jordan, and partly to establish itself as a major global player in the water technology industry. At a similar time but for very different geopolitical reasons, Singapore undertook a comparable techno-political-economic pivot, where desalination was positioned both as a way to reduce dependence on transboundary water resources and as a way of promoting the city state as a global leader in water technology R&D. This period also saw the growing use of desalination in industry and for specific manufacturing applications, particularly in China. Globally installed capacity rose sharply, particularly from the late 1990s, to over 35 million m^3/day by 2005 (IDRA, 2024).

4.1.3 Wave 3: 2000s–Present

This third wave of techno-optimism about desalination has been defined by several characteristics. While the first wave was closely associated with ideas of modernity and state-led development, desalination has now become more driven by private water companies and is connected to the neoliberal movement: to the commercialisation of water service delivery, the financialisation of water revenue, and the delivery of infrastructure through public–private partnerships (Swyngedouw, 2013; Loftus & March, 2016; Fragkou & Budds, 2020; Greer et al., 2021). Reverse osmosis membrane treatment has taken over distillation as the dominant technology. This, along with other innovations, has dramatically reduced the energy requirements and cost of desalination, making it more viable in regions outside the energy-rich Middle East. Although the Arabian Peninsula still dominates the sector, large-scale desalination projects have been developed in southern Europe (particularly Spain), the US West, Australia, North Africa, and Chile. The use of desalination in industry has continued to grow in China and elsewhere, & small- and medium-scale desalination has become increasingly used to support the tourist industry in places like the Caribbean, Indonesia, and the Mediterranean islands. The number of small and medium plants has proliferated rapidly, although the majority of global installed capacity comes from increasingly large mega-infrastructure projects. There is also a great deal of discussion around the potential of renewable energy to power new desalination programmes (as discussed below), although in reality, the sector is still dominated by fossil fuel power. Finally, in the last five years or so, we have increasingly seen the use of desalination to address water challenges in the Global South (for example, in South Africa, Ghana, Kenya, India, and Pakistan).

Global installed desalination capacity continued to rise during this period, to around 100 million m^3/day in 2024 (IDRA, 2024). According to the DesalData database there are now approximately 23,000 desalination plants worldwide, ranging from small-scale (that produce less than 1,000 m^3/day) to mega-plants (with capacities over 1 million m^3/day).

4.2 Problematising the Desalination 'Solution'

There is now a well-established body of research that pushes back against the idea that desalination is an unproblematic solution to water scarcity (see Williams & Swyngedouw, 2018 and Williams, 2022 for review), which is in turn part of a much longer tradition of problematising ideas of the 'technological fix' to complex social and environmental challenges. Nevertheless, having worked in this area for over a decade, I regularly encounter versions of the argument that 'the ocean is the biggest reservoir in the world, we just need to tap it'. In the remainder of this chapter, I consider some of the common claims made about desalination and explain why the picture is not as simple as many would believe.

4.2.1 Proposition One: The Oceans Are Inexhaustible, So the Potential for Desalination Is Unlimited

An astonishing number of the technical publications on desalination, as well as media reports and popular science on the topic, begin with some claim around the fact that 70% of the planet's water is salty and that desalination can reverse the hydrological cycle to transform these waters into a resource. However, the scope of desalination to supply new water is still limited by a number of complex factors. Energy and cost are discussed below. One of the other limiting factors is that desalination has been hotly contested in a number of places. For seawater desalination specifically, coastal land near areas with high demand for water is often very valuable, subject to different and competing uses (e.g., real estate) and is often ecologically sensitive. Constructing large, noisy, and ecologically destructive industrial infrastructure in such areas is, therefore, often fraught with challenges and impediments. While the oceans may be practically inexhaustible, coastal land is not.

This has been demonstrated clearly in southern California. Over the last 25 years there have been various attempts to develop municipal-scale desalination. These have all received sustained challenges by civil society organisations – some of them 'NIMBYist', but many of them citing issues around environmental sustainability, environmental justice, governance issues, and concern about water privatisation (see O'Neill, 2023; Patel, 2023). The high-profile rejection by California state authorities in 2022 of an application to construct a large plant at Huntington Beach – a project that had been in the permitting process for two decades – offers a cautionary tale of the kinds of politics that can limit the rollout of desalination.

Overall, while the desalting industry is certainly growing very rapidly and will continue to do so for the foreseeable future, its potential is far from unlimited.

4.2.2 Proposition Two: Desalination Is Needed to Fill the Gap Between Demand for Water and Dwindling Conventional Sources

The notion of a 'water gap' between demand and supply is often mobilised to demonstrate the need for unconventional water sources like desalination. Such claims tend to see water scarcity as an externally-imposed condition and ignore how conditions of scarcity are created by societies' water use and management practices. Indeed, in cases where large-scale desalination has been implemented, the evidence is that its effects are not simply to alleviate water stress but to create new water demand. Katz (2016), for example, shows how Israel's large desalination programme in the 2000s created perceptions of water abundance among the country's population, which in turn caused consumption and demand for water to increase. Moreover, large programmes to develop desalination tend to benefit particular groups, sectors, and water users, and so they can establish new water use regimes. For example, in Morocco, state agencies are leading the development of a series of mega-projects across the country, which are justified through reference to an ongoing multi-year drought and climate change projections. The first of these to be completed is in Agadir, which supplies water to both urban and agricultural users. Here, public money subsidises irrigation for large agribusiness, producing high-value crops that are mostly for export.

Overall, we should be critical of claims that desalination will fill gaps between demand and supply because of the ways in which new supply drives up demand.

4.2.3 Proposition Three: Desalination Used To Be Expensive, But It Keeps Getting Cheaper

There is a certain amount of truth in this claim, but it is often over-stated or exaggerated. Although the cost has been reduced in recent years to as little as US$0.30/m^3 (IDRA, 2024), this is very much the lower end on the scale of what is possible, and costs are typically more in the region of US$1–3/m^3. Cost depends on the technology used, plant size, the physical characteristics of the location, energy price, construction and labour costs, land value, regulation, and the amount of profit extracted. This means that in most places, desalination remains among the most expensive water source options available.

Returning to southern California to illustrate: The largest desalination plant currently operating in the state is at Carlsbad in San Diego County. A significant part of the financial justification for permitting this privately owned and operated plant in the early 2010s was a forecast that a combination of climate change and increased competition for the county's main water sources in northern California and the Colorado River would drive the cost of imported water higher than the price of desalted seawater at some stage in the plant's 30-year lifespan. A decade after the completion of the plant, however, this crossover point is nowhere near in sight. In fact, the situation has reversed, with the rising cost of energy resulting in an increase in the price of water desalted at Carlsbad from US$1.95/m³ in 2020 to US$2.76/m³ in 2024, while the price of (untreated) imported water to the county water authority has remained relatively low, at less than US$1.20/m³.

This is only an illustration, of course, and a lot depends on specific circumstances – indeed, in some places desalted water can be cost-competitive – but it demonstrates that we should be sceptical of overly-optimistic claims about the cost of desalination.

4.2.4 Proposition Four: Renewable Energy Solves a Lot of the Previous Problems with Desalination

One of the main constraining factors facing desalination has always been the amount of energy required for the process of separating dissolved salt from water. Granted, new technologies and efficiency measures have significantly reduced the energy intensity of desalination, particularly since the early 2000s, but energy requirements are still very high. The RO process requires between 2.5 and 4 kWh/m³ (Zarzo & Prats, 2018), compared to approximately 0.3 kWh/m³ for the treatment of conventional sources.[2] This is why techno-optimism about desalination is often accompanied by techno-triumphalism about energy supply.

A lot of this water-energy-abundance utopianism has coalesced around the potential of renewable energy, particularly solar, to power desalination. Scientific publications on the topic have burgeoned, in part because of the intuitive appeal in the idea that it 'makes use of the two most abundant resources on the planet – salt water and solar energy – to create fresh water' (Belmehdi et al., 2023, p. 1). While some of the optimism surrounding solar desalination is merited, the picture is not as simple as much of the narrative would suggest. Renewable energy doesn't solve the problems of desalination but simply transfers some of the issues in the water sector onto the energy sector. Just as nuclear-powered desalination in the 1960s embodied the contradictions of nuclear energy and fossil fuel-powered desalination in the 2000s embodied the contradictions of fossil fuels, so too does solar-powered desalination embody the issues and challenges of the green energy transition. These contradictions are well-documented and include the materiality of new energy infrastructures (e.g., lithium and other mineral extraction and associated social and environmental impacts), issues of environmental justice and the (re)creation of winners and losers associated with new energy regimes, and issues of how renewable energy infrastructures are financed. With solar power more specifically, land also becomes a significant political challenge. For example, Roth and Tal (2022) estimate that the solar panels required to power Israel's desalination plants would cover a land area more than half the size of Tel Aviv – and this in a region where land is already exceptionally politicised.

Overall, renewable energy offers the desalination industry lots of opportunity for innovation and growth, but it is far from unproblematic.

4.2.5 Proposition Five: Desalination Is Getting More Environmentally Friendly

Related to the above, another common claim is that, while old desalination plants from the 1970s and 80s were ecologically destructive, new technologies make the process environmentally sustainable. Similarly, while there is some truth in this claim, the impacts of desalination remain considerable (Elsaid et al., 2020). Of these, the issue of the brine waste product of desalination has received the most debate and attention (see Panagopoulos et al., 2019). All desalination processes produce highly saline brine, which must in some way be released back into the environment. For seawater desalination, the ratio is usually 50:50 freshwater to brine, and for brackish desalination the conversion rate is higher. The problem of brine disposal can be relatively easily addressed for coastal desalination, especially where there are strong currents to disperse the discharge. But for inland desalination, this creates significant issues and results either in environmentally destructive practices or expensive treatment technologies. Additionally, the desalination process requires the use of a number of chemicals, including various chlorines, anti-scalants, and corrosion inhibitors, that must also be discharged or dealt with.

Overall, the environmental impacts of desalination depend a lot on location, technology, and feedwater, but are never benign.

4.2.6 Proposition Six: Desalination Can Provide Safe Water to the 1 Billion People Who Currently Lack Access in the Global South

It is fairly common to hear claims that desalination can supply water where conventional sources have not, or that by increasing the overall supply of water, the most vulnerable in societies can be serviced. In some cases, new technologies do indeed offer opportunities to supply more vulnerable groups. In the context of the failure of centralised and 'modern' models of water delivery to provide universal access across much of the Global South, the emergence of small-to-medium-scale decentralised desalination units offers an intriguing alternative in certain situations (see Tarazona-Romero et al., 2022). In East Africa, for example, organisations like GivePower and Water Kiosk Africa are installing small-scale, solar-powered desalination systems in under-serviced areas. Both of these organisations, however, rely on donors and aid to keep the costs of water low enough for the communities, and even so, the water produced is generally too expensive for any use other than drinking, and even that is too costly for people in the poorest areas. In addition, many under-serviced areas are unsuitable for this model, and it remains unclear what the scope of this kind of technology really is.

In general, many of the claims made about the potential of desalination to reach un-serviced communities are naïve because they fail to consider the underlying structural causes of water inequality. The high costs of desalination, combined with the growing role of water corporations and public–private partnerships in the delivery of desalination infrastructures, means that desalination is usually only available to regions that can either afford the upfront costs and operational expenditure and/or have an attractive risk profile for investors. Considerable emphasis is placed on the role of the private sector and delivering projects through public–private partnerships, both within the industry and by governments and utilities in many of the countries where desalination is being deployed. An important impact of the emergence of desalination has been to accelerate ongoing – albeit highly uneven – trends towards the privatisation of water supply, or at least the growing influence of private actors in water resource management. This raises important questions around the governance of water (by and for whom?), equity, and affordability. The result is that desalination can accelerate uneven development, where profitable areas are supplied with high-quality water and unprofitable areas and communities are left un-serviced. Indeed, there is a growing body of research showing how desalination, depending on how it is deployed, can exacerbate social inequalities (e.g., Velásquez & Wachtendorf, 2023).

Overall, new and emerging desalination technologies can play a role in supplying safe, potable water to people who would otherwise not have access, but this entirely depends on context and how such technology is rolled out. Conversely, in some contexts, the use of desalination is making the situation worse.

4.3 Summary

The rapid growth of desalination is likely to continue for the foreseeable future. This technology is seen by many experts and non-experts as a 'gamechanger' in societies' responses to water crisis. There is, of course, an intuitive appeal in the idea that the largest untapped bodies of water in the world can become a resource through the application of technology. When we consider the severity of water challenges facing societies around the world, it is clear that desalination can and should play a part in some of those societies' responses. What we have shown in this chapter, however, is that utopian and techno-triumphalist appeals to future water abundance via desalination are highly problematic. This is demonstrated both in the history of the last 70 years and also in today's narratives around the potential for desalination powered by renewable energy. The reality in actual contexts is far more complex. In fact, the development of desalination usually reflects political desires and priorities other than simply addressing water stress, including the exercise of political power through control of water resources, modernist ideals around progress and growth, and neoliberal ideals around the transformation of water into a frontier of accumulation.

Acknowledgement of Funding

Joe Williams is supported by an ERC-assessed, UKRI-funded Starting Grant, 'Global development and the contradictions of new water – GloNeW' (EP/Y036530/1).

Further Readings

Cooley, H., Ajami, N., & Heberger, M. (2012–2013). *Key issues in seawater desalination in California: Volume I (Cost and financing), Volume II (Energy and greenhouse gas emissions) and Volume III (Marine impacts)*. The Pacific Institute. https://pacinst.org/publication/

Ibrahim, Y., Ismail, R. A., Ogungbenro, A., et al. (2021). The sociopolitical factors impacting the adoption and proliferation of desalination: A critical review. *Desalination, 498*, Article 114798.

Low, M. C. (2020). Desert dreams of drinking the sea, consumed by the Cold War: Transnational flows of desalination and energy from the Pacific to the Persian Gulf. *Environment and History, 26*(2), 145–174.

Williams, J. (2022). Desalination in the 21st century: A critical review of trends and debates. *Water Alternatives, 15*(2), 193–217.

Notes

1 There are many different techniques used to separate dissolved salt from water, which broadly fall into two categories: (1) thermal processes either heat water and collect evaporate (distillation) or separate salt by freezing; (2) membrane processes push or draw water across a physical barrier. Earlier desalination technologies typically used distillation, but the most common technique now is reverse osmosis membrane treatment.

2 This is also reflected in the cost of desalination, with around 50% of operating costs going to energy supply, although again, this varies considerably depending on technology type and feedwater.

Selected References

Al-Aghbari, A. (2021). *From abundance to scarcity: Exploring narratives and lock-in institutions around desalination in Bahrain* [Unpublished doctoral dissertation]. Institute of Development Studies, University of Sussex.

Belmehdi, F., Otmani, S., & Taha-Janan, M. (2023). Global trends of solar desalination research: A bibliometric analysis during 2010–2021 and focus on Morocco. *Desalination, 555*, 116490.

Borgomeo, E. (2024). A desalination research agenda fit for the 21st century. *Desalination, 586*, 117910.

Brannstrom, C., Jepson, W., Beckner, S., Sneegas, G., & Seghezzo, L. (2022). Not a silver bullet: Social perspectives on desalination and water reuse in Texas. *Urban Water Journal, 19*, 1025–1037.

Elsaid, K., Kamil, M., Sayed, E. T., Abdelkareem, M., Wilberforce, T., & Olabi, A. (2020). Environmental impact of desalination technologies: A review. *Science of the Total Environment, 748*, 141528.

Fragkou, M. C., & Budds, J. (2020). Desalination and the disarticulation of water resources: Stabilising the neoliberal model in Chile. *Transactions of the Institute of British Geographers, 45*(2), 448–463.

Greer, R. A., Lee, K., Fencl, A., et al. (2021). Public–private partnerships in the water sector: The case of desalination. *Water Resources Management, 35*(11), 3497–3511.

International Desalination and Reuse Association (IDRA). (2024). *IDRA desalination and reuse handbook 2023-24*. Media Analytics.

Katz, D. (2016). Undermining demand management with supply management: Moral hazard in Israeli water policies. *Water, 8*(4), 159.

Koch, N. (2023). *Arid empire: The entangled fates of Arizona and Arabia*. Verso.

Loftus, A., & March, H. (2016). Financializing desalination: Rethinking the returns of big infrastructure. *International Journal of Urban and Regional Research, 40*(1), 46–61.

McDonnell, R. A. (2014). Circulations and transformations of energy and water in Abu Dhabi's hydrosocial cycle. *Geoforum, 57*, 225–233.

Musk, E. (2024). Address at the 10th World Water Forum. www.youtube.com/watch?v=RY1gvV8Z5b8 (accessed 13 March 2026).

O'Neill, B. F. (2023). Water for whom? Desalination and the cooptation of the environmental justice frame in Southern California. *Environment and Planning E: Nature and Space, 6*(2), 1366–1390.

O'Neill, B. F., & Boyer, A.-L. (2023). 'Locking in' desalination in the U.S.-Mexico borderlands: Path dependency, techno-optimism and climate adaptation. *Water Alternatives, 16*(2), 480–508.

Panagopoulos, A., Haralambous, K.-J., & Loizidou, M. (2019). Desalination brine disposal methods and treatment technologies—A review. *Science of the Total Environment, 693*, 133545.

Patel, E. (2023). Debating desalination: Stakeholder participation and decision-making in southern California. *Water Alternatives, 16*(2), 509–540.

Ramey, J. T. (1968). Policy considerations in desalting and energy development and utilization. In *International conference on water for peace* (Vol. 2), i–ix. United States Department of the Interior.

Roth, M. B., & Tal, A. (2022). The ecological tradeoffs of desalination in land-constrained countries seeking to mitigate climate change. *Desalination, 529*, 115607.

Swyngedouw, E. (2013). Into the sea: Desalination as hydro-social fix in Spain. *Annals of the Association of American Geographers, 103*(2), 261–270.

Tarazona-Romero, B. E., Campos-Celador, A., & Maldonado-Muñoz, Y. A. (2022). Can solar desalination be small and beautiful? A critical review of existing technology under the appropriate technology paradigm. *Energy Research & Social Science, 88*, 102510.

Velásquez, C., & Wachtendorf, T. (2023). Desalination and the reproduction of water injustices in the San Andrés island water crisis. *Water Alternatives, 16*(2), 632–658.

Williams, J., & Swyngedouw, E. (Eds.). (2018). *Tapping the oceans: Seawater desalination and the political ecology of water*. Edward Elgar Publishing.

Zarzo, D., & Prats, D. (2018). Desalination and energy consumption. What can we expect in the near future? *Desalination, 427*, 1–9.

5 *REVISITING*: Large Dam Water Storage Is Unavoidable or an Anathema

Bruce Lankford, Matthew McCartney, and Florence Habets

5.1 Introduction

The debate surrounding large dams remains deeply contested, reflecting divergent perspectives on their development promises and socio-environmental costs. Proponents argue that large dams are a cornerstone of societal and economic progress, providing reliable water for irrigation, urban supply, industrial use, inland navigation, and hydroelectric power generation (Grey & Sadoff, 2006). Some go further to argue that they are indispensable tools for climate mitigation and adaptation (Biswas & Tortajada, 2021).

Critics, meanwhile, emphasise the significant environmental problems caused by large dams, such as the inundation of large areas of land, disrupted river flows, and fragmented hydrological connectivity. Societal issues include the displacement of communities, loss of cultural heritage, and degradation of ecosystem services, which have deepened inequalities and added barriers to development for millions of people globally (Richter et al., 2010). Some argue that when fully accounting for all environmental and societal costs, most large dams deliver net negative returns to national economies (Ansar et al., 2014).

Section 5.2 begins by summarising a well-documented debate on the 'myth' of the large dam that often obscures the externalities that come with an apparently good – even necessary – idea (Boelens et al., 2019). Section 5.3 navigates potential alternatives when thinking more widely and creatively about water storage. It asks whether storage can be rethought without succumbing to the 'solutionist' (Morozov, 2013) logic of 'scarcity = storage = dams'. Drawing on alternative ways of interrogating and integrating storage (Burke et al., 2023), it expands the debate about its place within water governance. Section 5.4 presents 10 points intended to broaden thinking about water storage in a time of climatic uncertainty and rapidly shifting societal demands. This exploratory framework is not a prescriptive model or multicriteria-decision tool, though such tools have value in storage planning.

DOI: 10.4324/9781003745532-5

5.2 On The Origins of Large Dam Myths

For much of the second half of the twentieth century, large dams were celebrated as emblems of modernity, national pride, and economic transformation. Promising secure water supplies, irrigation, hydropower, and flood control, they came to epitomise state-led development and technological prowess. Around the world, iconic projects, notably the multipurpose dams of the US Tennessee Valley Authority (TVA), reinforced the vision of dams as transformative instruments of economic development and state building. Two more examples illustrate this. In 1954, in India, Prime Minister Jawaharlal Nehru famously proclaimed dams to be the 'temples of modern India', capturing their symbolic power as monuments of national progress and postcolonial self-determination.

In the 1960s, President Gamal Abdel Nasser championed the Aswan High Dam as a monumental act of national liberation and modern statehood. Completed in 1970 and fully operational by 1971, the dam was designed to regulate the Nile River, expand irrigated agriculture and generate electricity, transforming Egypt's water management and enabling year-round cultivation. In 1964, Nasser called it 'the pyramid of the 20th century' and presented it as a triumph of Arab socialism and postcolonial self-determination. The Aswan High Dam delivered many tangible benefits. It stabilised water flows, protected Egypt from devastating floods and, in its early years, supplied over half of the country's electricity, fuelling industrial growth and improving livelihoods for millions. Yet it also exposed the limits of technocratic optimism. Lake Nasser, the reservoir created by the dam, displaced over 100,000 Nubian people and submerged irreplaceable archaeological sites. The trapping of nutrient-rich silt diminished the fertility of the downstream Nile Delta, increased dependence of farming on chemical fertilisers, accelerated coastal erosion and disrupted aquatic ecosystems and economically important fisheries.

This tension – between national ambition and long-term socio-environmental costs – is at the heart of what can be called the 'large dam myth'. It frames large dams as essential, straightforward, high-impact solutions to development challenges, while downplaying the profound trade-offs they entail, particularly for ecosystems and vulnerable populations. Versions of this story have played out globally – from Kariba on the Zambezi to the Three Gorges in China, the Narmada Valley in India, Belo Monte in Brazil, and many others. The success of the myth, if anything, can be gauged by the current number of large dams (higher than 15 m) in the world – around 60,000, notwithstanding the tens of millions of smaller reservoirs.

A particularly enduring myth is that large dams provide the sole or optimal hydraulic pathway to *modern* river basin management (Boelens et al., 2019). This techno-managerial vision – anchored in examples like the TVA – proposes large dams as multipurpose solutions that can simultaneously economically store very large volumes of water, tame floods, electrify economies, and irrigate deserts, thereby driving industrial development (Shi et al., 2019). The TVA, established in the 1930s as part of Franklin D. Roosevelt's New Deal, became one of the most influential models of integrated river basin development. It did not initiate the large dam myth, but by showcasing how a network of multipurpose dams could be used to promote regional development in one of America's poorest regions, it significantly amplified and institutionalised it (Ekbladh, 2002).

The 'TVA model' was elevated into a global template and exported widely as a development blueprint to countries including Jordan, India, Iran, and Turkey. Yet the failure to adapt the TVA model to local climatic, ecological, social, and political circumstances often led to misalignment and underperformance. In line with WCD (2000), the large dam myth tends to follow a recurring pattern:

1 Overestimation of benefits

- Promises of economic benefits from storage, irrigation, electricity generation, and flood control are frequently inflated.
- Projected gains often fall short due to siltation, operational inefficiencies, or maintenance challenges.

2 Underestimation of costs and risks

- Environmental costs, including biodiversity loss, altered river flows, and sediment trapping, are downplayed.
- Social impacts, such as community displacement, cultural loss, inequitable distribution of benefits, and loss of livelihoods, are inadequately addressed.
- Economic risks, including cost overruns, debt burdens, and underperformance, are common.
- Claims on upstream water needed to fill the reservoirs are rarely made explicit, and the potentially affected communities are seldom consulted during planning (van der Zaag & Gupta, 2008).

3 Neglect of alternatives

- More sustainable and cost-effective options – such as small-scale, decentralised or nature-based solutions – are often ignored.
- Demand-side interventions, such as improving water-use efficiency, receive less attention and funding.

4 Political and financial incentives

- Large dams often function as political trophies, symbolising state power, modernity, and development ambition.
- Their ability to attract international finance can sideline local perspectives and diminish the influence of affected communities (Merme et al., 2014).

- The hydropolitical tensions they exacerbate – especially in contested, transboundary river basins – are often ignored (Motta et al., 2024).
- The scale and visibility of the projects often overshadow more context-appropriate, decentralised, and socially inclusive alternatives.

The global failure to adequately assess the distribution of environmental and social impacts – especially on the most marginalised communities – has resulted in the displacement of millions of people, widespread degradation of freshwater ecosystems, and irreversible biodiversity loss. By altering natural flow regimes and fragmenting river systems, large dams have, for their host river basins, reduced fish stocks, weakened ecosystem services, and contributed to food and water insecurity. Dam outlets, often limited to downstream releases, make it physically difficult to equitably distribute water, often excluding poorer or more dispersed communities. Large dams help centralise decision-making over water management since a handful of gates or sluices determine how water is released and allocated. This not only reshapes river flows but also reinforces governance challenges in a sector marked by weak institutions. Globally, it is estimated that approximately 500 million people have been adversely affected by poorly planned or poorly managed dam projects (Richter et al., 2010).

By the late twentieth century, growing awareness of the adverse impacts led to a sharp decline in large dam construction. Civil society movements, indigenous resistance, and tighter funding constraints – particularly from institutions like the World Bank – curtailed new projects. The publication of the *World Commission on Dams* report (2000) marked a turning point, offering a rigorous global assessment and calling for better safeguards and for future dams to be more socially just, environmentally sound and economically viable. While the WCD report proposed a more balanced approach, to some, large dams of any type remain an anathema. The counter-myth is that all dams, and even storage in general, as exemplified by recent protests in France, are bad and undesirable.

In the twenty-first century, large dams are again being promoted, seemingly, without a balanced or cautionary approach (Gerlak et al., 2020). As well as responding to rising demand for water, food, and energy, climate change has added new urgency, with storage and hydropower reframed as tools for resilience and green growth. Governments are once again positioning large dams as essential to national development. Iconic examples include Ethiopia's Grand Ethiopian Renaissance Dam (GERD) and Tanzania's Julius Nyerere Hydropower Project. China's Belt and Road Initiative has supported dam construction, particularly in Africa and Southeast Asia, alongside growing participation from private investors and regional development banks. Accompanying these new challenges and framings are new myths – among them that large dams are the only viable and economic form of water storage that guarantee water security and that they are inherently climate-friendly.

Yet, evidence shows that large dams can fail under extreme conditions, are vulnerable to climate shocks, and, in some cases, emit significant greenhouse gases, particularly methane. As the world grapples with escalating water stress and a changing climate, the myth of the large dam – and the cautionary tale it provides – remains as relevant today as ever.

5.3 Navigating Alternatives

5.3.1 Multiple Types and Uses of Storage

To move beyond the biases embedded within traditional storage thinking – the myth of large dams and the knee-jerk counter-myth that all dams are bad – we must broaden the concept of storage. Drawing on Habets et al. (2018), storage is defined as the retention of water within the hydrological cycle, either naturally or artificially, where inflow-outflow conditions allow water to be slowed and held for immediate or future beneficial use. As shown in Figure 5.1,[1] storage manifests across landscapes in diverse forms: constructed dams, aquifers, soil moisture (between field capacity and permanent wilting point) and snow and ice – all interconnected in a network.

Globally, more than 99% of freshwater storage occurs in natural systems, yet these have been largely ignored and steadily degraded over the past 50 years due to climate change, environmental deterioration and over-exploitation. For example, the global extent of natural wetlands has declined

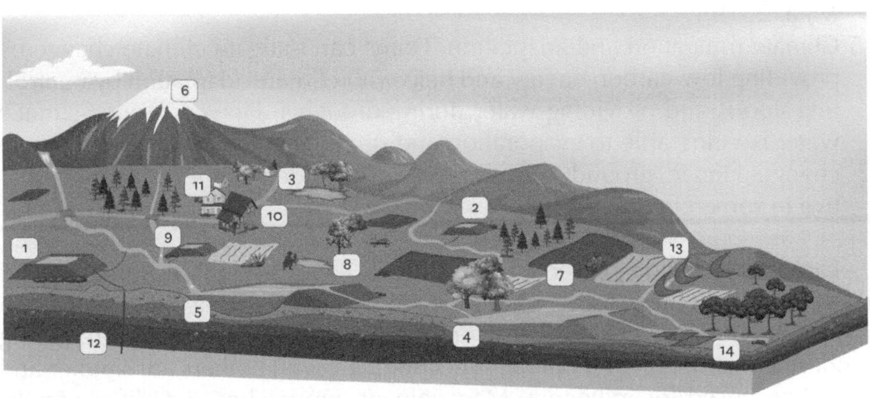

1. Artificial pond interlinked with groundwater
2. Artificial pond interlinked with river supply
3. Hillslope reservoirs, natural ponds and lakes
4. Diversion (off-stream) reservoir
5. Dam reservoir – usually interseasonal storage
6. Snow and glacier storage inc warming tundra
7. Soil moisture and shallow water table storage
8. Artificial and natural pond and wetland storage
9. Operational, canal, or night water storage for irrigation
10 Urban/rural/household on-site local tanks & network buffering
11. Storage within urban infrastructure, areas, drains,
12. Groundwater aquifer storage & subterranean reservoirs
13. Augmented natural / traditional harvesting-storage (sand dams, tanks)
14. Water storage in rice and paludiculture fields & polders

Figure 5.1 Different types of storage across a catchment and landscape.

significantly, with a total loss of around 3.4 million km^2 (21%) since 1700 (Fluet-Chouinard et al., 2023).

Whether natural or human-made, water storage serves multiple purposes. Classified into five categories below, these are presented in their idealised form. Yet, they contain within them sources of myths about storage and negative consequences if not managed well:

1 Hydrological regulation. Storage – either as a single store or in a network – slows and regulates water flows, thereby reducing flood peaks and mitigating impacts. For example, flood mitigation is provided by large dams or cumulative small reservoirs, rice fields and natural wetlands.
2 Meeting societal needs and supporting economic growth. Storage of all types and sizes, including aquifers, provides water for economic growth via irrigation, livestock, domestic use, and urban/industrial supply.
3 Facilitating operational management. Small operational reservoirs help manage irrigation scheduling, especially after a rainfall event when irrigation is suspended. For example, Sudan's Gezira system stores water in canals at night for daytime irrigation.
4 Supporting water management and governance. Examples include sustaining environmental flows; building resilience to drought via sufficient stored volume and multiple storage types; reconciling allocations during water-scarce periods (Yu et al., 2021); and complementing other technologies and institutions for water governance (Xie & Zilberman, 2018). Conversely, storage can exacerbate governance challenges when poor design and management lead to water depletion and conflict (Mahmood & Jia, 2019).
5 Climate mitigation and adaptation. Dams can mitigate climate change by providing low-carbon energy and help with climate adaptation by controlling floods and providing water during droughts. However, open-surface water is vulnerable to evaporation and quality degradation under climate change. Hence, groundwater storage can provide a more resilient alternative in some circumstances.

Each storage type presented in Figure 5.1 has specific advantages and disadvantages. How and by whom the water is accessed is different in each case. Access and governance vary: some options are highly technical, requiring modern construction and operation techniques, while others are simpler and have been used for millennia. Management approaches also differ greatly. In some cases, decision-making and responsibility rest with individuals or local communities, while in others, more complex and sometimes commercial institutional arrangements are required. Therefore, each type of storage has its niche. Overall effectiveness depends on seeing storage as a mosaic at the landscape scale with component parts understood through their technical feasibility, socioeconomic context, institutional fit and environmental impact.

5.3.2 Multiple Pitfalls of Different Types of Storage

The typologies in Figure 5.1 reveal not only a diversity of storage options but also how prevailing narratives become overly narrow. Commentary often fixates on the positives or negatives of large dams, treating them as symbols of modern water control (Flaminio, 2021) or merely as favouring elite interests (Hommes & Boelens, 2018) or appealing to political constituencies like large-scale farmers (Beg, 2019). Yet, paying excessive 'either-or' attention to storage types is to obscure mixed hybrid systems, uses and benefits and ignore the interplay of grey and green infrastructure.

Bias also leads to blind spots. Maintenance and sustainability issues plague all forms of storage. Small farm ponds may be dismissed as minor, yet they come with their own challenges and inefficiencies (Habets et al., 2018). Furthermore, both large and small dams often suffer from siltation due to upstream erosion. While modern dams incorporate sediment management, the issue of sedimentation continues to be downplayed politically. It also sets up false opposition of technology (*either* grey concrete structures *or* green, nature-based solutions), rather than acknowledging that there can be complementarities in hybrid systems and that dams can fit a natural cultural landscape.

Blind spots also occur for other forms of storage. For example, proponents of nature-based solutions (NBS) for water storage argue that regenerative agriculture boosts soil physical fertility and stores water. For example, 100,000 ha of regenerated land with an added extra 10 mm of moisture-holding capacity stores 10 million m^3 of water. This considerable volume is easy to dismiss alongside highly visible reservoir storage, but the equivalent volume of 10 MCM would require a costly dam 20 m deep by 707 m long and wide. Yet another blind spot arises when NBS proponents forget soil is a living system, subject to shifts in farming practices and environmental conditions. Under heat and severe drought, for example, soil organic matter oxidises, soil biota dies and moisture-holding capacity drops, potentially reversing gains in soil water storage.

Perhaps the most significant pitfalls relate to environmental and social justice. Large dams often harm ecosystems and displace marginalised communities. Benefits such as irrigation and hydropower tend to be captured by wealthy or urban elites (Siciliano et al., 2018). These injustices might be less visible for other forms of storage, but they are present, for example, with small water-harvesting structures (see Chapter 6). Likewise, when watertables drop in over-abstracted aquifers, deeper boreholes for cities or irrigation often undermine shallower handpumps in villages. NBS are often promoted as win-win strategies in addressing climate, water and biodiversity challenges; however, they are not inherently socially just. Matters of equality can still arise, particularly when solutions are implemented without careful consideration of context, power dynamics or local needs, including: (i) conflicts over land

and resource access, (ii) inequitable distribution of benefits, (iii) inequities of labour and burden and (iv) procedural injustice and token participation.

A growing concern is the failure of storage bodies – especially large dams – to refill sufficiently. This is necessary in normal-to-wet periods so that stored water can be drawn on during droughts and drier periods. Put simply, if water storage cannot be fully or partially refilled in time for the next dry period or drought, its utility as a storage body for buffering a future drought approximates zero. Modelling in the Groot Letaba catchment in South Africa shows that, once irrigation exceeds 30,000 ha, water demand in normal-to-wet periods consumes nearly all available refilling flows, leaving little or no surplus for storage (Lankford et al., 2023). This is typically exacerbated over longer timescales when a decade of relative wetness leads to increased use that needs to be curtailed during a subsequent drier decade. Refilling is jeopardised by excessive use but also by climate change (more erratic supply and higher evaporation) and basin over-building, where multiple dams are constructed in the same basin (although more physical capacity does not create more water to store).

The issue of refilling underscores a key point: storage is not solely a matter of supply engineering. If it is not linked to demand management, additional storage can become stranded assets, particularly under shifting climate and hydrological regimes. Another problem, exposed by the risk of failing to refill existing and new storage, is the set of long-recognised and deeply entrenched flaws in dam cost-benefit and risk analyses. These include the systematic underestimation of total project costs, the overestimation of anticipated benefits and narrow stakeholder engagement.

5.3.3 Political Economy of Storage

The political economy of large dams reveals deep tensions between national aspirations and local realities (Huber et al., 2017). Governments often champion dam projects as engines of national development and tangible symbols of progress. This attracts substantial domestic and international investment from multilateral development banks and private investors, who view them as financially attractive or even lucrative infrastructure ventures (Gerlak et al., 2020). Yet these projects frequently spark resistance from local communities, civil society and environmental advocates due to displacement, loss of livelihoods, ecological harm and weak accountability. Such opposition can lead to protests, legal challenges and international scrutiny, challenging both the legitimacy and long-term sustainability of dam projects.

In contrast, decentralised, small-scale storage – including sand dams, ponds, tanks and recharge wells – often evolves informally in a piecemeal fashion, through private or community initiatives, usually with minimal planning. While these interventions can improve local water reliability, their effectiveness is too frequently undermined by poor design, weak maintenance and a lack of oversight. Failures are widespread; for example, of approximately 4,000 rainwater-harvesting ponds constructed between 2003 and 2008 in the Amhara region of Ethiopia, the majority were non-functional by 2009 due to poor site selection, technical issues (e.g. failure of lining materials leading to seepage) and insufficient community commitment to maintain them (see Chapter 6 on water harvesting). In some cases, uncoordinated interventions have even caused negative health impacts, such as increased malaria incidence in Tigray, Ethiopia (Hagos et al., 2006).

Critical information gaps amplify these difficulties. In many basins, the number, the capacity and the cumulative impacts of small reservoirs remain unrecorded. For example, in both the Volta (West Africa) and Olifants (South Africa) basins, there are thousands of small reservoirs, but neither the exact numbers nor volumes of water stored are documented (Johnston & McCartney, 2010). Similar gaps exist in the Ganges basin and other rivers in South Asia. Even when hydrological data are available, dispersed small reservoirs are often difficult to access, and the technical understanding of flow and sediment regimes, aquifer recharge and climate variability is frequently insufficient. These limitations mean benefits are often suboptimal, and poorly designed interventions can exacerbate rather than alleviate local development challenges. Thus, while large dams embody the myth of centralised, technocratic control, decentralised systems risk creating other problems, such as piecemeal development without integration. Together, these factors signal a need for governance that balances local benefits with basin-wide sustainability.

The already complicated political economy of large storage is only further hampered by fantasy projects or stark versions of the hydraulic mission, such as are seen in the Mekong (Middleton, 2022). In other words, the case for a more intelligent, myth-busting approach to large dams, involving a more holistic water storage framework (see Section 5.4), is not helped by the single-minded championing of large dams by powerful political forces. At the same time, the decommissioning and removal of dams can also be viewed via a political economy of different interests (e.g. engineering influences) acting over and across different scales (Drapier et al., 2023).

5.4 Towards a Broader Framework on Storage

Large-scale storage has either been framed as a paramount indispensable technofix or, more recently, as an evil to be avoided at any cost. To challenge this dichotomy, it is useful to foreground a more holistic, adaptive, and pluralistic perspective (Burke et al., 2023). The ten issues below are an attempt at this more integrative yet non-prescriptive approach. Recognising that this view of dams is not new and in practice rarely implemented, the framework has two functions: it summarises the chapter and focuses on landscape-scale water storage systems rather than large dams alone (see also Ebrahim et al. 2024).

1 *Storage as innovative alternatives*. Storage should be viewed through the lenses of innovation and adaptability. This more creative approach invites comparisons between different forms – such as large dams, floodplain soil storage, wetlands and groundwater – and encourages exploration of complementary solutions. These might include, for example, enhancing river connectivity, substituting hydropower with solar power, managing soil for different purposes (biodiversity, crops and water retention) or deploying floating solar panels to mitigate reservoir evaporation.
2 *Storage from a cross-disciplinary perspective*. Water storage serves as a boundary object to be interpreted through hydrological, engineering, ecological, social, cultural and political lenses. It invites contributions from diverse disciplines and stakeholders, and the perspectives of both critics and proponents of large dams deserve consideration in an inclusive dialogue.
3 *Storage for water management*. Well-designed and well-operated storage infrastructure can smooth out both inter- and intra-seasonal variability in water availability, improve predictability and enable the distribution of scarce supply to match demand on daily and weekly bases. Both natural and artificial/built storage contribute to timely water distribution, increasing the economic value of water and thus the benefit to humans and ecosystems. However, each storage type brings unique challenges (Habets et al., 2018) and poorly managed storage can undermine water security.
4 *Storage as a mosaic of different types*. Storage depicted in Figure 5.1 operates as a system of different types, spatial scales and levels of diffusion, from point-based infrastructure, such as individual large dams, to diffuse capacity, like soil moisture and aquifers. Recognising this mosaic of storage allows the total water availability to be assessed and enhanced without defaulting to large infrastructure solutions alone.
5 *Storage as interlinked hydrological cascades and transfers*. A storage mosaic sees water continuously moving from one type of storage to another, whether naturally, within the hydrological cycle, or artificially. For example, saturated soil will add to surface runoff that flows to a pond or reservoir, thus contributing to groundwater recharge before further sustaining river baseflows or being taken up by wells. Poorly designed systems can disrupt river connectivity and downstream flows.
6 *Resilience as a function of storage*. The full array of storage options illustrated in Figure 5.1 emphasises diversity and redundancy – key features of societal and ecological resilience. Whether through the attenuation of floods or buffering of droughts, connectivity across freshwater systems enhances adaptive capacity and builds resilience. Conversely, the over-fragmentation of river systems can reduce resilience, and introducing artificial perennial flows into intermittently dry ecosystems can cause ecological harm.

7 *Storage for ecosystem health*. Storage is critical to the disruption or restoration of environmental flows and flooding. How these should be managed in relation to climate change remains a complex question.
8 *Storage governance*. The governance of water storage is inherently complex, given the political, social, economic, and administrative systems at play. Each level involves multiple institutions, jurisdictions, and stakeholders. Moreover, decisions made outside the formal water sector (e.g. in energy and agriculture) influence how stored water is used, when it is supplied, and who has the right to access it and its related services.
9 *Integrated water governance*. Storage is most effective when integrated into broader water strategies, such as demand management, reuse, and treatment. This ensures it is available to address drought shortfalls rather than simply enabling increased consumption.
10 *Legal and institutional frameworks for storage*. Clarifying ownership, management rights, and regulatory oversight of different types of storage is essential, especially during periods of hydrological stress such as floods and drought. This legal question applies to both the stored water and its grey/green infrastructure. In addition, understanding storage as a systemic landscape mosaic implies a need for legal reform away from siloed, infrastructure-specific norms towards frameworks that can govern integrated and dynamic storage systems. A review of the literature indicates that this is a significant gap requiring further exploration in many contexts.

5.5 Conclusion

Water storage is a lens through which many of the complexities and voices of water management are magnified. When storage is fragmented into myth and counter-myth or into isolated components, or when one type is prioritised at the expense of others, critical opportunities for resilient, equitable, and environmentally sustainable economic progress are lost. By embracing storage as a mosaic of meanings, functions and consequences, water governance is better placed to meet the challenges of a changing climate and increasingly water-scarce world.

This chapter calls for greater inclusion of diverse voices, paradigms, and options in the ongoing debate on water storage. Challenging the prevailing myths on all sides, it argues for a broader conceptual framework that accepts storage not merely as grey infrastructure, but as a diffuse yet interconnected and dynamic landscape mosaic in an integrated legal-institutional, socio-hydrological system.

Further Readings

Boelens, R., Shah, E., & Bruins, B. (2019). Contested knowledges: Large dams and mega-hydraulic development. *Water, 11*(3), 416.

Gerlak, A. K., Saguier, M., Mills-Novoa, M., Fearnside, P. M., & Albrecht, T. R. (2020). Dams, Chinese investments, and EIAs: A race to the bottom in South America? *Ambio, 49*(1), 156–164.

World Commission on Dams. (2000). *Dams and development: A new framework for decision-making*. Earthscan. https://awsassets.panda.org/downloads/wcd_dams_final_report.pdf

Yu, W., Rex, W., McCartney, M., Uhlenbrook, S., von Gnechten, R., & Priscoli, J. (2021). *Storing water: A new integrated approach for resilient development*. Global Water Partnership; International Water Management Institute.

Note

1 Which builds on Figure 2 in Carluer et al. (2016).

Selected References

Ansar, A., Flyvbjerg, B., Budzier, A., & Lunn, D. (2014). Should we build more large dams? The actual costs of hydropower megaproject development. *Energy Policy, 69*, 43–56.

Beg, S. (2019). Favoritism and flooding: Political alignment and allocation of irrigation water. *World Development, 114*, 175–195.

Biswas, A., & Tortajada, C. (2021). Roles of hydropower and dams in society, including mitigation and adaption to climate change. *International Journal on Hydropower and Dams, 2021*, 8–9.

Burke, E. R., Tront, J. M., Lyon, K. N., Rex, W., Castera Errea, M. I., Varughese, M. C., Newton, J. T., Becker, A. N., & Vale, A. L. (2023). *What the future has in store: A new paradigm for water storage—Overview for policy makers*. World Bank Group. https://policycommons.net/artifacts/3448289/what-the-future-has-in-store/

Carluer, N., Babut, M., Belliard, J., Bernez, I., Burger Leenhardt, D., Dorioz, J., Douez, O., Dufour, S., Grimaldi, C., & Habets, F. (2016). Impact cumulé des retenues d'eau sur le milieu aquatique. *Expertise scientifique collective. Rapport de synthèse*. https://doi.org/10.17180/y2xp-r208

Drapier, L., Germaine, M.-A., Lespez, L., Magilligan, F. J., & Sneddon, C. (2023). Networks, coalitions and the contestation of dam removal across political and institutional scales in France and New England (USA). *Geographical Review, 113*(2), 246–267.

Ebrahim, G. Y., Lautze, J. F., McCartney, M., Matheswaran, K., Nyikadzino, B., & Tafesse, N. T. (2024). Beyond dams: Assessing integrated water storage in the Shashe catchment, Limpopo River Basin. *Journal of Hydrology: Regional Studies, 55*, 101913.

Ekbladh, D. (2002). "Mr. TVA": Grass-roots development, David Lilienthal, and the rise and fall of the Tennessee Valley Authority as a symbol for U.S. overseas development, 1933–1973. *Diplomatic History, 26*(3), 335–374. https://doi.org/10.1111/1467-7709.00314

Flaminio, S. (2021). Modern and nonmodern waters: Sociotechnical controversies, successful anti-dam movements and water ontologies. *Water Alternatives, 14*(1), 204–227.

Fluet-Chouinard, E., Stocker, B. D., Zhang, Z., Malhotra, A., Melton, J. R., Poulter, B., Kaplan, J. O., Goldewijk, K. K., Siebert, S., Minayeva, T., Hugelius, G., Joosten, H., Barthelmes, A., Prigent, C., Aires, F., Hoyt, A. M., Davidson, N., Finlayson, C. M., Lehner, B., ... McIntyre, P. B. (2023). Extensive global wetland loss over the past three centuries. *Nature, 614*(7947), 281–286.

Grey, D., & Sadoff, C. (2006). *Water for growth and development*. Comision Nacional del Agua.

Habets, F., Molénat, J., Carluer, N., Douez, O., & Leenhardt, D. (2018). The cumulative impacts of small reservoirs on hydrology: A review. *Science of the Total Environment, 643*, 850–867.

Hagos, F. G., Yohannes, M., Linderhof, V., Kruseman, G., Mulugeta, A., & Abreha, Z. (2006). *Micro water harvesting for climate change mitigation: Trade-offs between health and poverty reduction in Northern Ethiopia* (PREM Working Paper). Vrije Universiteit Amsterdam.

Hommes, L., & Boelens, R. (2018). From natural flow to 'working river': Hydropower development, modernity and socio-territorial transformations in Lima's Rímac watershed. *Journal of Historical Geography, 62*, 85–95.

Huber, A., Gorostiza, S., Kotsila, P., Beltrán, M. J., & Armiero, M. (2017). Beyond "socially constructed" disasters: Re-politicizing the debate on large dams through a political ecology of risk. *Capitalism Nature Socialism, 28*(3), 48–68.

Johnston, R. M., & McCartney, M. (2010). *Inventory of water storage types in the Blue Nile and Volta river basins*. IWMI. www.iwmi.cgiar.org/Publications/Working_Papers/working/WOR140.pdf

Lankford, B., Pringle, C., McCosh, J., Shabalala, M., Hess, T., & Knox, J. W. (2023). Irrigation area, efficiency and water storage mediate the drought resilience of irrigated agriculture in a semi-arid catchment. *Science of the Total Environment, 859*, 160263.

Mahmood, R., & Jia, S. (2019). Assessment of hydro-climatic trends and causes of dramatically declining stream flow to Lake Chad, Africa, using a hydrological approach. *Science of the Total Environment, 675*, 122–140.

Merme, V., Ahlers, R., & Gupta, J. (2014). Private equity, public affair: Hydropower financing in the Mekong Basin. *Global Environmental Change, 24*, 20–29.

Middleton, C. (2022). The political ecology of large hydropower dams in the Mekong Basin: A comprehensive review. *Water Alternatives, 15*(2), 251–289.

Morozov, E. (2013). *To save everything, click here: Technology, solutionism and the urge to fix problems that don't exist*. Penguin.

Motta, S., Bock, I., Koehler, J., Wolf, A. T., & Pattberg, P. (2024). The financialization of rivers: Clean Development Mechanism (CDM) subsidized hydropower in the Mekong Regions basins at risk. *Global Environmental Change, 90*. https://doi.org/10.1016/j.gloenvcha.2024.102962

Richter, B. D., Postel, S., Revenga, C., Scudder, T., Lehner, B., Churchill, A., & Chow, M. (2010). Lost in development's shadow: The downstream human consequences of dams. *Water Alternatives, 3*(2), 14–42.

Shi, H., Chen, J., Liu, S., & Sivakumar, B. (2019). The role of large dams in promoting economic development under the pressure of population growth. *Sustainability, 11*(10), 2965.

Siciliano, G., Urban, F., Tan-Mullins, M., & Mohan, G. (2018). Large dams, energy justice and the divergence between international, national and local developmental needs and priorities in the global South. *Energy Research & Social Science, 41,* 199–209.

STATEWATCH. (2023). French government steps up repression after bloody battle of Sainte-Soline. www.statewatch.org/news/2023/may/french-government-steps-up-repression-after-bloody-battle-of-sainte-soline/

van der Zaag, P., & Gupta, J. (2008). Scale issues in the governance of water storage projects. *Water Resources Research, 44*(10).

Xie, Y., & Zilberman, D. (2018). Water storage capacity versus water use efficiency: Substitutes or complements? *Journal of the Association of Environmental and Resource Economists, 5*(1), 265–299.

6 *REVISITING*: Water Harvesting Is Necessary to Enhance Local Supply

M. Dinesh Kumar

6.1 The Emergence and Spread of Rainwater Harvesting

The term 'rainwater harvesting' (RWH) refers to the collection of rainwater from rooftops, urban runoff and runoff from small, natural, rural catchments and agricultural fields, and either storing it on surface or underground reservoirs for direct use or diverting it into the groundwater system to replenish aquifers. Over the last four decades, RWH as a concept and practice has gained ground in many developing countries of Asia and Africa.

Its emergence as a global practice can be traced back to certain successful initiatives. In India, these include the community-driven effort to revive the traditional water-harvesting system, the *johad*, in Rajasthan's Alwar district by the NGO Tarun Bharat Sangh, artificial groundwater-recharge projects by another NGO in the Kachchh district of Gujarat, western India, and the 'dug-well recharge movement' in Gujarat's Saurashtra region. Meanwhile, in the 1990s and early twenty-first century, a chorus had been growing among academics in support of indigenous water-harvesting systems (Denison & Wotshela, 2009; Shah, 2000) and environmental groups (Agarwal & Narain, 1997).

The credibility of the paradigm was bolstered internationally with assertions such as those by Denison and Wotshela (2009),

> Indigenous rainwater harvesting and conservation practices are the product of accumulated knowledge, practices and traditions which have evolved over many generations of experimentation and adaptation. These practices thus have inherent sustainability and present a sound platform on which to develop new practices aimed at maximizing the benefits of 'runoff farming', as rainwater harvesting might best be summarized.

In India, the popularity of Agarwal and Narain's *Dying Wisdom: The Rise and Fall of Traditional Water Harvesting Systems in India* (1997) boosted the morale of Civil Society Organisations (CSOs) seeking an alternative to large-scale, centralised solutions.

DOI: 10.4324/9781003745532-6

The practice got institutionalised as the opposition to the massive multi-purpose Sardar Sarovar Narmada Dam project in Gujarat, western India, gained traction in India and abroad with the endorsement of international environmental groups, which, in turn, were backed by multi-lateral financial institutions, such as the World Bank. The latter ultimately withdrew its support for the project based on the findings of the independent Morse Committee it had appointed that questioned the project's very rationale. Despite being built (over a period of 56 years), the significant controversy over the dam gave rise to a new outlook on water management favoured by the proponents of this new paradigm of water management as part of the larger narrative to garner support for it.[1]

The paradigm's emergence was witnessed in many countries of Africa, the Middle East, and elsewhere. Examples of RWH having existed since ancient times in the Middle East (Oman and Israel), West Africa (Burkina Faso), North Africa (Tunisia), North Central Africa (Chad), Pakistan (Baluchistan), and China were deployed to further the narrative globally (Oweis et al., 1999). Several international organisations got behind it and, through their publications, tried to make the international development agencies treat it as 'the lifeline for human wellbeing' (Baron, 2009) and to turn it into the foremost feature of water-management initiatives by those agencies in these regions.

Nonetheless, the vigour and enthusiasm observed in India was not seen elsewhere. In India it became almost a 'movement', driven by the notion that it could solve the nation's water problems (Shah, 2000). The view that it could address issues such as groundwater depletion and the drying-up of rivers was strongly endorsed by community leaders, CSOs, social activists working on water, religious and spiritual organisations and government agencies.

6.2 Issues with Rainwater Harvesting

Water-harvesting techniques are remarkably diverse. They have evolved in the most hyper-arid regions of the Middle East and North Africa to the most humid and high-rainfall regions of Sri Lanka (Kumar et al., 2006; Oweis et al., 1999). Some are centuries old (Agarwal & Narain, 1997; Oweis et al., 1999) and some are modern (Kumar et al., 2006). The potential benefits of the systems, both traditional and modern, have been documented by an extensive literature. However, the issues with RWH, particularly the unintended consequences and problems with planning, design and maintenance, are often concealed by the proclaimed desirability of this 'traditional knowledge'. What follows is a discussion of the notable shortcomings that surfaced with the increasing prevalence of RWH, particularly in naturally water-scarce regions.

6.2.1 Hydrological Impacts

Local water-management interventions are often based on very little understanding of the local hydrological regimes and hydrological impacts at a larger scale (Kumar et al., 2008). Rather, they tend to be based on the deep-rooted belief that the greater the size of the water-impounding structure, the greater the hydrological benefit in terms of water storage and recharge. Part of the reason is the lack of availability of data on streamflow for small catchments (Kumar et al., 2008). As a result, there is a tendency to over-build over-sized water-harvesting structures. The fact is also neglected that runoff harvesting is most suited to catchments with a high 'runoff'-to-'run-on' ratio. This is illustrated by studies of Tunisian areas where the traditional RWH system, the *jessr*, is commonly built to provide supplementary irrigation to olive plantations. Results showed that the 'catchment area-terrace area' ratio (CCR) should be at least 7.4 to yield sufficient water for olive cultivation, based on an average annual precipitation of 235 mm. The encroachment of catchments of RWH systems for crop cultivation is often rampant, which reduces the runoff potential (Schiettecatte et al., 2005).

At the same time, government agencies and NGOs usually implement rainwater-harvesting and groundwater-recharge projects in areas where the availability of water from surface systems and wells falls far short of demand due to limited rainfall, high aridity, and the presence of vast tracts of arable land requiring irrigation for crop production (Kumar et al., 2006). Such regions are also marked by a high degree of basin development, with reservoirs, diversion systems and intensive groundwater use meaning these basins are 'closed'. In such conditions, the unscrupulous building of RWH structures only leads to a redistribution of the available runoff water in the basin and an eventual reduction in flow to the pre-existing downstream reservoirs (Calder et al., 2008) with negative social and ecological consequences.

The evidence from southern Rajasthan and Saurashtra in Gujarat of the dryingg-up of public reservoirs is an indication of the larger menace of failing to exercise necessary caution in implementing RWH. The Aji reservoir suffered significantly reduced inflow following the construction of several check dams in its upper-catchment watersheds, affecting irrigators in the reservoir's command area as well as urban water users in the city of Rajkot (Kumar et al., 2008).

In 2006, the government of Rajasthan threatened legal action against that of Madhya Pradesh for refusing to dismantle hundreds of illegally constructed check dams along the Chambal River in the Malwa region. The structures were said to be affecting the inflow of water into the Gandhi Sagar Dam.

A detailed analysis of hydrological processes in the Merguellil catchment in semi-arid central Tunisia revealed a 41-50% decrease in the runoff coefficients of rainfall events between 1989 and 2005 (most notably for those under 40 mm) following the introduction of small reservoirs and contour benches on up to 32% of the catchment area (Lacombe et al., 2008).

Likewise, studies in the high-rainfall areas of Ethiopia's Alaba district showed a significant reduction in the water available for crop production in downstream areas due to intensive water harvesting to improve agricultural productivity in the upper-catchment areas (Seka et al., 2016).

6.2.2 Maintenance of Water-Harvesting Systems

Structures for RWH, such as check dams, ponds, bunds, and terraces, require regular maintenance. When communities or local governments fail to keep up with it, it leads to the silting up of their tiny reservoirs, structural collapse, or inefficient water capture. This reduces the long-term benefit of the system, thereby wasting the initial investment. When a government implements such a scheme, it is sold as a participatory or community-based water-conservation project. This implies that the operation and maintenance of the structure would be the responsibility of local community institutions. However, these exist only on paper. Since the schemes are often built by provincial government agencies or their local counterparts, paying no heed to the needs or priorities of the local communities, the latter show little interest in maintaining them. There is also limited understanding of the maintenance needs of complex systems like recharge structures (percolation tanks, sub-surface dykes, recharge shafts, etc.).

Many of Tunisia's traditional RWH systems (*tabias*, *jessours*, *meskat*, contour-bench terraces, and water-spreading systems) are being abandoned due to the difficulty and hard work involved in their upkeep (Jebari et al., 2015). In the absence of proper maintenance, for example, *tabias* can be damaged by breaches, which automatically cause further breaches in downstream *tabias*. The succession of ruptures concentrates the runoff, leading to the formation of gullies. Thus, the level of erosion in unmaintained areas of soil and water-conservation works (SWC) is worse than under natural conditions (Le Goulven et al., 2009). Studies on small tanks in El Gouazine showed that eight years after their construction, the SWC works had lost 60% of their storage capacity. The runoff coefficient and sediment transport rate both returned to values corresponding to pre-SWC works (Nasri et al., 2004). Le Goulven et al. (2009) note that, under such conditions, there is little likelihood that the remaining population of the upper basin will undertake the maintenance of plantations and conservation works, particularly given that they were not consulted during their planning or construction and even opposed them, thus forcing the authorities to build them on public land.

6.2.3 Technical Inadequacy

Since RWH is considered a universal 'best practice', there is frequent borrowing of techniques that are proven to be unsuitable to a particular set of rainfall patterns, climate, soil type, or topography. Structures that are ideal for humid, high-rainfall, hilly regions with steep terrain and deep gorges (such as check dams that help reduce the speed of flow and check silt transport in the streams) are used on flat terrain with shallow embankments. This causes problems such as flooding the banks and high evaporation from the small volume of water stored in the reservoirs with a large water-spread area.

Implementing RWH works on a massive scale has posed major technical challenges, and field evidence suggests some spending has been unwise. For instance, ponds have been dug in areas with scanty rainfall, ignoring factors such as catchment area and sources of recharge.

Groundwater-recharge schemes are sometimes implemented without the necessary geohydrological investigations having been carried out. Recent landslides in the Kozhikode district of Kerala point to the impending catastrophe such a careless approach can invite. The unscrupulous digging of RWH pits on hill slopes in this high-rainfall area (around 3,000 mm, 80% during the monsoon months) was reported to have caused the destabilising of slopes, which eventually led to landslides.

Structures for groundwater recharge (such as check dams, percolation ponds, and recharge tube wells) often only work in alluvial and sedimentary formations. When used in hard-rock areas (with very limited hydraulic conductivity and porosity), the result is that, in high-rainfall years, the system cannot push the excess runoff water from the catchment into the aquifer as it is already saturated, even with limited natural recharge from the rainfall.

6.2.4 Poor Micro- and Macro-Level Planning

For the sound planning of RWH at both the local and watershed levels, more scientific input is needed than is usually appreciated. This includes topography, analysis of the rainfall intensity and pattern, reliable estimation of catchment runoff, analysis of the engineering properties of the soil and geohydrological data, including the geohydrological parameters of the formations, and the mapping of geological structures and groundwater-surface water interactions (Kumar et al., 2006).

Further, the large-scale implementation of RWH at the basin level requires careful hydrological and economic planning, including an assessment of the non-utilised runoff available for further harnessing. This would determine the optimal number of structures to build for positive hydrological and economic results and to minimise any negative social and ecological effects at larger scales.

Scale effects in hydrology are well documented (Sivapalan & Kalma, 1995), but micro- and meso-level hydrology is far more complex than tends to be recognised. In micro- and meso-catchments the marginal groundwater discharge, or base flow, does not enhance runoff, meaning the fraction of rainfall available as runoff will be relatively small. Hydrological data (runoff and sediment yield) are usually available at the basin, and often sub-basin, level. Applying the rainfall-runoff relationships of large basins to small upper catchments would result in an underestimation of runoff, since upper catchments normally have rocky terrain capable of producing high runoff. Likewise, applying the data of the sedimentation available to large basins to small catchments can lead to an underestimation of the siltation rate, as this is generally high for hilly upper catchments (Kumar et al., 2006).

These essential hydrological data are rarely available for water-management planning of micro-catchments, which are often implemented in a decentralised and uncoordinated way by local arms of the government or NGOs, with very limited budgets and technical expertise.

6.2.5 Unreliable Local Water Supplies

A fact which is often ignored by proponents of RWH is that naturally water-scarce regions also experience high year-to-year variation in rainfall and high aridity along with low rainfall conditions. If an area's rainfall is too unreliable, the uses associated with the harvesting will also be precarious. Although small water-harvesting structures are meant to meet local demand, they will fall short in times of significant rainfall deficit when the catchment inflow is low, given the resultant increase in the need for water for irrigation and local domestic supply (Kumar et al., 2006).

A study of an RWH tank in Telangana (south India) over a 14-year period, for example, showed extreme variability in tank-filling, with the annual maximum volume ranging from 8,650 m^3 to approximately 200,000 m^3. It reached its maximum capacity only once, and for one-third of the time, its yearly maximum replenishment was below 15% of its capacity. The surface water availability remained limited since the tank dried up every year except for two (Boisson et al., 2022). The implication of such irregularity is a precarious domestic water supply.

6.2.6 Equity and Social Concerns

As per the legal provisions of most countries, water from natural catchments, which are on common land such as forest, pasture, and revenue wasteland, is a common-property resource. Private individuals are not allowed to harness the runoff from these catchments, except on their farm through bunds and ponds. In India, this was the situation for several decades. Recently, however, the governments of many water-scarce states started promoting the decentralised, community-led construction of small structures such as check dams, anicuts, weirs,[2] and ponds by providing funds to the local authorities. Although the entire work is supposed to be carried out with community participation, in several villages, influential rural elites took it upon themselves to plan and build RWH systems using public funds.

In these cases, structures such as check dams and anicuts are built on sites chosen by the village elites to the benefit of their private farmland, either directly through access to the small reservoirs or indirectly through the recharge of their wells. Often the poor villagers are excluded from such decisions and get only wage labour to work at the construction sites. In a study of externally funded water projects in two Indonesian villages, Lucas (2016) illustrated how elite 'capture of' and 'control over' the resources allocated for water and forestry projects in one village and complete 'elite capture' of resources in another resulted in inequitable outcomes in terms of the distribution of benefits from the project. In a study of dryland areas of Egypt, Hoffet et al. (2012) showed the influence of power differentials in local water management, resulting in water allocation being subject to 'elite capture', thereby impeding national priorities and compromising the equity and needs of local communities.

Meanwhile, in various water-rich states of India, wealthy individuals owning large tracts of land in rural areas have started building small dams and other RWH structures on rivers and streams to impound and divert water for private uses, such as irrigating crops. While there are no government funds for such initiatives, it is done under the guise of water conservation. Yet, such practices often threaten drinking water supply sources downstream, as witnessed in Kerala.

In the recent past, government subsidy schemes have also helped farmers over-exploit common-property groundwater resources in the name of water conservation. In Maharashtra, public subsidies for the construction of farm ponds on private land enabled resource-rich farmers owning large wells to pump groundwater during monsoon months to store in these ponds for use during the summer (Kale, 2017). Pumping during the monsoon induces recharge by creating more storage space in the aquifer, with the result of reduced runoff into the reservoir.

6.2.7 Economic and Financial Dimensions

There is growing concern over the economic viability of RWH systems with their high costs and uncertain benefits. Plans for local water-harvesting/groundwater-recharge schemes are not backed by proper hydrological or economic analysis. Thus, there is hardly any knowledge of the cost per cubic metre of the water harvested through such schemes. This becomes a far more serious issue in semi-arid and arid regions with extremely limited runoff potential and high inter-annual variability.

Studies in India show that RWH has limited potential to reduce the demand-supply imbalance or provide a reliable supply (see, for instance, Kumar et al., 2006) and find it economically unviable in water-scarce regions, given the high direct cost per unit volume of water harvested (Kumar et al., 2008). The above-mentioned poor design and maintenance of the structures reduce their lifespan to 2–3 years rather than the planned 10–25 years due to high siltation, structural damage or overtopping of the body wall.

Further, in these water-scarce regions, runoff harvesting offers very limited potential for augmenting the supply at basin level and instead creates huge negative economic (Batchelor et al., 2003), social, and environmental externalities (Ray & Bijarnia, 2006) downstream. Bouma et al. (2011) analysed the downstream impacts of RWH in a semi-arid basin in southern India, focusing on the trade-offs that arise when the source is changed from a downstream surface irrigation system to well irrigation. The results indicated that the downstream impacts were considerable and that net benefits were insufficient to pay back investment costs.

Another major issue is the lack of finances to cover the cost of maintaining the structures. With benefits that are non-measurable and uncertain, the agencies that execute RWH schemes have not been able to convert them into revenue-generating models, and therefore, no funds are earmarked for their upkeep. The assumption is that the local communities, which derive direct economic benefits from the structures, would maintain them. Yes, in many cases, these communities have very limited financial capacity and show low buy-in, given the uncertain benefits of the schemes for the community at large (Jebari et al., 2015).

6.2.8 Environmental Impact of Rainwater Harvesting

RWH poses several environmental concerns, depending on the type of system and the context. Building too many systems in catchments of naturally water-scarce regions with 'closed' basins can lead to desiccation and the drying-up of even low-order streams. This can damage the ecological health of the streams, affecting aquatic life and soil biota. Negative environmental impacts can also be induced by groundwater-recharge works (such as dug-well recharging or the construction of recharge shafts to direct runoff from agricultural watersheds into the aquifers), as the runoff can contain hazardous substances originating from fertiliser and pesticide residues.

Rooftop RWH systems that divert water directly, without proper filtration, into aquifers or tanks can also pose a serious hazard to human health if the roof catchments are contaminated with carbon from vehicular emissions, lubricants, or traces of heavy metals and microbes. Field studies have shown the incidence of chemical (Magyar et al., 2007) and microbial contamination (Evans et al., 2006) of the rainwater collected from rooftops, making it unfit. Consuming the chemically and biologically contaminated water can pose serious health risks, as experience in rural Auckland suggested. Simple slow sand filters currently promoted for the treatment of water from roof catchments at the household level are inadequate to remove chemical pollutants, while sophisticated treatment options are simply unviable.

6.2.9 The Ideological and/or Self-Serving Bias

When it comes to promoting RWH, NGOs and donor agencies usually display a certain degree of 'ideological blindness'. Since the structures they endorse are closer to traditional systems, they are found to be more appealing to their constituencies (Agarwal & Narain, 1997). Moreover, these techniques were globally accepted as low-cost, 'environmentally sound' and 'socially acceptable' solutions seen as unlikely to do any harm. This reputation helps them to provide 'quick fixes' for problems whose only other solutions require long-term planning, large-scale investment, recognition of the finite nature of the resource and a fundamental change in the way the water sector is run and administered. It helps them put their constituencies together.

For the government agencies, it has provided an easy way to show that they are prepared to address water problems with great sensitivity. This counteracts the severe criticism they often face from NGOs, environmental activists, and donor agencies while trying to implement large-scale water projects involving massive reservoirs and canal networks. Over the years, a whole industry has developed around this, with a mushrooming of private organisations providing consultancy services on water harvesting in urban areas and making tall promises about its benefits to their clients. Thus, many of the promoters of RWH become even more blinded by self-interest.

6.3 Conclusion

Over the past three decades or so a large body of work has covered the subjects of RWH and artificial groundwater recharge across the developing world, particularly in India and certain African countries. Efforts to revive traditional RWH systems and build other small-scale water-harvesting structures increased in these countries, initially with the support of donor agencies and later with government funds. Most paint a rosy picture of their economic and social benefits, while overlooking hydrological limitations and cross-scale and environmental impacts. In the process, water-harvesting projects had gained a high degree of legitimacy and global acceptance, with several international development agencies and financing institutions endorsing them as an alternative to large water projects. Nonetheless, the problems associated with this practice include: (i) inadequate technical inputs for planning and execution, (ii) poor consideration of hydrology in planning, (iii) lack of maintenance, and (iv) poor financial and economic viability. The unintended effects include: (i) impacts on the local hydrological regime in terms of reduced flows downstream, (ii) damage to river ecosystems due to environmental water stress or groundwater pollution and contamination of water supplies, (iii) the inability of the system to supply water at critical times and (iv) negative equity and social impact at the local level. While water harvesting can deliver in specific contexts, the ideological blindness of its proponents may obscure the many possible shortcomings this chapter has sought to identify.

Further Readings

Calder, I., Gosain, A., Rao, M. R. M., Batchelor, C., Snehalatha, M., & Bishop, E. (2008). Watershed development in India. 1. Biophysical and societal impacts. *Environment, Development and Sustainability*, *10*, 537–557.

Citchley, W., & Gowing, J. (2012). *Water harvesting in Sub-Saharan Africa*. Routledge.

Jebari, S., Berndtsson, R., & Bahri, A. (2015). Challenge of traditional water harvesting in Tunisia. *Middle East Critique, 24*(3), 289–306. https://doi.org/10.1080/19436149.2015.1046707

Kumar, M. D., Patel, A., Ravindranath, R., & Singh, O. P. (2008). Chasing a mirage: Water harvesting and artificial recharge in naturally water-scarce regions. *Economic and Political Weekly, 43*(35), 61–71. www.jstor.org/stable/40278725

Notes

1 The following views characterise this outlook: (i) any large project that involves human displacement and the submergence of forests should be completely avoided; (ii) rather than augmenting supply, efficiency should be improved in irrigated agriculture to manage the sector's demand and to allocate more water for other sectors, including the environment; (iii) sufficient flows should be maintained for the environment in all rivers; (iv) new schemes should be assessed on their ability to improve equity in access rather than augmenting supply; and (v) if required, new irrigation schemes should meet the growing economic needs of the farming enterprise rather than contributing to the country's grain basket (Kumar, 2018).

2 A dam built across a stream or river in order to supply water to the fields in the neighbouring areas.

Selected References

Agarwal, A., & Narain, S. (1997). Dying wisdom: *R*ise and fall of traditional water harvesting systems. Centre for Science and Environment.

Baron, J. (Ed.). (2009). *Rainwater harvesting: A lifeline for human wellbeing*. Stockholm Environment Institute.

Batchelor, C. H., Rao, M. S. R. M., & Rao, S. M. (2003). Watershed development: A solution to water shortages in semi-arid India or a part of the problem. *Land Use & Water Resources Research, 3*, 1–10.

Boisson, A., Villesseche, D., Selles, A., Alazard, M., Chandra, S., Ferrant, S., & Maréchal, J.-C. (2022). Long term monitoring of rainwater harvesting tanks: Is multi-years management possible in crystalline South Indian aquifers? *Hydrological Processes*. https://doi.org/10.1002/hyp.14759

Bouma, J. A., Biggs, T. W., & Bouwer, L. M. (2011). The downstream externalities of harvesting rainwater in semi-arid watersheds: An Indian case study. *Agricultural Water Management, 98*(7), 1162–1170.

Denison, J., & Wotshela, L. (2009). *Indigenous water harvesting and conservation practices: Historical context, cases and implications* (Report No. TT 392/09). Water Research Commission.

Evans, C. A., Coombes, P. J., & Dunstan, R. H. (2006). Wind, rain and bacteria: The effect of weather on the microbial composition of roof-harvested rainwater. *Water Research, 40*(1), 37–44.

Hoffet, N., Daoud, I., Alary, V., Tourrand, J. F., & Moselhy, N. (2012). *Participation, power and sustainable water resource management, a case study of the rainfed desert region of Matruh, Egypt*. Paper presented at the 10th European IFSA Symposium. Retrieved from https://agritrop.cirad.fr/566285

Kale, E. (2017). Problematic uses and practices of farm ponds in Maharashtra. *Economic and Political Weekly*, *52*(3), 20–22.

Kumar, M. D. (2018). Future water management: Myths in Indian agriculture. In A. K. Biswas, C. Tortajada, & P. Rohner (Eds.), *Assessing global water megatrends* (pp. 187–209). Springer.

Kumar, M. D., Ghosh, S., Patel, A., Singh, O. P., & Ravindranath, R. (2006). Rainwater harvesting in India: Some critical issues for basin planning and research. *Land Use and Water Resources Research*, *6*, 1–17.

Lacombe, G., Cappelaere, B., & Leduc, C. (2008). Hydrological impact of water and soil conservation works in the Merguellil catchment of central Tunisia. *Journal of Hydrology*, *359*(3–4), 210–224.

Le Goulven, P., Leduc, C., Batcha, M. S., & Poussin, J.-C. (2009). Sharing scarce resources in a Mediterranean river basin: Wadi Merguellil in Central Tunisia. In F. Molle & P. Webster (Eds.), *River basin trajectories: Societies, environments and development* (pp. 147–170). Comprehensive Assessment of Water Management in Agriculture, CABI Publishing.

Lucas, A. (2016). Elite capture and corruption in two villages in Bengkulu province, Sumatra. *Human Ecology*, *44*, 287–300.

Magyar, M. I., Mitchell, V. G., Ladson, A. R., & Diaper, C. (2007). An investigation of rainwater tanks quality and sediment dynamics. *Water Science and Technology*, *56*(9), 21–28.

Nasri, S., Albergel, J., Berndtsson, R., & Lamachère, J.-M. (2004). Impact des banquettes sur le ruissellement d'un petit bassin versant. *Revue des Sciences de l'Eau*, *17*(2), 265–289.

Oweis, T., Hachum, A., & Kijne, J. (1999). *Water harvesting and supplementary irrigation for improved water use efficiency in dry areas* (SWIM Paper 7). IWMI.

Ray, S., & Bijarnia, M. (2006). Upstream vs downstream: Groundwater management and rainwater harvesting. *Economic & Political Weekly*, *41*(23), 2375–2383.

Schiettecatte, W., Gabriëls, D., Cornelis, W., & Hofman, G. (2005). Impact of water harvesting techniques on soil and water conservation: a case study on a micro catchment in southeastern Tunisia. *Journal of Arid Environments*, *62*(2), 297–313.

Seka, A. M., Awass, A. A., Melesse, A. M., Ayele, G. T., & Demissie, S. S. (2016). Evaluation of the effects of water harvesting on downstream water availability using SWAT. In A. M. Melesse & W. Abtew (Eds.), *Landscape dynamics, soils and hydrological processes in varied climates* (pp. 763–787). Springer. https://doi.org/10.1007/978-3-319-18787-7_34

Shah, T. (2000). Mobilising social energy against environmental challenge: understanding the groundwater recharge movement in western India. *Natural Resources Forum*, *24*(3), 197–209.

Sivapalan, M., & Kalma, J. D. (1995). Scale problems in hydrology: Contributions of the Robertson workshop. *Hydrological Processes*, *9*(3–4), 234–250.

7 *REVISITING*: Abstracting Groundwater Is Safe as Long as You Pump Less Than the Natural Recharge

Sylvain Massuel and François Molle

7.1 Introduction

The interconnection between surface water and groundwater has long been hydrogeologists' bread and butter. Yet, a large percentage of hydrological studies or river basin development plans are still made on the implicit basis that they are distinct. Part of the problem lies in the difficulty of grasping the dynamics of a resource that is out of sight, that fluctuates in space and time, and that is partly, if not largely, unknown. These shortcomings result in erroneous water balances, double-accounting, overestimation of available resources, and surreptitious spatial reallocation (Zhou, 2009; Molle, 2023).

Groundwater resources are often implicitly conceived as a stock, akin to a tank, bathtub, or bank account, where the resource is available in proportion to the flow that has replenished or 'recharged' it (Molle, 2023). This deceptive view of aquifers is illustrated by the numerous statements linking 'aquifer overdraft' or overexploitation to 'exploiting groundwater resources at a faster rate than recharge' or 'extracting groundwater faster than it is being replenished' and ambiguous statements, such as 'we need to align use with recharge' or 'restore the balance between abstraction and recharge'. In other words, abstracting less or as much as the aquifer recharge would be the right and 'safe' thing to do to ensure sustainability.

This chapter begins by briefly recalling where the fallacy lies and how it has been addressed by scientists over the past few decades. It then illustrates why equating the 'safe yield' to the natural recharge – hereafter the 'recharge-is-the-limit' (RITL) rule – can be so problematic. While most of what follows is familiar terrain for hydro(geo)logists, it is important to point out that many studies fail to account for surface-water/groundwater interactions and therefore adopt misleading language. Water management and allocation schemes, but also legislation, usually fail to account for this complexity, even in places like the over-studied Colorado River basin (Perry et al., 2024). The pervasive and far-reaching consequences of these simplifications and misconceptions warrant reinstating some key clarifications.

DOI: 10.4324/9781003745532-7

7.2 The Water-Budget Myth and the 'Safe Yield' Debate

The power of the myth around safe yield is a vexing issue. As early as 1940, Theis pointed out the error of basing groundwater use on the estimated aquifer recharge (Theis, 1940). In the 1980s, John Bredehoeft and colleagues picked up the issue again and called it a 'water-budget myth' (Bredehoeft et al., 1982; Bredehoeft, 2002). Despite subsequent and recurring efforts to expose its consequences, including calls for 'better communication', the myth, mystifyingly, endures (Sophocleous, 1997; Alley & Leake, 2004; Zhou, 2009; Konikow & Bredehoeft, 2020; Molle, 2023).

The key entry point to understanding the appealing but unfortunately misconceived idea that pumping is safe as long as the recharge rate is not exceeded is to emphasise that aquifers *are not only replenished but also drained*: in (average) natural conditions, to state it plainly, as much water goes out as comes in. It is tempting to equate what comes in with natural recharge and what goes out with pumping. However, aquifers also naturally drain or leak their water to springs, riverbeds, wetlands, lakes, other aquifers, phreatic vegetation (that extracts groundwater through its roots), and oceans. When wells pump groundwater, the (net) abstracted volume is dynamically balanced by some combination of (1) storage depletion, (2) reduced discharge, and (possibly) (3) increased recharge (the drawdown of the water table may enhance the seepage from connected streams and aquifers but also leave more storage capacity for natural recharge by rainfall). The sum of the last two terms is called 'capture' (the fluxes the wells have captured by lowering the water table). When a new equilibrium is reached, with a time lag that varies significantly according to the geometry and characteristics of the aquifer, the water table is stabilised meaning that abstraction is equal to capture.

Hence, all groundwater use comes at the expense of *both* storage and drainage fluxes. Focusing on discharge readily makes it clear who stands to lose when an aquifer is depleted by pumping, that is, groundwater-dependent ecosystems and humans who depend on the springs, streams, and wetlands fed by the natural drainage of the aquifer. For some types of aquifers, it is also the 'next generations', i.e. those who will be deprived of water due to the long-term depletion of the 'stock'. Other major negative impacts include salinity intrusion (from deeper layers or from the sea) and land subsidence (Custodio, 2002; Zhou, 2009).

Obviously, the magnitude and significance of these impacts depend on the location and the intensity of abstraction. They are also mediated by the time lags and spatial heterogeneity specific to the hydrodynamic properties of a given aquifer. At one end of the range, modest pumping will go unnoticed; at the other, well-owners' benefits will come fully at the expense of those relying on natural groundwater drainage fluxes.

Ironically, there is one scenario (aside from isolated systems) where the RITL principle might apply: when pumping has been so severe that the sum of the aquifer's outflows has finally been zeroed due to the lowering of the water table. The RITL rule, then, is tantamount to stabilising groundwater at a level that cancels out all drainage outflows and associated benefits. Furthermore, the dropping of the water level below that of the river means not only that the aquifer is no longer contributing to river flows and downstream uses, but also that groundwater abstractors now *capture* part of the river flow coming from upstream.

Mistaking the recharge rate for the maximum abstraction level goes hand in hand with the misconception of groundwater as an additional resource instead of a major component of surface water. Globally, this contribution is estimated by Xie et al. (2024) at 59% ± 7%. In France, 60% of surface water comes from the ground (springs or baseflow). Allowing the full aquifer recharge to be pumped amounts to allowing the abstraction of the entire natural discharge (once a new equilibrium has been found), including the river baseflow – unfortunately a frequent phenomenon. The 'double-accounting' of the same groundwater unit, first as part of the recharge of the groundwater stock and then as part of the surface water flow sustained by aquifer drainage, perversely also results in the 'double allocation' of water in river basin development plans.

This unfortunate separation is often legally etched in separate pieces of legislation that ignore the interconnectedness of the resource and inadvertently promotes double-accounting. In California, for example, it was only with the 2014 Sustainable Groundwater Management Act (SGMA) that these links were legally acknowledged. In the Murray-Darling basin, Australia, it took until the early 2000s before the management of hydraulically connected surface and groundwater was fully addressed so as to amend double allocation.

7.3 The Fallacy of the Ritl Rule Illustrated

The commonplace misconceptions involved in the RITL rule have significant implications for the (mis)management of groundwater resources. Examples can be found for any type of aquifer (confined and unconfined) and for any type of porosity (matrix, fracture, conduit) under any climatic conditions. While the dynamics will vary, the principle is the same overall, where there are hydraulic interactions between the aquifer and surface and other water bodies. The RITL approach obfuscates the reality that any pumping rate, even well below the recharge rate, affects the ecosystems or people dependent on the drainage pathways back to the surface. An abstraction rate equal to (RITL), or higher than, the recharge will of course have an impact, but so will a lower value, if to a lesser extent, as shown by numerous case studies.

7.3.1 Falling Groundwater Levels and the Drying Up of Rivers

One invisible and therefore surreptitious impact of groundwater abstraction is the diminution in river baseflow, that is, the groundwater that seeps to riverbeds (Barlow & Leake, 2012). Indeed, incised riverbeds act as drains. Reduced baseflow lessens the river flow itself, potentially turning it into an intermittent stream or even drying it up entirely. The management of the Ogallala aquifer in Kansas, described by Sophocleus (2000), is a textbook example of streams and rivers that historically flowed year-round becoming intermittent due to lowered groundwater tables.

In the 1970s, local Groundwater Management Districts were created to tackle a general decline in the unconfined aquifer of the High Plains. In three districts, the strategy was to allow no more than 40% of the aquifer's storage to be depleted over a period of 20–25 years, because limiting pumping outright to the average recharge rate would have threatened the regional economy. The timescale matched a period of investment in irrigation infrastructure, after which pumping would be limited. In 1991, the districts sought to arrest depletion by limiting pumping to the estimated average natural recharge. However, groundwater levels continued to fall, albeit at a slower pace, while the perennial streams of the 1960s showed a marked reduction in length, and numerous dead trees were observed in the landscape. State authorities finally decided to close some areas to new groundwater development.

The aquifer was initially developed with the assumption that extraction could match recharge. However, this was found to be unsustainable, although not because of groundwater depletion but because of the drop in the natural aquifer discharge to the riverbeds. The RITL approach failed not only to stabilise groundwater levels but also to account for natural groundwater discharge, with consequences for baseflow to rivers and in the visible degradation of riparian vegetation.

A similar phenomenon was observed in the Indo-Gangetic plains in northern India, where groundwater began to be used for irrigation after the Green Revolution. Mukherjee and colleagues (2018) evidenced a long-term downward trend (1999–2013) in the river's water level of several centimetres per year in the dry season, despite an upward trend in annual precipitation. They estimated that the main contribution to the river's baseflow in the dry season was its drainage of groundwater by the river (gaining stream). The river depletion was explained by a reduction in the contribution from groundwater (~−59% in 2016), with some reaches ending up being drained by the aquifers (losing stream). Scattered groundwater withdrawals from adjacent aquifers had gradually lowered the water table, enough to reduce the baseflow in the pre-monsoon season and even to capture the river streamflow in some reaches. The reduction in the Ganges River's pre-monsoon streamflow had severe consequences on domestic water supply, ecological sustainability, and surface-water irrigation.

Prior to this study, the implications of groundwater use for surface-water resources during the pre-monsoon season were not taken into consideration in water accounting, as if the aquifer and the river were separate compartments and issues. Water-allocation plans were missing part of the equation, with likely double-accounting: some of the apparent reduction in aquifer depletion was wrongly attributed to successful policies aimed at curbing pumping, while in fact it was the consequence of water (unnoticeably) missing in the river in the dry season. This common phenomenon was also observed in Spain's Mancha Oriental System, where 406 million m^3 (Mm^3) are pumped annually in an aquifer whose recharge is estimated at 320 Mm^3/year. Water tables dropped less than expected because the aquifer came to capture part of the Jucar River flow (Sanz et al., 2011).

In the US, the issue of wells abstracting groundwater flows to rivers, or even indirectly sucking up the river itself, has challenged the inter-state compacts that govern the sharing of surface-water resources. Kansas sued Nebraska regarding the sharing of the Republican River, claiming that wells were actually subtracting part of the water that would have entered the compact, had it been allowed to reach the river as baseflow. A similar dispute involved New Mexico and Texas against Colorado for violating the Rio Grande compact (Molle & Closas, 2017).

In all similar main alluvial plains or sedimentary plateaus, the invisible yet essential exchanges between streams and shallow aquifers (see detailed overviews in Konikow & Bredehoeft, 2020, and Barlow & Leake, 2012) are modified by groundwater abstraction even where it does not reach or exceed recharge. Globally, but most prominently in (semi)arid environments like the Mediterranean basin or the US Southwest, the most frequent and visible consequence is streams shifting from perennial to intermittent (de Graaf et al., 2019).

7.3.2 Impacting Wetlands and Lakes

Capturing the natural groundwater discharge affects not only streams but wetlands, lakes, and ponds with consequences just as disastrous. In the mouth of the Guadalquivir River in Spain, the Doñana wetlands are illustrative of the complexity of groundwater discharge processes and interactions between water bodies (Green et al., 2024). Abstraction from the deep aquifer, mainly for urban, tourism, and agriculture use, led to water levels dropping up to 20 m between 1972 and 1992. As a result, the connected shallow aquifers started feeding deeper aquifers and the water table dropped, harming wetlands and terrestrial habitats. Some ponds and shallow lakes are now almost permanently dry, and many wetlands visible in historical documents have disappeared. The Court of Justice of the European Union condemned Spain in 2021 for the lack of monitoring and action to prevent damage to Doñana's habitats. Century-old cork oaks have been dying in Doñana National Park (8% loss since 2009) and scrub has now colonised the basins of many dried-up ponds (59% between 1985 and 2018). The diving ducks that once bred in the marsh have disappeared.

In France, the Conie River (Beauce Region) is actually a groundwater resurgence, formerly consisting of marshy and unhealthy land drained by a canal dug at the beginning of the twentieth century. In the 1990s, the pumping of groundwater for agricultural purposes led to a drop in the Beauce water table, causing the Conie River to dry up with severe ecological consequences. Following this, extensive consultation between the government and the agricultural sector led to the establishment of annual pumping quotas varying from 150 to 420 Mm^3/year (200 on average) on top of 120 Mm^3 for domestic and industrial use. The irrigation quota is determined each year based on actual piezometric levels. The fact that withdrawals have to be limited to around 17% of the recharge flux to protect wetlands and river baseflow shows how unsuitable the RITL rule would be (Flipo et al., 2012).

There are countless examples of lakes and wetlands drying up or being severely impacted by unchecked groundwater abstraction in their vicinity (among other stressors), whether following the RITL rule or ignoring the interconnection between surface water and groundwater. These include iconic locations, such as Azraq (Jordan), the Everglades (Florida), Lake Baiyangdian in the Hai river basin (China), Lake Haramaya (Ethiopia), Owens Lake (USA), and the Tablas de Daimiel marshes (Spain). Applying the RITL rule dramatically decreases surface flows and baseflow to the lakes/wetland – to devastating effect. Between 1965 and 2019, for example, the groundwater stored in the Baiyangdian Lake watershed decreased by 49 billion m^3 (Bm3) (Ma et al., 2023). In 2010, annual abstraction was just under 1.1 Bm3, against an aquifer recharge estimated at 1.8 Bm3, yet this did not prevent the lake from turning from being fed by the aquifer to feeding it.

7.3.3 Impacting Groundwater-Dependent Vegetation

Groundwater abstraction following the RITL rule also neglects how the vegetation that relies on a dropping aquifer may be affected by dieback (Nevill et al., 2010), since the depth of the groundwater determines its type and distribution. Deep water tables tend to support drought-adapted shrubs and trees, while herbaceous species often thrive where groundwater is shallow. Interactions can be more complex, as in savannah ecosystems, where groundwater gives trees a competitive advantage over grasses during dry seasons (Rapiya & Ramoelo, 2025). The decline in native plant species in response to groundwater decline may also be followed by an invasion of non-phreatophytic species and an increase in the cover of exotic or invasive plants, possibly creating a new ecosystem. In the Mancha Occidental (Spain), for example, graminoid and herbaceous groundwater-dependent vegetation transitioned to shrubs. In the lower Tarim basin, China, aquifer decline has driven the gradual disappearance of herbaceous taxa, followed by shrubs and then trees (Liu et al., 2005).

It is not only the riparian vegetation along streams or around wetlands that is affected but also that on plains above shallow aquifers. In Australia, floodplain forest dieback between 1990 and 2006 was estimated to have increased from 45% to 70% of the floodplain in the part of the Murray River floodplain in Victoria due to less frequent flooding, dropping water tables, and increasing groundwater salinity (Cunningham et al., 2011). In southwestern Australia, overextraction lowered groundwater levels and resulted in extensive stress and death of Banksia vegetation. In the western US, phreatophytic shrubs and grasses have been greatly reduced in the Owens Valley, as with mesquite woodlands in much of Arizona (Zektser et al., 2005).

A further (little-known) result of retreating groundwater is the weakening of trees, making them vulnerable to insects and pathogens. Resulting tree death was observed in Europe after recent dry spells (e.g. in Germany and the Landes, France).

The quality of the groundwater itself is also affected, as in Western Australia's sandy, unconfined aquifer in the pine plantations of the Gnangara Mound. The water table dropped by 5 m in 20 years and has become acidic (pH < 5) and more mineralised due to soil leaching.

7.3.4 Impacting Water Quality in Coastal Areas

The capture of groundwater discharge can have a direct effect on water quality in coastal aquifers and, therefore on, all those tapping this resource. This is a graphic example of a situation where the RITL rule is tantamount to a disaster: the total salinisation of the aquifer. This is not uncommon. In the Mediterranean basin, a substantial number of coastal aquifers have become saline after groundwater was extracted even at rates lower than the recharge (Mastrocicco & Colombani, 2021).

In the Salinas Valley, California, a large marshland area was reclaimed through massive well development. As early as the 1940s, wells were abandoned due to excessive salinity. Between 2004 and 2018, abstraction rates amounted to 56% of the total recharge, with 78% of the water sources coming from storage depletion (SVBGSA, 2022). Although far below the RITL rule, this drawdown created a reversal in the groundwater gradient, allowing seawater to creep into the aquifer. The injustice of the situation was that pumping in the northern part of the valley, near the sea, was found to be responsible for the saltwater intrusion, yet both the north and south were required to contribute to the cost of resolving the issue. A similar situation was observed in the Pajaro Valley, further south.

As well as the intrusion of marine water, salinisation can be a result of leakage from adjacent or deeper salty aquifers. The detrital coastal aquifer of the Castellón plain, Spain, suffered excessive pumping that led to a drop in the water table of up to 8 m in the late 1970s – enough to cause the intrusion of seawater and sulphate-rich water from the underlying aquifer. The coastal aquifer of the Gaza Strip has witnessed a similar process, with abstraction causing salinity and boron contamination.

In each of these examples, the RITL rule would be self-defeating. Achieving sustainability means observing piezometric levels and/or outflow to the sea in order to avoid degrading water quality. This means reducing the allowable pumping to far below the natural recharge, unless the latter is augmented artificially, which is often the solution that is advanced.

7.4 Time Lags and Spatial Variations Are Not Captured by Water Budgeting

Establishing a water balance involves comparing the available resource against its use and requires the selection of a specific geographical area (usually the approximated extent of an idealised aquifer) and a timeframe (usually a year). Hence, the RITL rule (or any abstraction target) will be based on gross annual 'recharge' or 'renewable resources'. However, this obscures the spatial and temporal variability and complexity of water resources. Fine-tuning abstraction by month may provide room to increase use by distributing the effect across the year.

The Lez River, in the South of France, is fed by one of the largest perennially flowing (in undeveloped conditions) karstic resurgences in Europe. It has supplied the city of Montpellier since 1854, diverting a discharge that reached 400 L/s in 1931. In response to population growth, large pumps were installed in 1965 with the capacity to draw up to 1,000 L/s from the overflow basin. This intensive abstraction meant the Lez River became intermittent, the resurgence overflowing only in the wettest months. In 1983, a new pumping station allowed the abstraction of water directly from a deep conduit 48 m below the outlet sill level, with a maximum allowed flowrate of 1,700 L/s and a maximum drawdown of 30 m below the sill (Maréchal et al., 2013).

During the dry season (spring to summer), pumping temporarily exceeds natural discharge, drawing water from the karst aquifer's storage and lowering water levels below the outlet level. Following the summer, the water table rebounds rapidly in response to rainfall, reaching overflow levels and indicating that withdrawals are lower than actual discharge. In winter, pumping rates remain well below natural discharge, and the Lez Spring flows freely. This seasonal cycle means the karst reservoir functions like a regulated dam: it overflows during periods of high recharge but is depleted by abstraction during drier months. Despite the optimisation of abstraction, the average annual withdrawal is around 1.1 m^3/s against an average natural flow of around 2 m^3/s, and thus remains far below the RITL rule (Maréchal et al., 2013).

However, this optimisation, or so-called 'active management', is only possible if the spring and upper reach of the river are allowed to be dry for several months (63% of the year on average). In order to preserve biodiversity and meet the legal environmental flow in the river upper reach, 180 L/s is abstracted from the pumping conduit and released into the river. And when the –30 m drawdown limit is reached, pumping must stop. At such times, domestic water demand is met by sourcing water from the Rhone River through a 73 km-long inter-basin transfer. However, as releases to the river in summer generally fail to meet the 650 L/s environmental flow required in its downstream reach, or the amenity of the river flowing through the city, it is necessary to 'inject' 500–1000 L/s, also transferred from the Rhone River... This optimised management (with pumping at 55% of recharge) is only possible because the externalities created – flow rates not meeting environmental or amenity requirements – can be offset by an alternative water supply.

Examples such as this underline the importance of analysing seasonal dynamics, particularly in systems where they are slow. When the effect on the aquifer is delayed, abstraction may impact a system over decades. Indeed, local pumping and recharge dynamics are rarely synchronised, causing systemic lags that potentially mask resource depletion when seen through simple budgets focused on storage. This is where the specific nature of the aquifer comes into play, with a wide range of aquifer behaviours dependent on how easily water can move through the formation (via conduits, fractures, or matrix).

Sustainable groundwater use also requires an understanding of how pumping interacts with the aquifer's physical and spatial characteristics. Exploiting the full recharge will be further constrained by the aquifer geometry, hydraulic conductivity, as well as the spatial distribution of abstraction. Romano and Preziosi (2010) illustrated this point in the central Italian aquifer of Petrignano d'Assisi. A well located near a river will have a disproportionate impact on its flow, regardless of the overall recharge. Accounting for the location and respective impact of wells is thus key to the management of water markets in the Lower Murrumbidgee basin (Australia), or to environmental flows in the Jucar River (Spain) (Molle & Closas, 2017).

7.5 Conclusion

Evidence from around the world shows, together with the basics of hydrogeology, that an aquifer's recharge rate does not represent a convenient line in the sand below which abstraction can be considered safe or sustainable. The fact is that abstracting the volume of the recharge will eventually draw all outflows to zero. This will not only have a dramatic effect on surface water (baseflows) and aquatic ecosystems but will often cause a spatial/social reallocation of the resource. Indeed, where the phreatic vegetation, salinity risk, or environmental flow constraints were considered, as illustrated in this chapter, abstraction had to remain much under the recharge value. None of this is news to hydrogeologists, but the puzzling persistence of the RITL rule and the confusion and fallacies it generates warrant the reiteration of the evidence.

Although each situation with regard to the type of aquifer, ecosystems and human use has to be analysed in its own right, the appealing but fallacious 'the recharge-is-the-limit' narrative must be replaced by a trade-off approach, wherein what is gained from pumping is set against an analysis of what is lost and by whom. The result must acknowledge the social process by which a middle ground is deemed 'acceptable' but also the inevitability of any groundwater pumping having an impact on the environment as well as other users. The burden of proof, therefore, is to show that the abstraction of water is safe in all respects.

Further Readings

Barlow, P. M., & Leake, S. A. (2012). *Streamflow depletion by wells—Understanding and managing the effects of groundwater pumping on streamflow* (Circular 1376). U.S. Geological Survey.

Konikow, L. F., & Bredehoeft, J. D. (2020). *Groundwater resource development: Effects and sustainability*. The Groundwater Project. https://gw-project.org/books/groundwater-resource-development/

Sophocleous, M. (2000). From safe yield to sustainable development of water resources—The Kansas experience. *Journal of Hydrology, 235*, 27–43.

Zhou, Y. (2009). A critical review of groundwater budget myth, safe yield and sustainability. *Journal of Hydrology, 370*(1–4), 207–213.

Selected References

Alley, W. M., & Leake, S. A. (2004). The journey from safe yield to sustainability. *Ground Water, 42*, 12–16.

Bredehoeft, J. D. (2002). The water budget myth revisited: Why hydrogeologists model. *Ground Water, 40*(4), 340–345.

Bredehoeft, J. D., Papadopulos, S. S., & Cooper Jr., H. H. (1982). Groundwater: The water budget myth. In *Scientific basis of water resource management* (pp. 51–57). National Academy Press. www.nationalacademies.org/read/19530/chapter/1#viii

Cunningham, S. C., Thomson, J. R., Mac Nally, R., Read, J., & Baker, P. J. (2011). Groundwater change forecasts widespread forest dieback across an extensive floodplain system. *Freshwater Biology, 56*(8), 1494–1508.

Custodio, E. (2002). Aquifer overexploitation: What does it mean? *Hydrogeology Journal, 10*(2), 254–277.

de Graaf, I. E., Gleeson, T., van Beek, L. P. H., Sutanudjaja, E. H., & Bierkens, M. F. (2019). Environmental flow limits to global groundwater pumping. *Nature, 574*(7776), 90–94.

Flipo, N., Monteil, C., Poulin, M., de Fouquet, C., & Krimissa, M. (2012). Hybrid fitting of a hydrosystem model: Long-term insight into the Beauce aquifer functioning (France). *Water Resources Research, 48*(5).

Green, A. J., Guardiola-Albert, C., Bravo-Utrera, M. A., Bustamante, Camacho, A., Camacho, C., …, & Díaz-Delgado, R. (2024). Groundwater abstraction has caused extensive ecological damage to the Doñana World Heritage Site, Spain. *Wetlands, 44*(2).

Liu, J. Z., Chen, Y. N., Chen, Y. J., Zhang, N., & Li, W. H. (2005). Degradation of Populus euphratica community in the lower reaches of the Tarim River, Xinjiang, China. *Journal of Environmental Sciences, 17*(5), 740–747.

Ma, Z., Wang, W., Hou, X., Wang, J., Duan, L., Wang, Y., …, & Li, L. (2023). Examining the change in groundwater flow patterns: A case study from the plain area of the Baiyangdian Lake Watershed, North China. *Journal of Hydrology, 625*, 130160.

Maréchal, J.-C., Vestier, A., Dörfliger, H., & Dörfliger, N. (2013). L'hydrosystème du Lez: une gestion active pour un karst à enjeux. *Karstologia, 62*(2), 1–6.

Mastrocicco, M., & Colombani, N. (2021). The issue of groundwater salinization in coastal areas of the Mediterranean region: A review. *Water, 13*(1), 90.

Molle, F. (2023). Aquifer recharge and overexploitation: The need for a new storyline. *Groundwater, 61*(3), 293–294.

Molle, F., & Closas, A. (2017). *Groundwater governance: A synthesis*. IWMI. http://gw-mena.iwmi.org/outputs/

Mukherjee, A., Bhanja, S. N., & Wada, Y. (2018). Groundwater depletion causing reduction of baseflow triggering Ganges river summer drying. *Scientific Reports, 8*(1), 12049.

Nevill, J. C., Hancock, P. J., Murray, B. R., Ponder, W. F., Humphreys, W. F., Phillips, M. L., & Groom, P. K. (2010). Groundwater-dependent ecosystems and the dangers of groundwater overdraft: A review and an Australian perspective. *Pacific Conservation Biology, 16*(3), 187–208.

Perry, D., Swanson, R. K., & Springer, A. E. (2024). Policy deficiencies and contingency plans: Groundwater management implications for baseflow contributions to the Colorado River. *Frontiers in Environmental Science, 12*, 1444015.

Rapiya, M., & Ramoelo, A. (2025). Groundwater-vegetation interactions in rangeland ecosystems: A review. *Water, 17*(8). www.mdpi.com/2073-4441/17/8/1174

Romano, E., & Preziosi, E. (2010). The sustainable pumping rate concept: Lessons from a case study in central Italy. *Ground Water, 48*(2), 217–226.

Salinas Valley Basin Groundwater Sustainability Agency (SVBGSA). (2022). *Salinas Valley Groundwater Basin, Monterey Subbasin Groundwater Sustainability Plan*. SVBGSA.

Sanz, D., Castaño, S., Cassiraga, E., Sahuquillo, A., Gómez-Alday, J. J., Peña, S., & Calera, A. (2011). Modeling aquifer-river interactions under the influence of groundwater abstraction in the Mancha Oriental System (SE Spain). *Hydrogeology Journal, 19*(2), 475–487.

Sophocleous, M. (1997). Managing water resources systems – Why "safe yield" is not sustainable. *Ground Water, 35*(4), 561.

Theis, C. V. (1940). The source of water derived from wells: Essential factors controlling the response of an aquifer to development. *Civil Engineer, 10*, 277–280.

Xie, J., Liu, X., Jasechko, S., Berghuijs, W. R., Wang, K., Liu, C., ..., & Koirala, S. (2024). Majority of global river flow sustained by groundwater. *Nature Geoscience, 17*(8), 770–777.

Zektser, S., Loáiciga, H. A., & Wolf, J. T. (2005). Environmental impacts of groundwater overdraft: Selected case studies in the southwestern United States. *Environmental Geology, 47*(3), 396–404.

8 *REVISITING*: Planting Trees Will Sustain Springs and Streamflow

Vazken Andréassian

8.1 Introduction

The relationship between trees and water is an age-old question that has been written about since the great Roman naturalist Pliny the Elder. For most of the nineteenth century, a passionate debate raged around the role of forests on streamflow, pitting foresters and water engineers against each other and only calming at the beginning of the twentieth century with the introduction of quantitative water-balance (afforestation and deforestation) experiments. By the 1960s, a wide consensus had been reached on the issue of forest influences: reforestation was agreed to increase evapotranspiration and decrease groundwater recharge and streamflow (including low flows).

Yet, the opportunity for dispute remained since the role of trees in the water cycle contains a fundamental duality. On one hand, they use a lot of water (intercepting some on their leaves or needles, transpiring some through their stomata), which is then unavailable to local streamflow and groundwater recharge. On the other hand, trees return vapour to the atmosphere, which will precipitate somewhere, sometime in the future, so that the continental water cycle only actually loses that vapour which precipitates on the oceans. The difficulty of the debate lies in this duality with characteristic space and time scales that are fundamentally different.[1]

The last 25 years witnessed an unexpected revival of the debate on forest influences as development agencies and non-governmental organisations (NGOs) began to advocate measures that contradicted the established consensus. The idea was introduced that reforestation could not only boost the terrestrial carbon sink but also enhance water resources in general and low flows in particular. They described forests as sponges, storing water in the wet period of the year and releasing it during the dry period. A variety of theories have been offered in support of this view, and their enthusiastic proponents are particularly active on social media. This chapter examines the arguments on either side of the reinvigorated debate.

DOI: 10.4324/9781003745532-8

8.2 A Breakdown in Consensus

8.2.1 What the Established *Science Says*

It is widely admitted by hydrologists that reforestation increases evapotranspiration and thus decreases streamflow. Indeed, it has been demonstrated by numerous controlled catchment-scale experiments (in deforestation and reforestation), which in turn have been extensively reviewed by Hibbert (1967), Bosch and Hewlett (1982), Andréassian (2004), Brown et al. (2005) and Calder (2007). More recent reviews are available in Goeking and Tarboton (2020), van Meerveld and Seibert (2024), and Bruijnzeel et al. (2025).

This established science is based on evidence collected from small catchments (mostly under 1 km^2), using a specific experimental design, the *paired-catchment design*, first implemented in the Colorado mountains of the United States from 1910 to 1926 (Bates & Henry, 1928). This approach gained recognition among foresters and hydrologists and was subsequently adopted in many countries around the world during the twentieth century. The paired-catchment design consists of selecting two neighbouring catchments covered with the same type of forest, as similar as possible, and monitoring them jointly for a number of years. Ideally, this preliminary calibration period is varied enough (i.e. includes both wet and dry years) to describe as completely as possible the hydrology of each catchment (indeed, two catchments might behave similarly in average years but differ in very dry years). At the end of the initial study, forest cover is modified on one of the catchments (the 'treated' catchment), while the other (the 'reference' or 'control' catchment) remains untouched. As long as hydrologic monitoring continues, the relationship established prior to treatment is used to reconstitute the streamflow of the treated catchment (to assess what it would have been without the treatment) and thus to quantify the impact of the intervention directly in mm runoff or m^3/s of flow.

The main hypotheses underlying this experimental design are the following:

- first, it is assumed that the two catchments behave similarly so that the uncertainties in streamflow reconstitution are small (this assumption can be tested during the calibration phase of the experiment);
- second, the catchments are assumed to be subject to the same weather variations (precipitation and potential evaporation) – the reason for choosing neighbouring catchments;
- last, the hydrological behaviour of the reference catchment must be stable throughout the period of study (in particular, its forest cover should neither increase nor decline, and there should not be changes in the species composition).

The paired-catchment design has the important advantage of avoiding the key problems encountered in uncontrolled experiments: climate and inter-catchment variability. Indeed, an experiment relying on a single catchment would be difficult to interpret due to natural climate variability. A design could also be conceived that compared two catchments with different forest cover, but that too would be difficult, if not impossible, to interpret, since without prior calibration no distinction could be made between land-cover impact and the natural variability of catchment behaviour (caused by differences in topography, soil type, or hydrogeological characteristics).

The main conclusions drawn from the paired-catchment experiments of the twentieth century are the following:

- *annual streamflow*: this was found to increase following deforestation and decrease with afforestation. However, there is large variability in the actual amounts, depending on climatic, ecologic, geologic, and soil conditions (see Figure 8.1: although the sign of the changes is clear, there is no simple relationship between the outcome of the experiment and the percentage of catchment that was treated). Also, the impact of the treatment does not last when a forest is allowed to recover after felling, and it can even change sign as young trees regrow rapidly.
- *low flows*: these were shown to increase with felling and decrease with reforestation. A South African study (Scott & Lesch, 1997) provided a complete illustration by following a reforestation experiment with one on deforestation: two catchments were monitored for 16 years after afforestation with eucalypts or pines. The trees were then felled, and the catchments allowed to return to their former state (grassland). Afforestation caused a complete drying of the small intermittent water courses, which stopped flowing completely after nine and 12 years respectively (eucalypts were faster to reduce streamflow). The felling of the eucalypts did allow the low flows to recover to their previous levels, but complete recovery was only obtained after five years;
- *floods*: deforestation was shown generally to increase flood peaks and volumes, although with far more variability than the effect on total flow and even an inversion of the trend in certain years or seasons. Also, the available reforestation experiments show a very limited effect on floods in general and none at all on larger floods. As Cosandey (1995) and Fritsch (1990) note, the increased floods observed in deforestation studies seem to reveal the impact of felling on the soil rather than that of the forest cover itself.

Naturally, the experiments conducted on small, mostly temperate headwater catchments have their limitations, which we will discuss below.

8.2.2 The Alternative *Viewpoints*

Alternative viewpoints are actively espoused by communities opposing (or simply unaware of) the established hydrologic science and focusing on the

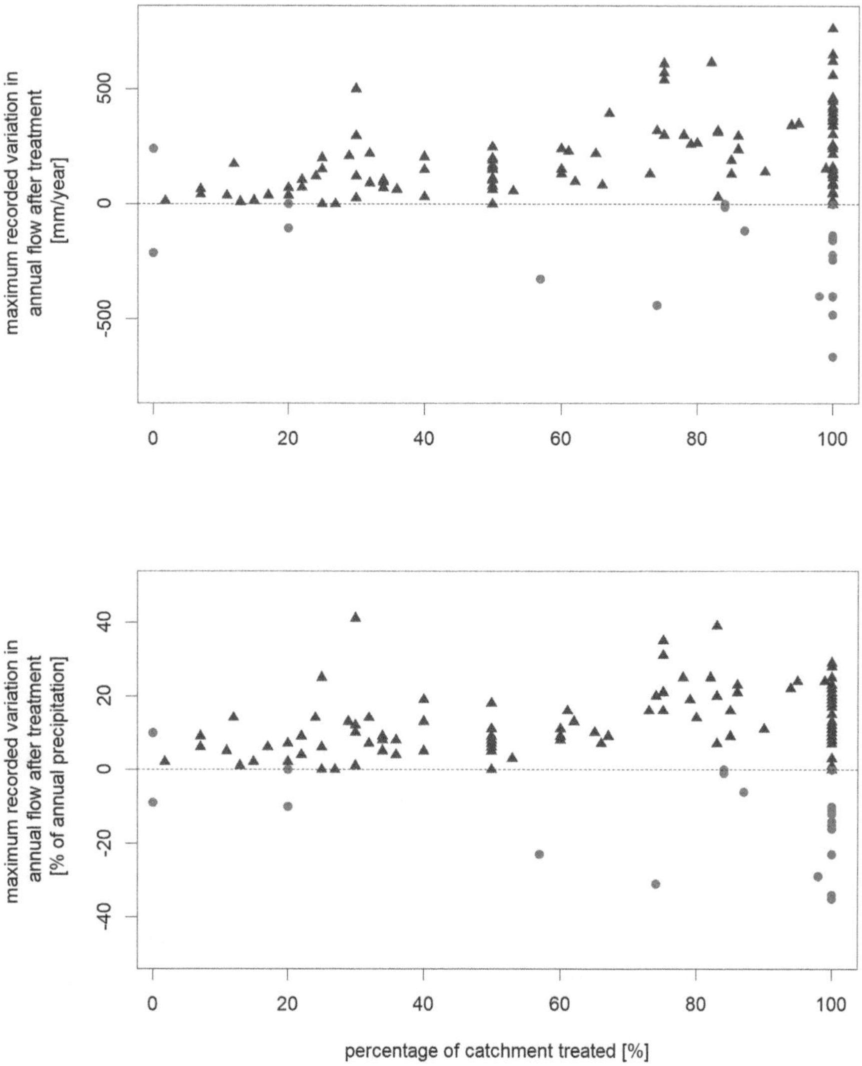

Figure 8.1 Maximum variation in annual flow following treatment as a function of the percentage of catchment subjected to treatment: data from Andréassian (2004) covering 21 reforested catchments (circles) and 115 deforested catchments (triangles).

forest's climatic impact. Since no experimentation is possible at this scale, they rely on theories, historical narratives, anecdotes, or 'Earth system models' (planet-scale computations that are difficult if not impossible to validate). The following are the main alternative claims:

- **Forests 'attract rain', 'promote the formation of clouds' and 'pump moist winds from the sea'**

As living plants, forest trees, shrubs, and grasses are all part of the global water cycle through the interception-transpiration-evaporation process chain. However, it is sometimes claimed that trees hold the special power to 'attract' rainfall and hence offset the higher water consumption of forests. For example, a group called 'L'autoroute de la pluie'[2] (the rainfall highway) promotes the idea of 'increasing rainfall by developing vegetation by adapting the landscapes to increase climatic connectivity and generate coolness and humidity'. Their website states that 'beyond technical considerations, one of our ambitions is to nurture an imaginary world in which great changes are once again possible'. They advocate the planting of a continuous belt with approximately 40 trees/hectare – a density that they relate to the Dogon agroforestry system in Mali.

Attempts to lend scientific justification to the above-mentioned rain-attraction theory have included the argument that forests favour cloud formation. The physical reason given is that forest trees are dark, meaning their albedo is low, and they therefore receive more shortwave energy from the sun. The additional heating makes turbulent currents ascend over the forest. Also, the presence of dust particles in the air is known to aid the formation of droplets in ascending moist air (meteorologists talk about cloud-condensation nuclei).[3] Because forest trees have been shown to release organic compounds in the atmosphere, some forest advocates say they promote the formation of clouds by emitting 'rain seeds': the role of a class of volatile organic compounds known as sesquiterpenes has recently been examined (e.g. Dada et al., 2023).

For the sake of completeness, it should be noted that it is not only forest trees that release sesquiterpenes (among cultivated plants, hops, for example, release significant amounts, but one does not hear of the rain-attraction impact of beer production). Also, there are many particles in the atmosphere that can act as cloud-condensation nuclei, for example, all the atmospheric pollutants released by traffic in large cities (Tiszenkel et al., 2025), but no one rejoices at the rain-attraction impact of urban air pollution. There exists, nonetheless, a relative consensus[4] that it rains more in cities: see the review of Lalonde et al. (2023). Nonetheless, the forest-rain-seeds theory is actively defended by some forest activists.[5]

Almost 20 years ago, the rain-attraction theory received the backing of two nuclear physicists, who published a 'biotic pump theory' (Makarieva & Gorshkov, 2007). According to their analysis, actively evaporating inland forests create a continuously ascending air motion due to 'previously unstudied properties of atmospheric water vapour, which can be either in or out of aerostatic equilibrium depending on the lapse rate of air temperature'. According to the authors, the ascending air above forests 'sucks in' moist air from the ocean, hence the coining of 'biotic pump'.

Although the atmospheric phenomena described in their paper attracted interest, few meteorologists agree with Makarieva and Gorshkov that they can be sufficient to modify atmospheric circulation patterns. Classical meteorology states that these patterns are created by large-scale pressure fields and not by local vegetation. There are further issues with this demonstration. For example, Meesters et al. (2009) write that:

> the fact that regions with natural forest vegetation are also regions of high precipitation is not necessarily proof of a complete redirection of atmospheric circulation by the forest (although there may be some influence). Rather, it is the atmospheric circulation that determines where forests can grow [...] [I]t would follow from consequent application of the 'evaporative force' theory that the oceans, being mostly stronger evaporators than the continents, should themselves 'suck' moisture from the air, and this would cause horizontal flows of moisture at the expense of the moisture supply of the land, directly opposing the biotic pump theory.

To conclude, the issue of the biotic pump should be put in the perspective of the other atmospheric phenomena involving the concentrated transport of moisture – so-called 'atmospheric rivers' (Newell et al., 1992). Described with particular reference to the west coast of the United States, these narrow bands of enhanced tropospheric water vapour transport provide a framework for understanding extreme rainfall and flooding. Atmospheric researchers tend to endorse the concept of atmospheric rivers[6] while rejecting the biotic pump theory.

- **'Forests promote the recycling of atmospheric moisture'**

No one would reject the statement that, *at the global scale*, evaporation and precipitation play a zero-sum game. Indeed, moisture cannot accumulate indefinitely in the atmosphere. Based on this observation, some argue that forests promote the recycling of atmospheric moisture thanks to their larger transpiration. It should be added, however, that irrigated agriculture does the same, and that the huge evaporation of Nile water stored in Lake Nasser (ca 13 km^3/year) also 'recycles' river flow into atmospheric moisture (however, this has not modified the rainfall in the surrounding desert landscape[7]).

To understand the debate on the recycling of atmospheric moisture it is essential to have a clear notion of the scales of time and space involved. Oak trees, sugar cane shrubs and maize plants all produce vapour that ends up as rain somewhere. If the evaporating vegetation is in the middle of Siberia, the evaporated water will very probably fall as rain on the continent. Elsewhere it may well fall on the ocean. The distance involved is typically hundreds of km (as an order of magnitude), and for large regions (e.g., the Amazon or the Congo basin) this effect is certainly relevant.[8] But even for a catchment as large as the Nile basin (cf. the example of Lake Nasser), recycling probably plays a minor or even undetectable role.

- **'Forests promote low flows'**

When working to preserve aquatic ecosystems, the timing of river flow is an important factor to consider, and forests are sometimes hailed by NGOs for their assumed power to increase low flows. Yet this is not supported by small catchment experiments: everywhere a long-term follow-up was available, afforestation had the effect of reducing low-water flows (cf. the study by Scott & Lesch, 1997 and the review by van Meerveld & Seibert, 2024), making it difficult to see where the idea came from. It might be related to the better infiltration capacity of forest soil (although experience has shown that the additional infiltration is usually counterbalanced by additional evaporation during the growing season). However, as Bruijnzeel (1989) and Ilstedt et al. (2016) noted, there must be an optimum tree density that would lead to higher groundwater recharge than with more or fewer trees. Where this optimum lies, and how it can be maintained, remains unclear.

8.3 The Danger of Over-Simplistic Views

8.3.1 A Surprising Disregard for Issues of Scale

The fact is that the established science and the alternative viewpoint on forests disagree on almost everything. To understand the difference in the opposing points of view it is helpful to acknowledge the limits of catchment experiments, particularly the issues they were never meant to address. Providing evidence at the small catchment scale, they do not (and cannot) include continental or planetary processes. They focus on stream- or groundwater-flow and, while it is obvious that the water used by trees leaves the catchment as vapour (either by transpiration or interception), neither catchment experiments nor the established science addresses the question of what happens to the evaporated water.

Obviously, water vapour cannot remain up in the atmosphere indefinitely and will convert to precipitation somewhere sometime, its residence time being on average eight to ten days (Gimeno et al., 2021). Catchment experiments cannot tell us where this precipitation will fall, nor when, nor at what rate, but a growing field of research deals with continental moisture recycling. Closely related to the alternative viewpoint, this new research has shifted the focus of the debate – consciously or otherwise – from local hydrology to the global climate.

The net effect of additional evaporation on precipitation and water resources can only be evaluated by means of numerical models. One such example is that of Hoek van Dijke et al. (2022) who aimed to predict the hydrological impact of a planetary-scale tree restoration project of 900 million ha. Their modelling experiment yielded a complex pattern of water availability shifts, with an increase of up to 6% in some regions… but a decrease of up to 38% in others. Another numerical experiment was that of Asselin et al. (2024) who modelled the effects of large-scale forestation in Europe, showing that it could enhance the recycling of precipitation over continental Europe but not enough to offset the concurrent increase in evapotranspiration. Streamflow was reduced as a consequence. They conclude that this continental-scale reforestation would turn blue water green, bringing heat relief but compromising water availability in some regions. With the current generation of Earth system models, there is no consensus concerning the net effect of large-scale deforestation on regional water resources (e.g. Boysen et al., 2020).

8.3.2 A Parallel with Economic Theory

It could be argued that the established viewpoint misses the global scale… but even if forest hydrology in the twentieth century was shortsighted, this should not allow forest supporters to ignore local effects in the twenty-first century. Here, a parallel can be drawn with an economic theory: Most high-school students have heard of Keynesian theories, and of the *Keynesian multiplicr*[9] that refers to the effect of an increase in public spending causing an even greater increase in private spending within the economic system. In other words, the final change (in output) is more than proportional to the initial change (in input). Yet which of Keynes's admirers would dare to start spending recklessly, alone or in a small group, with the conviction that they would soon benefit from the multiplier effect of their spending?

8.3.3 Over-Enthusiastic Solutions

Several concerns have been raised on the topic of tree planting. The late Ian Calder, an internationally renowned forest hydrologist, warned against the over-enthusiastic view that 'more trees is always better'. He stressed that, while forests may produce water benefits and play an important role in the ecosystem, the world's water resources cannot be managed based on this over-simplistic claim, and stressed that forest policy should aim to ensure that 'forest benefits outweigh water costs' (Calder, 2007) since forests cannot be held good for the water environment in all circumstances (i.e., always increasing rainfall and runoff, reducing erosion and floods, regulating flows and improving water quality).

Several examples could be cited to sustain Calder's worries. He discussed the actual impact of China's very large-scale Sloping Land Conversion Programme, India's plantation forestry programme meant to increase groundwater recharge, and the massive reforestation programme in the Panama Canal catchment. Douville et al. (2024)[10] also raise concern about the current trend for touting large-scale afforestation programmes as the most effective way to increase the terrestrial carbon sink, with an almost complete disregard for the hydrological consequences. They mention in particular the surprising fact that forestry is identified as a priority in terms of adaptation and mitigation in achieving the Paris Agreement goals in 50% of the national contributions. This, they point out, concerns not only forest restoration programmes but also includes large-scale tree planting. Similarly, a review paper on the forest and low-flow controversy by van Meerveld and Seibert (2024) cites various NGOs that seem unaware of the results presented by established science. They show, for example, how the European Commission endorsed unscientific beliefs by stating on the European Environment Agency website that forests are not only effective in 'reducing the damage from flooding' but 'can also help mitigate the effects of droughts' thanks to their 'water-retention potential' (EEA, 2015). The only counter example we identified comes from South Africa, where after decades of encouraging the planting of fast-growing tree species, the country reversed course with its 1998 Water Act, introducing a tax on the planting of trees that are recognised as net water consumers. Also in South Africa, the so-called 'alien plant control project' oversees the removal of non-native species (van Wilgen et al., 2012).

8.4 Conclusion

The present debate feels like a dialogue of the deaf. While enthusiastic defenders of the cause of trees look with faith at the sky and discuss climatic issues at the continental and planetary scales, pragmatic water engineers worry about plans to extend plantation forests, fearing decreased groundwater recharge or inflows to their local reservoirs. It is, in fact, reminiscent of the debate on water and forests in late-nineteenth-century France (Andréassian, 2004), a debate about which the first director of the US Forest Service, Gifford Pinchot, once wrote that it had been 'unfortunate' because 'the friends and the enemies of the forest had both said more than they could prove'.

What remains uncontested is the fact that all plants, wild and cultivated, ligneous and herbaceous, tall and short, are part of the global water cycle through their transpiration and the water they intercept. Healthy catchments need healthy vegetative cover protecting healthy soil. Indeed, the soil's hydraulic properties are essential to catchment functioning, and the hydrogeologic properties of the underlying aquifer are essential to sustaining streamflow in dry periods. As for planet-scale computations of where and when a gram of water vapour will fall as a raindrop, these are still a matter of research. Establishing the actual cost-benefit balance of forests, which directly enhance evaporation and indirectly enhance precipitation, remains a challenge for science.

The degree of uncertainty calls for more nuance in the debate on forest influences. Enthusiasm need not forgive simplistic views. In order to promote sustainable forestry and catchment conservation, it is important to identify, as Calder put it, where and when *forest benefits may outweigh water costs*.

A context-sensitive vision based on this trade-off must replace the reductive narrative that 'more trees are always better'.

Acknowledgements

The author gratefully acknowledges the reviews of Hervé Douville, Julien Lerat, François Molle, Charles Perrin, Jan Seibert, Ilya van Meerveld, and the great editing work of Sophie Bradford.

Further Readings

Andréassian, V. (2004). Waters and forests: From historical controversy to scientific debate. *Journal of Hydrology, 291*(1–2), 1–27. https://doi.org/10.1016/j.jhydrol.2003.12.015

Calder, I. R. (2007). Forests and water—Ensuring forest benefits outweigh water costs. *Forest Ecology and Management, 251*, 110–120. https://doi.org/10.1016/j.foreco.2007.06.015

Laval, K. (2024). *Les pouvoirs de la forêt: De l'eau et des arbres*. Odile Jacob.

van Meerveld, I., & Seibert, J. (2024). Reforestation effects on low flows: Review of public perceptions and scientific evidence. *WIREs Water*, e1760. https://doi.org/10.1002/wat2.1760

Notes

1 Hydrologists would still point out that the presence/absence of trees also affects soil properties, and meteorologists would probably add that trees modify the albedo, roughness length, and Bowen ratio at the land surface. Thus, they alter the interactions between the atmospheric boundary layer and the free troposphere and may affect the precipitation regime, depending on location, season, and place. Obviously, the debate can easily be made even more complex.

2 L'autoroute de la pluie, 2022. www.autoroutedelapluie.org/.

3 Atmospheric physicists have long conducted experiments to 'seed' clouds with silver iodide to increase rainfall. But after decades of research, the atmospheric science community has not been able to conclude on the effectiveness of cloud-seeding experiments. A report released in 2003 by the US National Research Council states: 'science is unable to say with assurance which, if any, seeding techniques produce positive effects. In the 55 years following the first cloud-seeding demonstrations, substantial progress has been made in understanding the natural processes that account for our daily weather. Yet scientifically acceptable proof for significant seeding effects has not been achieved'. National Research Council. 2003. *Critical Issues in Weather Modification Research*. Washington, DC: The National Academies Press. https://doi.org/10.17226/10829. See also Rauber et al. (2019).

4 With naturally a few opposed opinions, e.g., Givati and Rosenfeld (2004).

5 See, for example, the beautifully illustrated video by permaculture advocate Andrew Millison, who argues poetically that 'trees and the atmosphere have a partnership in creating rain': www.youtube.com/watch?v=oY8ds4BiG1A.

6 See the collection of articles available at: https://agupubs.onlinelibrary.wiley.com/doi/toc/10.1002/(ISSN)1944-8007.ATMOS_RIVERS1.
7 Howard Penman, author of the renowned Penman Potential Evaporation formula, discusses this topic in *Vegetation and Hydrology* (1963).
8 van der Ent et al. (2010) provide several examples of continent-wide impact: 'moisture evaporating from the Eurasian continent is responsible for 80% of China's water resources' or 'the main source of rainfall in the Congo basin is moisture evaporated over East Africa, particularly the Great Lakes region'.
9 The British economist John Maynard Keynes based this theory on a mechanism of wealth increase following a rise in the employment rate. Simple in its basic principle, the multiplier effect remains the subject of debate among various schools of economic thought as to its actual repercussions.
10 In an article entitled 'Call for caution regarding the efficacy of large-scale afforestation and its hydrological effects' (Douville et al., 2024).

Selected References

Asselin, O., Leduc, M., Paquin, D., de Noblet-Ducoudré, N., Rechid, D., & Ludwig, R. (2024). Blue in green: Forestation turns blue water green, mitigating heat at the expense of water availability. *Environmental Research Letters, 19*(11), 114003. https://doi.org/10.1088/1748-9326/ad796c

Bates, C. G., & Henry, A. J. (1928). Forest and streamflow experiment at Wagon Wheel Gap, Colorado. *Monthly Weather Review Supplement, 30*, 1–79.

Bosch, J. M., & Hewlett, J. D. (1982). A review of catchment experiments to determine the effect of vegetation changes on water yield and evapotranspiration. *Journal of Hydrology, 55*, 3–23. https://doi.org/10.1016/0022-1694(82)90117-2

Boysen, L. R., Brovkin, V., Pongratz, J., Lawrence, D. M., Lawrence, P., Vuichard, N., Peylin, P., Liddicoat, S., Hajima, T., Zhang, Y., Rocher, M., Delire, C., Séférian, R., Arora, V. K., Nieradzik, L., Anthoni, P., Thiery, W., Laguë, M. M., Lawrence, D., & Lo, M.-H. (2020). Global climate response to idealized deforestation in CMIP6 models. *Biogeosciences, 17*, 5615–5638. https://doi.org/10.5194/bg-17-5615-2020

Brown, A. E., Zhang, L., McMahon, T., Western, A. W., & Vertessy, R. (2005). A review of paired catchment studies for determining changes in water yield resulting from alterations in vegetation. *Journal of Hydrology, 310*, 28–61. https://doi.org/10.1016/j.jhydrol.2004.12.010

Bruijnzeel, L. A. (1989). (De)forestation and dry season flow in the tropics: A closer look. *Journal of Tropical Forest Science, 1*(3), 229–243. https://jtfs.frim.gov.my/jtfs/article/view/2163

Bruijnzeel, L. A., Peña-Arancibia, J. L., Sheil, D., Ziegler, A. D., Zhang, J., Zwartendijk, B. W., Birkel, C., Sun, G., Wang, Y., & Zhang, X. (2025). Potential for improved groundwater recharge and dry-season flows through forest landscape restoration on degraded lands in the tropics. *Forest Ecosystems, 14*, 100376. https://doi.org/10.1016/j.fecs.2025.100376

Calder, I. R. (2007). Forests and water—Ensuring forest benefits outweigh water costs. *Forest Ecology and Management, 251*, 110–120. https://doi.org/10.1016/j.foreco.2007.06.015

Cosandey, C. (1995). La forêt réduit-elle l'écoulement annuel? *Annales de Géographie, 104*(581–582), 7–25. www.persee.fr/doc/geo_0003-4010_1995_num_104_581_13866

Dada, L., Stolzenburg, D., Simon, M., Fischer, L., Kürten, T., Wimmer, D., Rissanen, M. P., Zaidan, M., Lacher, L., Rondo, L., Lóránt, M., Lehtipalo, K., Keronen, P., Tsuru, F., Sipilä, M., Schobesberger, S., Zorn, M., Ehn, M., Kerminen, V.-M., … Worsnop, D. R. (2023). Role of sesquiterpenes in biogenic new particle formation. *Science Advances, 9*, eadi5297. https://doi.org/10.1126/sciadv.adi5297

Douville, H., Allan, R. P., Arias, P. A., & Fisher, R. A. (2024). Call for caution regarding the efficacy of large-scale afforestation and its hydrological effects. *Science of the Total Environment, 950*, 175299. https://doi.org/10.1016/j.scitotenv.2024.175299

EEA (European Environmental Agency) (2015). Water-retention potential of Europe's forests. Technical report 13/2015. EEA.

Fritsch, J.-M. (1990). *Les effets du défrichement de la forêt amazonienne et de la mise en culture sur l'hydrologie des petits bassins versants* [Doctoral dissertation, Université des Sciences et Techniques du Languedoc]. Montpellier. https://horizon.documentation.ird.fr/exl-doc/pleins_textes/pleins_textes_2/etudes_theses/35059.pdf

Gimeno, L., Eiras-Barca, J., Durán-Quesada, A. M., Dominguez, F., van der Ent, R., Sodemann, H., Sanchez-Murilo, R., Nieto, R., & Kirchner, J. W. (2021). The residence time of water vapour in the atmosphere. *Nature Reviews Earth & Environment, 2*, 558–569. https://doi.org/10.1038/s43017-021-00181-9

Givati, A., & Rosenfeld, D. (2004). Quantifying precipitation suppression due to air pollution. *Journal of Applied Meteorology and Climatology, 43*, 1038–1056. https://doi.org/10.1175/1520-0450(2004)043<1038:QPSDTA>2.0.CO;2

Goeking, S. A., & Tarboton, D. G. (2020). Forests and water yield: A synthesis of disturbance effects on streamflow and snowpack in western coniferous forests. *Journal of Forestry, 118*, 172–192. https://doi.org/10.1093/jofore/fvz069

Hibbert, A. R. (1967). Forest treatment effects on water yield. In W. E. Sopper & H. W. Lull (Eds.), *Forest Hydrology, Proceeding of a National Science Foundation Advanced Science Seminar* (pp. 527–543). Pergamon Press.

Hoek van Dijke, A. J., Herold, M., Mallick, K., Benedict, I., Machwitz, M., Schlerf, M., Pranindita, A., Theeuwen, J. J. E., Bastin, J. F., & Teuling, A. J. (2022). Shifts in regional water availability due to global tree restoration. *Nature Geoscience, 15*, 363–368. https://doi.org/10.1038/s41561-022-00935-0

Ilstedt, U., Bargués Tobella, A., Bazié, H. R., Bayala, J., Verbeeten, E., Nyberg, G., Sanou, J., Benegas, L., Murdiyarso, D., Laudon, H., Sheil, D., & Malmer, A. (2016). Intermediate tree cover can maximize groundwater recharge in the seasonally dry tropics. *Scientific Reports, 6*, 21930. https://doi.org/10.1038/srep21930

Lalonde, M., Oudin, L., & Bastin, S. (2023). Urban effects on precipitation: Do the diversity of research strategies and urban characteristics preclude general conclusions? *Urban Climate, 51*, 101605. https://doi.org/10.1016/j.uclim.2023.101605

Makarieva, A. M., & Gorshkov, V. (2007). Biotic pump of atmospheric moisture as driver of the hydrological cycle on land. *Hydrology and Earth System Sciences (HESS), 11*, 1013–1033. www.hydrol-earth-syst-sci.net/11/1013/2007/

Meesters, A. G. C. A., Dolman, A. J., & Bruijnzeel, L. A. (2009). Comment on "Biotic pump of atmospheric moisture as driver of the hydrological cycle on land" by A. M. Makarieva and V. G. Gorshkov. *Hydrology and Earth System Sciences (HESS), 13*, 1299–1305. https://doi.org/10.5194/hess-13-1299-2009

Newell, R. E., Newell, N. E., Zhu, Y., & Scott, C. (1992). Tropospheric rivers? – A pilot study. *Geophysical Research Letters, 19,* 2401–2404. https://doi.org/10.1029/92GL02916

Penman, H. L. (1963). *Vegetation and Hydrology* (Technical Communication No. 53). Commonwealth Bureau of Soils.

Rauber, R. M., Geerts, B., Xue, L., French, J., Friedrich, K., Rasmussen, R. M., Tessendorf, S. A., Blestrud, D. R., Kunkel, M. L., & Parkinson, S. (2019). Wintertime orographic cloud seeding—A review. *Journal of Applied Meteorology and Climatology, 58,* 2117–2140. https://doi.org/10.1175/JAMC-D-18-0341.1

Scott, D. F., & Lesch, W. (1997). Streamflow responses to afforestation with *Eucalyptus grandis* and *Pinus patula* and to felling in the Mokobulaan experimental catchments, South Africa. *Journal of Hydrology, 199,* 360–377. https://doi.org/10.1016/S0022-1694(96)03336-7

Tiszenkel, L., Flynn, J. H., Dodero, A., & Lee, S. H. (2025). Sulfuric acid, base, and low-volatility organics contribute to aerosol nucleation in urban Houston. *Communications Earth & Environment, 6,* 364. https://doi.org/10.1038/s43247-025-02310-4

van der Ent, R. J., Savenije, H. H. G., Schaefli, B., & Steele-Dunne, S. C. (2010). Origin and fate of atmospheric moisture over continents. *Water Resources Research, 46,* W09525. https://doi.org/10.1029/2010WR009127

van Wilgen, B. W., Forsyth, G. G., Le Maitre, D. C., Wannenburgh, A., Kotzé, J. D. F., van den Berg, E., & Henderson, L. (2012). An assessment of the effectiveness of a large, national-scale invasive alien plant control strategy in South Africa. *Biological Conservation, 148,* 28–38. https://doi.org/10.1016/j.biocon.2011.12.035

9 *REVISITING*: Pricing Irrigation Water Will Reduce Its Use

François Molle and Chris Perry

9.1 Irrigation Cost Recovery and the Advent of Efficiency Pricing

Supplying water for irrigation always entails a measure of human labour and capital investment. Concern, therefore, over who should shoulder the financial burden and how it could be recouped was already manifest in colonial settings as well as in the (failed) private investments in the US and Australia in the late nineteenth century. It was to fall away in the four decades following WWII, as irrigation and dams became pivotal in the drive for economic development, food self-sufficiency, and state legitimacy. The prevailing view in the 1970s was that investment costs might be too high for (often poor) beneficiaries to repay and that a 'reasonable' share would be acceptable. In practice, even operation and maintenance (O&M) costs proved difficult to recover, and there was virtually no capital cost recovery (Duane, 1986).

In 1976, a World Bank discussion paper acknowledged that 'efficiency pricing of irrigation water is usually not possible' yet asserted that 'even a nominal price for water would offer users some incentive to eliminate at least some of the conspicuous waste and overwatering [...] which occurs when water is treated as a free good' (Ray et al., 1976). Although a further World Bank report (1981) could not establish a causality between charges and field-level water use efficiency, and neither did a similar study by the Asian Development Bank, the narrative stating otherwise was set in motion.

A further review of loan conditionality and cost recovery in 1986 confirmed that recovery rates of O&M costs ranged from zero to 100%, with most in the 15%–45% bracket (World Bank, 1986). The limited adherence to loan covenants was explained by (a) the lack of government commitment, (b) unreliable water supply due to poor O&M of irrigation systems, and (c) the often-heavy burden of direct and indirect taxes already imposed on the farming sector (World Bank, 1986). The lack of relation between O&M performance and cost recovery questioned the Bank's emphasis on the latter, with Duane (1986) considering the Bank's approach to be 'heavily influenced by its thinking about authorities supplying public utilities such as electricity, water

DOI: 10.4324/9781003745532-9

for domestic use, etc., which were expected to be self-sustained by commercial revenues'.

In retrospect, 1986 was notable for the crystallisation of a consensus on the role of irrigation service fees and their use alongside other mechanisms to improve irrigation performance. As captured by Small and Carruthers (1991), reports by the World Bank, ADB, USAID, FAO, and others all broadly agreed that water charges alone were inadequate to tackle wastage and that the focus should be on controlling water supply and enhancing distribution. The most critical factor was found to be how closely an irrigation agency's operating budget was tied to the revenue generated by irrigation fees. This was described as providing 'an incentive for cost-effective, goal-oriented performance that is otherwise often weak or lacking' (Duane, 1986). In terms of cost recovery, the principle of charging for water was to be taken in the context of the ability to pay and the overall taxation of agriculture.

The 1990s represented a watershed in terms of economic paradigms. Economic instruments and the economic value of water found legitimacy in the 4th Dublin Principle, and Agenda 21 of the UN's Rio Declaration on Environment and Development supported the 'implementation of allocation decisions through demand management, pricing mechanisms and regulatory measures'. Water pricing and economic incentives more generally came to be seen as key policy tools with the potential to achieve multiple objectives. Beyond ensuring the financial autonomy of the irrigation sector and cutting state expenditure, 'efficiency pricing' was expected to encourage water savings and agricultural use with a higher return to water (typically high-value crops). It was even believed that water pricing would maximise the economic efficiency of water use across sectors (favouring uses yielding 'more dollars per m^3 of water'). Such promises inevitably attracted the attention of policymakers, academics, environmental non-governmental organisations (NGOs), development agencies, and banks. They all started to see it as an elegant solution to irrigation's long-standing problems, changing behaviour through incentives while avoiding the painstaking technical, social, and political intricacies of irrigation management.

The economic approach soon percolated into global water policy. The World Bank's Water Resource Management Policy Paper of 1993 observed that 'waste and inefficiencies have resulted from the frequent failure to use prices and other instruments to manage demand and guide allocation'. This helped to establish a powerful narrative around the causal link between water crises, water waste, and underpricing. Indeed, the decade would see a massive upsurge in academic literature, books, reports, and conferences on the subject, as well as mainstream media articles and bold statements. Wolfensohn, the then president of the World Bank, claimed that 'The biggest problem with water is the waste of water through lack of charging'. The pricing principle began to be incorporated into policies, not least the European Water Framework Directive (2000) or water laws in countries such as South

Africa, Brazil and China. The centrality of pricing found its acme, perhaps, with the World Water Commission's (2000) report proclaiming that 'the single most immediate and important measure that we can recommend is the systematic adoption of full-cost pricing for water services'.

Notwithstanding the enthusiasm and hopes vested in the pricing-based policies of the 1990s, a number of findings gradually meant a reassessment was necessary. A review of literature by Bosworth et al. (2002) confirmed earlier studies which had found no relation between water charging and scheme efficiency. While examples of shifts in cropping patterns and technology were more numerous, these were typically explained by a host of inter-related factors, of which water pricing was seldom of more than marginal significance. It thus became apparent that on-the-ground evidence of the impact of economic tools fell well short of expectations and promises. In 2003, the World Bank issued a new strategy for the water resources sector (World Bank, 2003) as an update to the document issued ten years earlier. Stating that 'it is not enough to just extol the virtues of pricing', it acknowledged the 'yawning gap between simple economic principles [...] and on-the-ground reality'. Indeed, the theme of that report, 'Principled Pragmatism', encapsulated the tension between the pricing theorists and the practitioners with field experience. Reflecting this shift in tone, the four 'urgent action[s] needed for water security' issued from World Water Week 2002 in Stockholm made no reference to economic instruments in water management.

Further publications focused on the practical difficulties as well as inconsistencies in some of the theoretical assumptions (Cornish & Perry, 2003; Cornish et al., 2004; Hellegers & Perry, 2006; Molle & Berkoff, 2007; etc.). They stressed the need to distinguish between policy objectives (resource use, financial sustainability, economic benefits, etc.) and pointed out that volumetric pricing was uncommon. They provided insight into why the relationship between 'demand' and volumetric pricing was not as straightforward as it appeared. Dinar and Saleth (2005) acknowledged that

> efficient water pricing schemes are rare, if not completely absent, even in economically advanced regions with extreme water scarcity levels, [which] provides sufficient evidence for the persistence of a vast gap between the development of pricing theory and its practical application.

Dinar and Mody (2004) observed that, though becoming more common, financial cost recovery was hard to implement, noting that pricing rarely elicited more efficient on-farm water use and that, when it did (often through crop shift or technological change), it did not automatically translate into overall water savings.

In other words, a new consensus emerged in the mid-2000s, largely echoing the conclusions established 20 years earlier: water pricing is primarily designed to recover operational and maintenance costs and is heavily

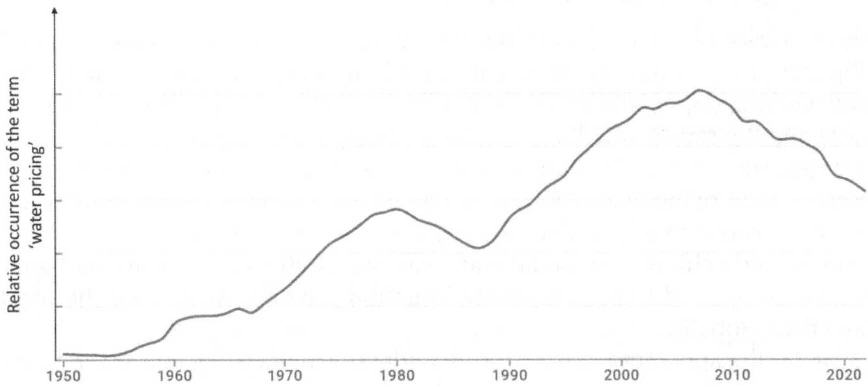

Figure 9.1 Evolution of the occurrence of 'water pricing' in (digital) literature.

influenced by users' willingness and ability to pay. Although conservation and allocation remain key objectives, the ability of water pricing to advance these goals is restricted to very specific conditions.

The issue has resurfaced periodically over the past ten years. There are steps back to orthodoxy (Kochhar et al., 2015) but also continued academic efforts to explore contexts (Dinar et al., 2015) and understand why 'the price of water almost never equals its value and rarely covers its costs' (Grafton et al., 2020).

The fate of 'water pricing' is illustrated in a diagram of the occurrence of the term over the past seven decades (Google Ngram viewer). It confirms the rise and burst of the policy bubble in the mid-2000s but also points to a downward trend that is quite exceptional, since 'best practices' or buzzwords generally keep growing or plateau but seldom decline (Figure 9.1).

9.2 Irrigation Water Pricing and Its Pitfalls

While economic tools are an important lever in water policy, it is becoming clear that they have often been presented as universal principles and endorsed without the necessary qualifications or caveats. Parallels with land rights provide limited guidance in addressing water rights, and comparisons between the water and energy sectors can also be misleading. Further, the debate on water pricing has been obscured by the indiscriminate conflating of cases, from domestic supply to irrigation water, classical, large-scale surface irrigation to pumped irrigation, public schemes to farmer-managed operations, low- to high-tech distribution systems, staple to cash-crop production, or high- to low-income countries. The following section examines these crucial differences.

9.2.1 Large-Scale Public Gravity Irrigation Schemes

The recovery of O&M costs in public irrigation schemes is usually a central objective in national policy, yet it is seldom achieved. This is chiefly due to defaulting (especially in irrigation systems with unpredictable/uneven service) and the state's unwillingness and/or incapacity to impose or levy water charges. This chapter, though, focuses on the narrative that inefficient irrigation (in terms of the ratio of water consumed in crop production to total water supplied) is due to a lack of, or inadequate, water charging and that charging more would achieve substantial water savings. Both causalities are problematic when applied to public gravity irrigation schemes, which are the most prevalent globally.

Firstly, the apparent 'wastage' used to identify inefficiency in irrigation may not be what it seems. Not only do abundant returns flow to drains, rivers, or aquifers go back to the system, but they also recharge aquifers and feed aquatic ecosystems. In the context of monsoons, for example, pricing to reduce demand would be nonsensical. Low efficiency where there is continuous full supply can also be rational, since water is a substitute for management inputs (monitoring, adjusting gates, imposing rotations, etc.).

It can make sense to reduce overall water diversions if excess delivery is lost to productive use (where water percolates to an underlying aquifer that is saline, for example). Yet, the cause of the inefficiency may not lie within the end users' control. Farmers can do little to prevent the system-level losses that typically constitute up to half of the total. System wastage (and shortage) is often due to unpredictable supply to the scheme and poor internal management and/or design, rather than farmer behaviour, and none of these would be affected by increasing the price of water. Unlike in urban, on-demand piped systems, where users normally access water at will, farmers in non-technified systems – common in low-income countries – often only use water when (and if) it is delivered to them and have little influence on this distribution. Inefficient surface irrigation at the plot level is more likely to be due to poor plot levelling than underpricing, and when system management improves, 'wastage' declines, thus lowering the potential for savings from user-level water pricing.

It is understandable that pricing is most often proposed where there is an overall shortage or scarcity of water, and thus minimising real losses becomes a priority. Yet, this is precisely when farmers face the crisis of supply not meeting the requirements of their crops. Blaming them for 'wastage' in such a scenario is illogical and does not do justice to the extraordinary inventiveness they are forced to deploy in times of shortage. Driven by the very high value of the scarce water, they will either access it at their own (additional) expense, for example by tapping alternative sources through pumping, or adapt their farming practices.

Even when water can be considered to be wasted at the farm level, raising the price can only have an impact on demand if use is monitored and charged volumetrically. In other words, the individual farmer can opt for more or less water at the agreed marginal price. This is rare in gravity irrigation systems; farmers are more commonly charged a flat fee based on the area, or sometimes crop, to be irrigated. Therefore, raising the cost would not affect behaviour. It is often asserted that even a nominal charge would provide an incentive to users to reduce 'conspicuous waste'. In practice, however, raising flat fees actually makes farmers feel entitled to use all the water they can access since they are paying for it (Moore, 1989).

9.2.2 Individual Volumetric Use in Collective Irrigation Schemes

It is difficult to pass on efficiency incentives to end users where there is no volumetric monitoring of individual use. An innovative approach here is bulk allocation. This works where a group of users shares, for example, a secondary canal or communal irrigation system supplied by a regulated river system. This can make the operation simpler for the agency and improve the reliability of the bulk supply, which in turn may allow farmers to negotiate allocations among themselves. With the quota effect reducing overall supply (see below), it is this improvement in management that increases productivity and minimises waste rather than pricing, in addition to the fact that charges are generally area-based or too low. Reaching individual farmers with incentives requires well-organised collectives or mid-level managers who can organise collective action for better management (possibly against a financial reward shared with the farmers, as trialled in China). The infrastructure and social organisation needed for this to work are rare.

There are cases where the water abstracted for *gravity irrigation* in collective schemes is metered or estimated at the farm level (typically based on the duration of delivery at a roughly known flowrate). This tends to be found in private water rights systems (e.g. Chile, Western US, Australia), where a fixed, plus a variable, fee is charged to cover O&M costs, but also where hydraulic structures allow such monitoring (e.g. some schemes in California, Spain, Morocco, Mexico). Even where water use could be charged by the volume diverted, it is still often by area because the priority remains the recovery of O&M costs, which may not warrant the substantial administrative/transaction costs of volumetric billing (Cornish et al., 2004). Exceptions include the Chancay-Lambayeque system in Peru, which combines quotas in times of scarcity and individual volumetric pricing (Vos & Vincent, 2011) and some schemes in Morocco where farmers pay by their requested water turn ('arranged demand') and have the option to forego supply.

China, however, stands out for experimenting with a wide range of water pricing schemes (Zhang & Oki, 2023). These initiatives combine infrastructure upgrades to enable metering, crop- and irrigation-specific quotas, targeted subsidies, and gradual price increases to recover O&M costs. This suggests that flexibility, appropriate technology, and robust policy enforcement can create the conditions necessary for quotas and pricing to effectively reduce water diversions.

The gradual modernisation of gravity irrigation schemes, notably in more developed countries, means it is more common to find a network of canals from which collective pump stations distribute water to farm/plot hydrants. Water is more expensive, partly due to water being lifted and pressurised, and lends itself to the possibility of individual pricing. It is worth noting, however, that a parallel with domestic 'on demand' supply might be deceptive, since the water may only be available in turns and networks are often beset by the same problems of uneven pressure and distribution (e.g. the Jordan Valley).

It is notable that where water can be metered, scarcity is actually often managed by quotas (whether individual or collective), with prices serving to discourage, yet permit, marginal use beyond the quota. This configuration resembles a block tariff system, with a first block functioning as a quota and reflecting O&M costs and a second block providing a degree of flexibility when water has a high marginal value in crop production. This is observed for example in Spain, Italy, the south of France, or in Israel, where water used beyond the quota (set annually based on projected water availability) is charged at three or four times the flat rate. The prevalence of quotas can be explained by their effectiveness in limiting demand to the available supply, as well as their simplicity and relative transparency; their relatively low infrastructural and transaction costs; and their limited impact on farm income compared with price-based regulation (Molle, 2009). However, reducing waste and managing supply and demand are distinct objectives. A quota system *always* does the latter and may do the former; pricing may do the former, but never (in the real world) does the latter.

In developed countries, the cost of water provision can be substantial, particularly when long-distance transfers and pumping are required. Charging users for O&M costs can strain the economic viability of certain crops or even of the farm, often prompting subsidies. However, whether these charges cover full O&M costs or only a fraction, they are typically too low to influence water use behaviour (Albiac et al., 2020). Farmers generally respond to pricing signals (demand is 'elastic') only for charges significantly exceeding O&M costs. Yet, setting prices at such levels solely to elicit behavioural change is impractical (Molle & Berkoff, 2007). Farmers are unlikely to accept being charged more than it is costing the state or agency to deliver the water (running costs), especially since that rate may be increased by poor management, overstaffing and, sometimes, corruption. Where greater participatory management or financial autonomy in public schemes is promoted, with users expected to meet O&M costs themselves, there is even less willingness to charge for externalities or opportunity costs. Indeed, Rosegrant and Cline (2002) point to the growing evidence in developing countries that 'farmer demand for irrigation water is price inelastic, and that prices high enough to induce significant changes in water allocation (or recover capital costs) will severely reduce farm incomes'.

Even where it is possible to charge enough for water to cover O&M costs (or more) and enter the 'elasticity range' (which prompts a response in users), it still has to be politically feasible. In economic jargon, elasticity occurs when the price is high enough to force the user to consider the trade-off between the benefit derived from the input and the price paid. Stated simply, the price is harming the farmer to the point that they can only reduce their loss by opting for a second-best alternative (that they had initially rejected as less profitable). In the CACG (Compagnie d'Aménagement des Coteaux de Gascogne) system in south-west France, experts were asked to design a pricing scheme that would cover O&M costs and might also generate water savings. This led to lobbying by large farmers who made sure that the pricing proposal was shelved. Likewise, the provision of costly desalinated water for irrigation in Murcia, Spain, is subsidised in order to make its price acceptable. In other words, water demand tend to be elastic at levels of charge that are politically unfeasible and/or economically unviable.

Farmers who benefit from a pressurised supply will generally employ sprinkler or drip irrigation, whether by choice or because it comes with their scheme's technological package. This means that the prospect for water savings is far lower. In addition, drip-irrigated vegetables or fruit trees generally command high added value and will be largely unaffected by changes in the price of water. In such cases, water can only be 'saved' by reducing allocations.

9.2.3 Prompting Shifts in Cropping Patterns and Irrigation Technology

In theory, high water prices can elicit shifts in cropping patterns and irrigation technology with the potential to lower water demand. However, empirical evidence suggests that such shifts are primarily determined by factors other than the cost of water, notably subsidies, labour shortages, technology, or market opportunity (Molle & Berkoff, 2007). There are many reasons why farmers might choose to irrigate low-value crops. These include environmental conditions (e.g. rice in Asian lowlands), limited market options, aversion to risk or drudgery, or a lack of capital, skill, or labour. They also explain why the farmers had not embraced high-value and less water-intensive crops in the first place. Indeed, decision-making models often fail to capture farmer responses, risks, and difficulties fully, with the implication that they are likely to overstate the mobility of farming systems and their response to pricing.

When water prices rise high enough to incentivise a shift to water-saving crops – or to reduce cultivated area – they often reach levels that severely dent farmers' incomes. Rising prices may crowd out crops with a low water use efficiency (€/m^3), but this transition can be economically devastating if the market lacks alternative crops with comparable returns. Politicians typically seek to avoid such outcomes due to their severe impact on the farming sector. Additionally, higher water prices can exacerbate equity and social challenges. Poorer farmers may be forced out of production, leading to the concentration of land and resources among a few capital-intensive farms.

Furthermore high-value crop patterns are not necessarily less water-intensive, and there is thus also a possible contradiction between aiming for water savings alongside economic efficiency. Drip-irrigated fields may accommodate successive vegetable cycles, fruit trees with longer irrigation periods, and even crops like sugar cane, which can all increase water consumption. High-value crops are also necessarily limited by market demand and account for no more than 7% of the cropped area worldwide. A rush to cash crops, as spurred by water policies and subsidies, frequently results in overproduction, depressed prices, and farm failure (e.g. tomato or citrus in Morocco in recent years).

Similarly, the spread of micro-irrigation is constrained by agricultural and economic factors (drip irrigation in the US, for instance, amounts to no more than 4% of the total irrigated area). A lower water bill can seldom offset the expense of adopting water-saving technology. It is quite natural, therefore, that high-value crops and efficient irrigation often coincide. They tend to occur where farm conditions (capital, skills, information, subsidies, etc.), water supply (secure, typically groundwater-based and on-demand), and market availability (high-value crops for cities or export) favour the development of intensive agriculture.

Even with a net reduction in the per-hectare crop consumption of water, the savings may be used to expand the irrigated area, thereby increasing the overall consumption (this is called 'the rebound effect', as discussed in Chapter 2). Aiming to reduce gross withdrawals, for example, to leave more water in the river, may need additional tools, such as quotas or strong licensing. In any case, technically optimised systems already minimise return flows, limiting the scope to achieve savings through pricing or otherwise.

9.2.4 Individual and Collective Groundwater Pumping

At face value, irrigation from individual or collective tubewells offers better conditions for water pricing to be effective. Rather than drawing on shallow wells or aquifers with a relatively low abstraction cost, water from tubewells is more or less on-demand, and pumping it is expensive. There are examples around the world (including Spain, Syria, and India) of farmers changing crops, reverting to rainfed irrigation, or abandoning fields altogether when the cost of pumping increases, whether due to a fall in the water table and/or energy price hikes. Farmers will generally revert to rainfed farming when no high-value crops are available as a substitute (e.g. those irrigating wheat in Syria). Shah et al. (2009). have documented the impact of increased pumping costs in the Indo-Gangetic basin and identified a series of mitigation strategies that go well beyond the conventional call for reducing alleged 'wastage'. Indeed, whether they adopt deficit irrigation or water-saving crops or technology, reduce the irrigated area, or even abandon agriculture altogether, it generally dents their income considerably. At the other end of the spectrum, in the case of commercial fruits and vegetables farmed with drip irrigation, it is very rare that the cost of water represents a significant share of the overall input expenditure. The elasticity here is nil.

The same is true of groundwater 'cooperatives'. Operating under different labels, modalities, and degrees of dependence from state authority, whether in Bolivia, Turkey, India, Spain, or Mexico, for example (Molle & Closas, 2017), they seek autonomy in terms of management and financial arrangements. This autonomy remains elusive, however, particularly regarding maintenance (e.g. in Tunisia), especially when subsidies are cut or energy services privatised (as in Turkey).

In both these cases, thus, pumping (O&M) *costs* obviously affect water use/savings and crop and technology choices, but this is not the same as a *managed water pricing* policy. In the case of the latter, a *price* would be set in line with the objective of limiting demand to reflect the available supply and allocating that supply to the most productive users. There is hardly any example of state administration imposing a volumetric fee on groundwater abstraction, on top of private O&M costs, with the aim and effect of limiting abstraction. Where this has been the case, like in Jordan, fees have been kept at politically acceptable, but ineffective, levels.

Energy fees, which are often subsidised or regulated by the state, may also offer an indirect way to manage water use through interventions in that sector, such as pricing, subsidies, quotas, and double power lines (domestic and agricultural uses). Independent farmers in Morocco using groundwater juggle the available energy sources (diesel, petrol, gas, solar) in their attempt to keep pumping and farming profitable. But state policies are predominantly driven by social and economic goals, to the detriment of water conservation, and price mechanisms are rarely brought to bear fully.

Two final caveats should be mentioned, where individual water use is constrained, whether by price or otherwise. First, it is commonly observed that users evade expensive or unreliable supply by turning to other sources, typically groundwater if they can find it below their feet (e.g. in California, northern China, or southern Italy). This means the pressure on resources merely shifts. Second, reducing gross withdrawals will mostly translate into reduced return flows to drains, rivers, and aquifers. Where water is scarce, it is very likely that these return flows are already tapped. Efficiency gains actually amount to reallocation (see Chapter 2).

9.3 Conclusion

This chapter examined the conditions under which water pricing in agriculture can enhance efficiency. It argued that pricing is effective only in specific contexts, i.e. those where supply is (quasi)on-demand and (quasi) volumetric, where prices fall within the elasticity range yet remain near operation and maintenance costs, and where the political conditions allow its implementation. In practice, such circumstances are rare and apply mainly to high-cost systems, dependent on pressurisation or tubewells, that are precisely where conveyance losses are already minimal and demand largely inelastic, as in intensive horticulture. While water pricing is essential for cost recovery and financial sustainability, charges rarely cover O&M costs and are hardly conducive to savings. Empirical evidence shows that where scarcity *is managed*, it is generally through quotas, rotations, or allocation rules rather than charging. In pressurised systems, consumption may be indirectly influenced by energy pricing or subsidies, yet social and political constraints discourage measures that significantly affect agricultural income. Ultimately, it is likely that the old, unglamorous issue of managing supply through infrastructure, allocation, and regulation will continue to prevail over that of managing demand through pricing.

Further Readings

Cornish, G., Bosworth, B., Perry, C., & Burke, J. (2004). *Water Charging in Irrigated Agriculture: An Analysis of International Experience* (FAO Waters Reports 28). FAO.

Dinar, A., Pochat, V., & Albiac-Murillo, J. (Eds.). (2015). *Water Pricing Experiences and Innovations* (Vol. 9). Springer International Publishing.

Molle, F., & Berkoff, J. (Eds.). (2007). *Irrigation Water Pricing: The Gap between Theory and Practice*. CABI. https://horizon.documentation.ird.fr/exl-doc/pleins_textes/divers16-09/010045237.pdf

Small, L. E., & Carruthers, I. (1991). *Farmer-financed Irrigation: The Economics of Reform*. Cambridge University Press.

Selected References

Albiac, J., Calvo, E., Kahil, T., & Esteban, E. (2020). The challenge of irrigation water pricing in the Water Framework Directive. *Water Alternatives, 13*(3), 674–690.

Bosworth, B., Cornish, G., Perry, C., & van Steenbergen, F. (2002). *Water Charging in Irrigated Agriculture. Lessons from the Literature* (Report OD 145). HR Wallingford.

Cornish, G. A., & Perry, C. J. (2003). *Water Charging in Irrigated Agriculture: Lessons from the Field* (Report OD 150). HR Wallingford.

Dinar, A., & Mody, J. (2004). Irrigation water management policies: Allocation and pricing principles and implementation experiences. *Natural Resources Forum, 28*(2004), 112–122.

Dinar, A., & Saleth, R. M. (2005). Issues in water pricing reforms: From getting correct prices to setting appropriate institutions. In H. Folmer & T. Tietenberg (Eds.), *The International Yearbook of Environmental and Resource Economics 2005/2006* (pp. 1–51). Edward Elgar.

Duane, P. (1986). Cost recovery in irrigation projects: Perceptions from World Bank operation evaluation. In FAO and USAID (Eds.), *Water Management Synthesis II Project. Technical Papers from the Expert Consultation on Irrigation Water Charges* (Vol. I, pp. 122–157). FAO.

Grafton, R. Q., Chu, L., & Wyrwoll, P. (2020). The paradox of water pricing: Dichotomies, dilemmas, and decisions. *Oxford Review of Economic Policy, 36*(1), 86–107. https://doi.org/10.1093/oxrep/grz030

Hellegers, P. J., & Perry, C. J. (2006). Can irrigation water use be guided by market forces? Theory and practice. *Water Resources Development, 22*(1), 79–86.

Kochhar, K., et al. (2015). *Is the Glass Half Empty or Half Full? Issues in Managing Water Challenges and Policy Instruments* (IMF Staff Discussion Note). IMF.

Molle, F. (2009). Water scarcity, prices and quotas: A review of evidence on irrigation volumetric pricing. *Irrigation and Drainage Systems, 23*(1), 43–58.
Molle, F., & Closas, A. (2017). *Groundwater Governance: A Synthesis* (Report submitted to USAID, Vol. 6). IWMI. http://gw-mena.iwmi.org/outputs/
Moore, M. (1989). The fruits and fallacies of neoliberalism: The case of irrigation policy. *World Development, 17*(11), 1733–1750.
Ray, A., Bruce, C., & Hotes, F. L. (1976). *Cost Recovery Policies for Irrigation Projects: Informal Guidelines*. Paper presented for informal guidance of persons involved in cost recovery policies for irrigation projects. World Bank.
Rosegrant, M. W., & Cline, S. (2002). The politics and economics of water pricing in developing countries. *Water Resources Impact, 4*(1), 6–8.
Shah, T., Hassan, M. U., Khattak, M. Z., Banerjee, P. S., Singh, O. P., & Rehman, S. U. (2009). Is irrigation water free? A reality check in the Indo-Gangetic Basin. *World Development, 37*(2), 422–434.
Vos, J., & Vincent, L. (2011). Volumetric water control in a large-scale open canal irrigation system with many smallholders: The case of Chancay-Lambayeque in Peru. *Agricultural Water Management, 98*(4), 705–714.
World Bank. (1981). *Water Management in Bank-supported Irrigation Project Systems: Analysis of Past Experiences* (Report No. 3421). Operations Evaluation Department, World Bank.
World Bank. (1986). *World Bank Lending Conditionality: A Review of Cost Recovery in Irrigation Projects* (Report No. 6283). Operations Evaluation Department, World Bank.
World Bank. (2003). *World Bank Water Resources Sector Strategy: Strategic Directions for World Bank Engagement*. World Bank.
World Water Commission. (2000). *A Water Secure World: Vision for Water, Life, and the Environment*. World Water Commission.
Zhang, C. Y., & Oki, T. (2023). Water pricing reform for sustainable water resources management in China's agricultural sector. *Agricultural Water Management, 275*, 108045.

10 *REVISITING*: Payment for Environmental Services Is a Win-Win

Jean Carlo Rodríguez-de-Francisco and Audrey Joslin

10.1 Introduction

Payments for Environmental Services (PES), also called Payments for Ecosystem Services, have generated tremendous interest among politicians, conservation practitioners, development agencies, and scholars over the past quarter of a century. Proponents claim that compromise between stakeholders in nature conservation is unnecessary; nature and everyone else will benefit, whether private companies, Indigenous peoples, local communities, urban consumers, governments, or conservation organisations. Everybody wins – or do they?

This chapter interrogates the myth of PES as a win-win and explores why such schemes can, in fact, lead to social injustice and even environmental degradation. While not all experiences with PES will be negative, new proposals continue to emphasise the idea of 'win-win' without much scrutiny. Taking a critical look at PES, this chapter discusses its origins, its application to water, its promises, and its popularity, before examining how the assumptions underlying the myth thwart the win-win theory in practice. Drawing on examples such as watershed PES in Latin America, it discusses issues of measurement, power relationships, and environmental outcomes.

10.2 Markets as the Solution to Environmental Degradation

The concept of 'ecosystem services' offers a utilitarian lens through which to interpret the relationships between humans and nature (Gómez-Baggethun et al., 2010). Ecological economists and conservation biologists initially used it to highlight nature's economic value so as to encourage funding for conservation initiatives (Daily et al., 2000). Notwithstanding the competing conceptualisations of human-environment relationships, the ecosystem services approach gained significant influence in shaping environmental policy worldwide from the late 1990s onward (Norgaard, 2010). The United Nations' Millennium Ecosystem Assessment (2005) solidified its influence by presenting a clear and accessible framework for understanding ecosystem

DOI: 10.4324/9781003745532-10

services. Linking nature to human life, wellbeing, and economic activity, it classified them into four types: provisioning (food, water, materials), regulating (storm mitigation, disease control, air purification), cultural (recreation, spiritual enrichment), and supporting (photosynthesis, nutrient cycling).

As the concept gained traction and scholars made efforts to attach monetary value to their ecosystem benefits, an economic interpretation of environmental degradation also gained popularity. From a neoclassical-economics perspective, environmental problems are recognised as a failure to value nature sufficiently. Many ecosystem services lack recognised monetary value and are treated as public goods. As such, people or businesses can benefit from them without directly paying for their use, but these economic actors can also harm ecosystem services through activities like pollution or deforestation without bearing full social or ecological costs. These costs, known as negative externalities, are a core source of environmental problems as markets do not account for them, and there is no motivation or incentive to avoid them. That said, the Coase theorem suggests that if property rights are clearly defined, transaction costs are low, and information is complete, externalities can be internalised into markets, allowing stakeholders to negotiate and reach mutually beneficial agreements.

PES as a conservation mechanism draws on Coasean logic. It works by assigning economic value to previously unvalued ecosystem services and connecting consumers of these services to those living in and around service-producing areas. In practice, landowners receive compensation – typically a portion of their opportunity costs – for protecting ecosystems that provide crucial services downstream. For example, a landowner may receive payment to conserve forests that mitigate erosion, purify water, and support biodiversity.

Compared to command-and-control approaches, like protected areas, proponents of PES argue that it offers better targeting, greater cost efficiency, and fairer distribution of conservation costs and benefits. As critical areas often overlap with regions of rural poverty (Pagiola et al., 2005), PES continues to be presented as a poverty-alleviation tool. The 'win-win-win' rhetoric promises environmental benefits, poverty reduction, and reduced environmental expenditure (see Chapter 15 on the discourse of conciliation). Advocates portray it as easily implementable and scalable, helping it gain widespread support.

International environmental NGOs (e.g. World Wide Fund for Nature [WWF], The Nature Conservancy [TNC], and Conservation International [CI]), alongside institutions like the World Bank, Global Environmental Facility, regional development banks and research centres (e.g. Center for International Forestry Research [CIFOR]), have been instrumental in promoting PES as a central conservation strategy. These organisations have advanced PES through lobbying efforts, encouraging governments to incorporate it into environmental and development agendas. The World Bank, in particular, shaped the global PES landscape by offering financial resources and technical assistance, enabling initiatives to scale up in low- and middle-income countries. The

NGOs, meanwhile, have spearheaded on-the-ground implementation through partnerships with governments, corporations, and local communities. They have also contributed through applied research, tool development, and capacity-building investments. Scholars, too, have advanced PES by producing analyses and success narratives that frame it as an ideal solution. Collectively, these actors form a robust, transnational epistemic community positioning PES as a mainstream conservation strategy, particularly for water-resource management (Rodríguez-de-Francisco & Boelens, 2015).

10.3 Unpacking Watershed PES: From Textbook Economics to Complex Ecologies

Though exact definitions vary (see Wunder, 2015), over 550 PES schemes currently operate worldwide. Those aimed at conserving hydrological services outnumber other types, with approximately 70% of global PES initiatives centred on watersheds across more than 60 countries (Salzman et al., 2018). Developing economies are a major focus, as PES is seen as offering stability for environmental protection where governmental environmental enforcement capacity is weak.

Watershed PES schemes target spatial areas that are important for water-resource production. They focus on maintaining or improving potable water quality and quantity, as well as regulating hydrological flows to recharge groundwater, mitigate flooding, and ensure steady streamflow. Common strategies include controlling nutrient runoff and erosion, reducing pollution and sediments, and absorbing excess precipitation. As such, participants in targeted areas may enter contracts for activities such as planting trees, removing livestock, abstaining from fire use, ceasing agriculture, embracing new agricultural technology, or guarding land against unsanctioned use (Joslin, 2022). In return, land managers receive direct cash transfers, in-kind transfers, or goods and materials. Governments, private companies, NGOs, or mixed entities typically fulfil the 'buyer' role.

An influential early example is the New York City watershed protection programme. Beginning in the 1990s, the city struggled to meet water regulatory standards. Over 90% of the city's water comes from the Catskill-Delaware watershed, and continued land development resulted in declining water quality. City officials determined they needed either $6 billion for water-filtration facilities or approximately $1 billion for watershed protection to maintain natural filtration (Postel & Thompson, 2005). TNC facilitated the watershed protection option, working with the city to secure land acquisitions and collaborating with local communities, government agencies, and private landowners to initiate programmes incentivising water-quality protection.

TNC used New York City as inspiration for the 'water fund' model, which channels urban user contributions into a trust financing conservation by upstream landholders (Goldman-Benner et al., 2012). The organisation has been championing this model across Latin America since the late 1990s and expanded it to Africa in 2018. Over two dozen Latin American, and more than a dozen African, water funds now exist (TNC, 2023), though few have reached complete operation, limiting robust impact evaluation.

Another prominent watershed PES model is the Reciprocal Agreement for Water, backed by the organisation Rare Conservation – also highly active in Latin America. This community-based model explicitly stresses mutual benefit, trust, and local ownership. Rather than direct cash payments, it focuses on in-kind exchange. While watershed PES models like these attract much attention, their implementation on the ground reveals a more complex reality shaped by diverse ecological, social, and institutional contexts that often diverge from the assumptions of textbook economics. This chapter focuses on these shortcomings, providing only selected key references.

10.3.1 Messy Metrics: Questioning the Economic Foundations of Watershed PES in Unruly Landscapes

PES schemes appeal to policymakers because of their conceptual simplicity. Ecosystems are generally viewed as self-regulating and generating predictable outputs tied to specific land-cover types. Within this logic, paying someone to conserve an ecosystem should produce predictable ecosystem service flows. Consequently, measuring and valuing ecosystem services and conservation labour is highly important. Ecosystem service payers want assurance that their contributions will yield quantifiable environmental outcomes, equal to or greater than their investment.

However, ecosystems are inherently complex and do not conform to the financial logic of consistent, discrete measurement (Robertson, 2004). Numerous factors interacting across space shape hydrological services: land-management practices, precipitation patterns, soil types, vegetation, topography, geology, and water bodies. Every watershed is unique, so streamflow drivers in one area may not apply elsewhere. Interconnected and nested watersheds make defining clear boundaries difficult, as streamflow may connect to subsurface processes in adjacent watersheds. These spatial interdependencies challenge the assumption that ecosystem services can be neatly bounded for economic valuation.

Even if spatial complexity for measuring hydrological ecosystem services could be resolved, temporal variation adds difficulty for establishing reliable baselines. Hydrological conditions vary significantly by season and year-to-year due to climate variability, land-use change, and ecological feedback. Features, such as rivers, may naturally change course and impact surrounding hydrology. Multiyear studies may capture these dynamics better, but they are often too costly and time-consuming for PES implementation timeframes. Climate change further undermines the notion of fixed baselines against which land-management impacts can be evaluated.

Yet, many watershed PES schemes demand immediate intervention to demonstrate progress, secure investors, or meet funder expectations. Because of this pressure, interventions are implemented before their effects are fully understood. Instead, they rely heavily on simplified causal assumptions and proxies to represent ecosystem services, such as land-cover types or management practices, reflecting what is deemed desirable within dominant conservation paradigms rather than what is ecologically verified in specific contexts.

Accurately attributing water quality or quantity changes to specific land-management actions is notoriously difficult. Commonplace myths persist despite limited empirical support: forests increase total water yield (though higher evapotranspiration often reduces it), prevent large-scale flooding (which they rarely do), or always reduce erosion (which depends on deforested land use). These assumptions shape PES interventions, often overlooking ecological nuance or contradicting local knowledge.

When land-management interventions are proposed by outside experts lacking local ecological knowledge, they may have unintended consequences. For example, early Quito Water Fund initiatives centred on afforestation, assuming that tree planting would improve hydrology (see Chapter 8). Communities were incentivised to plant Polylepis, believed to be native. However, locals observed that the particular subspecies grew larger and consumed more water than their local variety, potentially drying nearby streams relied upon for household needs and agriculture. After the trees were planted and concerns raised, a fire swept through and destroyed the plantings. Though unconfirmed, some participants speculated the fire may have been intentionally set in response to the project's unintended impact.

Furthermore, when expected outcomes fail to materialise, or monitoring proves inadequate, PES schemes may terminate, as payers no longer perceive value in their investment (Cremanschi et al., 2012). Others may persist despite limited or no clear evidence of benefit. Rather than being sustained by demonstrated success, many continue due to ideological commitments, institutional inertia, or ongoing external funding.

Far from being offered well-defined and measured, easily tradeable ecosystem services, buyers often lack a clear understanding of what they are purchasing, as the relationship between ecosystem management and resulting services is complex, uncertain and plagued by imperfect knowledge. Given the difficulty of measurement, it is unsurprising that much scholarship questions the market-based characterisation of PES. Kaiser et al. (2023), for example, find that most PES programmes rely on government funding and oversight rather than market transactions, reflecting that ecosystem services resist commodification; they cannot be easily quantified, stored, or traded like conventional goods. The heavy dependence on public support undermines the original vision of PES as an efficient, cost-effective market mechanism intended to reduce reliance on the state, and suggests that the idea of a fully functional, market-based PES system remains largely theoretical.

10.3.2 Social Impact: The Uneven Terrain of Power in Watershed PES

PES is often portrayed as an equitable framework in which all actors collaborate to negotiate responsibilities and compensation. In practice, however, watershed PES schemes operate within entrenched power structures. Downstream actors, such as municipalities and agribusinesses, often hold disproportionate political and economic influence, allowing them to impose environmental policies on upstream communities. This structural imbalance pressures marginalised groups to participate in a system misaligned with their interests. The Coasean assumptions underlying PES fail to account for these power asymmetries, framing upstream communities and downstream users as equals, while obscuring the political dynamics of resource control. The case of Pimampiro, Ecuador – initially promoted as an exemplary model of watershed protection (Wunder & Alban, 2008) – in fact illustrates the political tensions that appear in watershed PES schemes and is useful in showing how inequities materialise in practice.

Coercion masked as free will can characterise PES schemes. Proponents may argue that PES is voluntary; however, in many rural and Indigenous contexts, so-called 'choices' are shaped by limited economic alternatives, power asymmetries, and unclear tenure rights, making participation less a choice than a survival strategy. The Pimampiro scheme imposed limitations on land use for upstream residents while failing to compensate them fully. Local authorities pressured landowners to enrol by invoking threats of legal enforcement or land seizure (Rodríguez-de-Francisco et al., 2013). As such, many participants reported feeling coerced to participate rather than freely choosing to engage.

In addition to coercion from outside the community, power imbalances shaped intra-communal relationships through elite capture (when a small, influential group disproportionately benefits from initiatives intended to serve broader communities). In Pimampiro, a small sub-group of prominent upstream landowners with extensive forested plots was able to influence decisions over PES (i.e., community involvement, distribution of benefits), while other community members faced increased restrictions. The largest landholding family benefitted significantly despite minimal reliance on upstream land, weakening community cohesion and exacerbating inequality (Rodríguez-de-Francisco & Boelens, 2014, 2016). Similar dynamics have been observed elsewhere. In the Philippines and China, PES payments are channelled through municipalities or local groups, and actual individual amounts received are often unclear, raising concerns about transparency and fairness (Chen et al., 2015; Cremanschi et al., 2012).

Beyond inequality within communities, PES can perpetuate long-term regional patterns. In many Global South countries, PES operates in contexts already marked by deep disparities in resource access and control. In Ecuador, for instance, water has long been a contested resource, with historical irrigation rights favouring large estates and excluding small-scale peasant landholders. Pimampiro's case reflects this broader pattern: although upstream communities argued their conservation efforts should entitle them to water rights, these claims were largely ignored. Instead of promoting fairness, the Pimampiro watershed PES model helped preserve the long-standing privileges of wealthier downstream users, leaving smallholders with restricted land and water access and deeper socioeconomic disparities.

Framing PES schemes as technical solutions to environmental problems often conceals the influence of powerful actors in defining what those 'problems' are. This dynamic is evident across many contexts, where the dominant narrative blames poor upstream farmers for water scarcity while overlooking growing downstream commercial demand: greenhouse and rose production in Pimampiro, sugarcane cultivation in the Nima River basin, and soft-drink manufacturing that outcompetes human consumption in Bogotá (Rodríguez-de-Francisco & Budds, 2015; Duarte-Abadía et al., 2023). Water funds in Bogotá and Tungurahua similarly perpetuate blame-shifting by disregarding rural communities' conservation work and framing the inhabitants of the páramo as environmental degraders, obscuring how conservation projects channel rural waters toward capital accumulation sites (Duarte-Abadía et al., 2023). Such selective framing allows powerful interests to deflect responsibility onto marginalised groups.

Broadly, PES can undermine autonomy and self-determination. It frequently targets the territories of Indigenous and local communities – land once dismissed as unproductive but now valued for ecosystem service provision. These hydrosocial territories are complex, socio-historical and geopolitical spaces where economic, environmental and cultural activities intersect. Historically, Indigenous peoples and small, local communities have claimed and defended these territories through legal and political struggles to establish rights. However, PES schemes often function as mechanisms to consolidate external control, reinforcing social hierarchies and perpetuating historical inequities in land and water governance. As a result, they can undermine community self-determination and restrict access to traditionally stewarded resources. In Mexico, PES forest-conservation initiatives have privileged existing power structures and formal landowners, resulting in unequal benefit distribution and limited local autonomy (McAfee & Shapiro, 2010). In Indonesia, PES undermined Indigenous governance systems, especially when customary tenure rights are not formally recognised (Myers et al., 2018). In Tanzania, community-based PES schemes show how elite capture and unclear land rights can limit equitable participation, often benefitting local elites or external actors more than the intended community members (Corbera et al., 2007). Rather than empowering local actors, PES can deepen social inequality.

PES is thus a highly political mechanism that can consolidate control for those already possessing the most political and financial power. It may do little to improve the overall welfare of upstream communities, focusing instead on downstream water security at their expense. While appearing to reward conservation efforts, it often ends up transferring control of the resource to economically powerful actors, disregarding the rights and needs of communities that have long protected these ecosystems. Numerous studies have stressed the urgent need for more social context in future PES research (Kolinjivadi et al., 2023). However, much of the literature in support of PES remains disconnected from the political histories shaping local social and ecological relations.

10.3.3 PES Undermining Environmental Stewardship

Environmental goals may not be best served by PES and may even be made harder to attain. A key aspect of PES arrangements is using incentives to influence land managers to engage in conservation activities that promote ecosystem services. There is an underlying assumption that human engagement with the environment is driven by rational economic logic and that individuals will pursue land-management strategies that maximise financial gain while minimising cost. This is to presume that degrading the environment is the default human behaviour and that conservation requires external financial motivation, ignoring the intrinsic values and practices that may already sustain ecosystems. Predicated on narrow assumptions about value and decision-making, this logic may contribute to environmental harm rather than reduce it.

A fundamental challenge is spatial displacement through 'leakage', whereby deforestation prevented in one location is simply shifted elsewhere. Land managers may preserve ecosystems in areas targeted for payments while continuing or relocating degrading activities to adjacent areas (Le Velly et al., 2017). This reveals a fundamental limitation: PES schemes cannot comprehensively cover all vulnerable areas, and even expanded coverage cannot prevent land managers from shifting activities beyond programme boundaries or to alternative resource uses.

A further problem is that incentives used in PES schemes can end up proving counterproductive and actively undermine conservation. Payments may enable an intensification of environmental degradation when land managers use those funds to expand harmful activities in unprotected areas, for example by purchasing improved equipment, hiring additional labour or acquiring machinery to facilitate operations (Wunder, 2008). The threat of PES restrictions can also trigger defensive responses from those resisting enrolment: in Pimampiro, upstream farmers opposed to participation began cultivating fields previously left fallow for over 20 years to demonstrate productive use and prevent them from being classified as conservation areas (Rodriguez-de-Francisco & Boelens, 2014). In these scenarios, PES directly or indirectly induces environmental harm rather than preventing it.

Beyond spatial and behavioural effects, PES can erode intrinsic environmental stewardship by introducing economic logic into communities where diverse relational values – tied to co-management, spiritual beliefs, local customs and identity – shape human-environment relations. The introduction of external economic incentives risks 'crowding out' these intrinsic motivators (Rode et al., 2015). Individuals may modify previously sustainable behaviours to capture economic incentives, intensify grazing or abandon crop rotation to demonstrate productive land use for PES eligibility, or refuse to perform stewardship previously undertaken without compensation. PES regulatory frameworks can compound this problem by inadvertently disrupting traditional sustainable practices. In Pimampiro, participating farmers maintained continuous cultivation to prevent fallow land from triggering Forestry Law restrictions that would preclude future agricultural use (Rodriguez-de-Francisco & Boelens, 2014). This effectively displaced a traditional rotational fallow system – a practice rooted in intrinsic land stewardship – with continuous cultivation driven by compliance requirements rather than ecological or cultural rationale.

Perhaps most problematic, PES frequently fails to internalise the externalities produced by ecosystem service-buyers themselves, instead redirecting attention from environmental degradation caused by powerful actors toward the activities of rural, often economically marginalised communities (Rodríguez-de-Francisco & Budds, 2015; Duarte-Abadía et al., 2023). Buyers may use PES schemes as tools for greenwashing, deflecting responsibility for their environmental impact onto less powerful actors, while exerting control over upstream territories through PES mechanisms (Kolinjivadi et al., 2019). The Quito Water Fund exemplifies this dynamic: early TNC reports acknowledged that páramo degradation was also driven by large-scale infrastructure projects, including Quito's reservoirs, pipelines, and road construction. Yet the water fund addresses only threats attributed to smallholders, ignoring broader, systemic impacts from powerful actors, while portraying these actors as environmental stewards (Joslin & Jepson, 2018). Similarly, Mexico's national PES programme (PSA-H) frequently favours wealthier, politically connected landowners capable of navigating application requirements, while small, marginalised communities face participation barriers. The scheme emphasises visible, localised forest use while neglecting large-scale deforestation drivers, such as logging concessions or infrastructure projects, thereby displacing responsibility onto rural actors while more powerful contributors are left untouched (Shapiro-Garza, 2013).

10.4 Debunking the PES Myth

While PES offers a theoretically appealing framework for incentivising conservation and sustainable resource management, its 'win-win' promise remains largely unachievable on the ground. The belief that downstream water consumers, local communities providing ecosystem services, economic growth, and ecosystems can simultaneously benefit is an optimistic simplification. Rather, PES schemes may produce unanticipated environmental impacts and social injustices rather than universally beneficial outcomes.

This failure stems from multiple shortcomings. Complex natural processes and local ecological knowledge are frequently sidelined by simplified models ignoring the diversity and intricacies of ecosystem services. Power dynamics skew supposedly apolitical PES schemes, disproportionately benefitting powerful downstream stakeholders at the expense of marginalised upstream communities. These communities may be coerced into restrictive agreements that compromise traditional resource management, exacerbate inequality, and contradict calls for human rights-centred conservation. Meanwhile, the 'leakage' effect means conservation in one area causes environmental degradation in another, while 'crowding out' erodes intrinsic motivations for environmental stewardship. Furthermore, PES schemes can unintentionally promote greenwashing, allowing large organisations to shift blame onto less powerful groups while obscuring their own environmental impacts. Thus, PES is far from the panacea it is often presented to be.

PES supporters might respond, 'If PES is so flawed, why does it continue to grow?', which is a fair question. In part, it reflects strong donor interest, institutional momentum, and the appeal of market-based solutions to global environmental problems. However, growth does not equate to effectiveness or fairness. Many programmes persist due to ideological commitments, sustained external funding, and the political attractiveness of seemingly technical, non-confrontational approaches to conservation, rather than demonstrable success. Evidence of long-term ecological and social benefits remains limited, and few schemes survive once donor support ends. The growth of PES should not be mistaken for proof of its fairness or functionality. Achieving true sustainability requires approaches that genuinely address ecosystem complexity, social equality, and local engagement.

Further Readings

Joslin, A. (2023). Struggles for recognitional justice through payments for ecosystem services contracts in Ecuador's Andes. *Geoforum, 145*, 103832.

Kolinjivadi, V., van Hecken, G., & Merlet, P. (2023). Fifteen years of research on payments for ecosystem services: Piercing the bubble of success as defined by a Northern-driven agenda. *Global Environmental Change, 83*, 102758.

McAfee, K., & Shapiro, E. N. (2010). Payments for ecosystem services in Mexico: Nature, neoliberalism, social movements, and the state. *Annals of the Association of American Geographers, 100*(3), 579–599.

Rodríguez-de-Francisco, J. C., & Boelens, R. (2016). PES hydrosocial territories: De-territorialization and re-patterning of water control arenas in the Andean highlands. *Water International, 4*, 140–156.

Selected References

Chen, C., König, H. J., Matzdorf, B., & Zhen, L. (2015). The institutional challenges of payment for ecosystem service program in China: A review of the effectiveness and implementation of sloping land conversion program. *Sustainability, 7*(5), 5564–5591.

Cremanschi, D. G., Lasco, R. D., & Delfino, R. J. P. (2012). Payments for watershed protection services: Emerging lessons from the Philippines. *Journal of Sustainable Development*, *6*, 90–103.

Corbera, E., Kosoy, N., & Tuna, M. M. (2007). Equity implications of marketing ecosystem services in protected areas and rural communities: Case studies from Meso-America. *Global Environmental Change*, *17*, 365–380.

Daily, G. C., Söderqvist, T., Aniyar, S., Arrow, K., Dasgupta, P., Ehrlich, P. R., … Walker, B. (2000). The value of nature and the nature of value. *Science*, *289*(5478), 395–396.

Duarte-Abadía, B., Galarza Suárez, L., & Hidalgo-Bastidas, J. (2023). ¿Seguridad hídrica urbano-rural en los fondos de agua? Un análisis desde las relaciones de poder, la participación y la co-creación de conocimientos. *Journal of Political Ecology*, *30*(1), 391–400.

Goldman-Benner, R. L., Benitez, S., & Boucher, T. (2012). Water funds and payments for ecosystem services: Practice learns from theory and theory can learn from practice. *Oryx*, *46*, 55–63.

Gómez-Baggethun, E., De Groot, R., Lomas, P. L., & Montes, C. (2010). The history of ecosystem services in economic theory and practice: From early notions to markets and payment schemes. *Ecological Economics*, *69*, 1209–1218.

Joslin, A. (2022). Labor as a linchpin in ecosystem services conservation: Appropriating value from collective institutions? *Capitalism Nature Socialism*, *33*, 90–110.

Joslin, A. J., & Jepson, W. E. (2018). Territory and authority of water fund payments for ecosystem services in Ecuador's Andes. *Geoforum*, *91*, 10–20.

Kaiser, J., Krueger, T., & Haase, D. (2023). Global patterns of collective payments for ecosystem services and their degrees of commodification. *Ecological Economics*, *209*, 107816.

Kolinjivadi, V., van Hecken, G., Almeida, D. V., Dupras, J., & Kosoy, N. (2019). Neoliberal performatives and the 'making' of payments for ecosystem services. *Progress in Human Geography*, *43*, 3–25.

Le Velly, G., Sauquet, A., & Cortina-Villar, S. (2017). PES impact and leakages over several cohorts: The case of the PSA-H in Yucatan, Mexico. *Land Economics*, *93*(2), 230–257.

Millennium Ecosystem Assessment. (2005). *Ecosystems and human well-being*. Island Press.

Myers, R., Larson, A. M., Ravikumar, A., Kowler, L. F., Yang, A., & Trench, T. (2018). Messiness of forest governance: How technical approaches suppress politics in REDD+ and conservation projects. *Global Environmental Change*, *50*, 314–324.

Norgaard, R. B. (2010). Ecosystem services: From eye-opening metaphor to complexity blinder. *Ecological Economics*, *69*, 1219–1227.

Pagiola, S., Arcenas, A., & Platais, G. (2005). Can payments for environmental services help reduce poverty? An exploration of the issues and the evidence to date from Latin America. *World Development*, *33*, 237–253.

Postel, S. L., & Thompson, B. H. (2005). Watershed protection: Capturing the benefits of nature's water supply services. *Natural Resources Forum*, *29*, 98–108.

Robertson, M. M. (2004). The neoliberalization of ecosystem services: Wetland mitigation banking and problems in environmental governance. *Geoforum*, *35*, 361–373.

Rode, J., Gómez-Baggethun, E., & Krause, T. (2015). Motivation crowding by economic incentives in conservation policy: A review of the empirical evidence. *Ecological Economics*, *117*, 270–282.

Rodríguez-de-Francisco, J. C., & Boelens, R. (2014). Payment for environmental services and power in the Chamachán Watershed, Ecuador. *Human Organization, 73*, 351–362.

Rodríguez-de-Francisco, J. C., & Boelens, R. (2015). Payment for environmental services: Mobilising an epistemic community to construct dominant policy. *Environmental Politics, 24*, 481–500.

Rodríguez-de-Francisco, J. C., & Budds, J. (2015). Payments for environmental services and control over conservation of natural resources: The role of public and private sectors in the conservation of the Nima watershed, Colombia. *Ecological Economics, 117*, 295–302.

Rodríguez-de-Francisco, J. C., Budds, J., & Boelens, R. (2013). Payment for environmental services and unequal resource control in Pimampiro, Ecuador. *Society & Natural Resources, 26*, 1217–1233.

Salzman, J., Bennett, G., Carroll, N., Goldstein, A., & Jenkins, M. (2018). The global status and trends of payments for ecosystem services. *Nature Sustainability, 1*, 136–144.

Shapiro-Garza, E. (2013). Contesting the market-based nature of Mexico's national payments for ecosystem services programs: Four sites of articulation and hybridization. *Geoforum, 46*, 5–15.

The Nature Conservancy. (2023). *The journey of Africa's water funds*. The Nature Conservancy. www.nature.org/en-us/about-us/where-we-work/africa/stories-in-africa/water-funds-overview/

Wunder, S. (2008). How do we deal with leakage? In A. Angelsen (Ed.), *Moving ahead with REDD: Issues, options and implications* (pp. 65–75). CIFOR.

Wunder, S. (2015). Revisiting the concept of payments for environmental services. *Ecological Economics, 117*, 234–243.

Wunder, S., & Albán, M. (2008). Decentralized payments for environmental services: The cases of Pimampiro and PROFAFOR in Ecuador. *Ecological Economics, 65*, 685–698.

11 *REVISITING*: Making Space for Water

Jeroen Warner, Dik Roth, and Rens de Man

11.1 The Genesis of 'Making Space' in Flood-Risk Governance

River and sea floods are a major potential threat to societies all over the world. Floods come at increasing human and economic costs due to growing population densities, settlement, and economic infrastructure in vulnerable urban areas, as well as the more erratic weather patterns of a changing climate. In Europe and elsewhere, the 1990s were a decade of growing flood-risk awareness and of rethinking established flood-risk management policies.

Since this time, the concepts of river restoration and the greening of (peri-) urban spaces gained popularity with their use of natural dynamics to reduce flood risk by enhancing storage capacity and restoring natural values, biodiversity, and water quality. Emerging from opposition to fixed and costly 'modern', engineering-based flood-protection measures, the ecological approach takes various forms depending on each country's context in terms of hydrology, socioeconomic and political dynamics, societal values, cultures of participation, modes of thinking about risk, protection and resilience, objectives and priorities, and relationships between spatial and infrastructural measures. Policies may also aim to regenerate landscapes and landforms that strengthen 'natural resilience' or create 'new nature' or opportunities for 'rewilding', tourism, and recreation. Overall, 'creating more space' became a firmly established tenet of flood-risk governance, and with it came attempts towards reconstructing rivers' original courses and floodplains, rehabilitating the vernacular land- and river-management practices that had fallen out of favour since the nineteenth century, such as controlled flooding and living on mounds. 'Making space' initiatives thus tend to hark back to an idealised past of flexible, amphibious living with and on water, implemented in highly engineered, 'fixed' coastal, delta and riverine systems.

In the 1990s, France began developing *Plans de Prévention des Risques* (Risk Prevention Plans: PPR). Germany's 2009 Federal Water Act led to the establishment of flood-generation zones in headwater areas with a high probability of heavy precipitation and a landscape conducive to rapid runoff. Along the lower Rhine River, structural measures (e.g. 'technical retention') and ecological measures (e.g. increasing buffer capacity and natural retention) have

DOI: 10.4324/9781003745532-11

been implemented since 1990. Zevenbergen et al. (2013) identify projects on the Mississippi (1928) and Sacramento rivers (1933) in the US as precursors of the Dutch 'Room for the River' (RftR) policy. Like more recent projects in Greece and Hungary, these were essentially technical interventions, while in Belgium, France and Sweden more participatory processes are found in RftR-related projects. Thailand has its 'monkey cheeks' (*kaem ling*): water retention areas established to protect vulnerable cities like Bangkok. Since the early 2000s, China has sought alternatives to 'hard' infrastructure to create so-called 'sponge cities', while Bangladesh, having enclosed its open wetland structure with embankments and Dutch-style polders, is returning to a (controlled) spatial approach to flood-risk governance known as 'tidal river management'.

A window of opportunity opened in the Netherlands when unexpectedly high river discharge in 1993 and 1995 necessitated the evacuation of 200,000 citizens and, post-event, a round of fast-tracked dyke reinforcement. Policymakers used the opportunity to redirect river management to a national green engineering plan of over 30 local RftR interventions, aiming to make the four main rivers of the central Netherlands (Rhine, Meuse, Waal, and IJssel) flood-proof by 2015 (see e.g. Klijn et al., 2018). Like many innovations, the time was right for the idea of 'green rivers' to take hold in many places at the same time; In the UK, for example, the Environment Agency stopped dredging the Thames in the mid-1990s in favour of a green flood-risk policy, also developing 'green rivers' in close dialogue with like-minded Dutch and other 'green' engineers on the continent.

But the emboldened Dutch went a step further. While implementing Room for the River at home (officially terminated in 2018 but in practice until 2023), they operated an international promotional campaign abroad, encouraging other countries to adopt the Dutch model. The Dutch Enterprise Agency actively promotes it within the toolbox of its Disaster Risk Reduction and Surge Support, recently for Durban, South Africa, and the Brazilian state of Rio Grande do Sul. The low-lying Indian state of Kerala and also New Zealand proved intent on replicating RftR, as did the 'Respecting our River' policy adopted by the Canadian province of Alberta. As in the nineteenth century when the Dutch were invited to France and the UK to share their knowledge on dyke building, their consultants were welcomed in the US following Hurricane Katrina. Indeed, *The New York Times* (e.g. NYT, 2017) regularly celebrates their water solutions, even when the Dutch towns of Valkenburg and Maastricht flooded in summer 2021. The present first author was contracted by the World Bank in 2014 to advise on the replicability of RftR in Bangladesh. A positive answer is not self-evident: while a RftR plan for Colombia's River Cauca involved Dutch consultants, it underestimated the degree of resistance from sugarcane growers wary of hard infrastructural solutions. Its viability was further hampered by weak state control and the lack of a compensation scheme (Moreno-Quintero & Selfa, 2018).

What, then, was the attraction of the Dutch RftR programme? Its two key objectives were enhancing flood protection by creating more space for the main rivers and improving 'spatial quality' (Rijkswaterstaat, 2012). This meant doing away with both the exclusive focus on hard infrastructure and the rigid separation of water and land. While the former objective was approached through the 'hard' numbers of river hydrology, the latter hinged on an extremely elastic neologism that could cover anything from hard infrastructure and housing projects to creating 'new nature', recreational sites, or restoring the river in the name of improving the (environmental, economic, and in some cases social) attractiveness of the landscape. It also entailed a change in river governance. While the previously all-powerful Public Works Department only set the target for reducing peak flood levels at each project site, the regional and local authorities and their networks were given the scope to devise their own plans to reach those targets, albeit in close cooperation with the national authority (Rijke et al., 2012).

Each project utilised a mix of measures, such as relocating dykes, creating bypasses, and deepening the riverbed, often in combination with 'traditional' hard infrastructural measures (see Figure 11.1). RftR produced 'iconic' projects: a bypass to relieve a bottleneck in the River Waal at Lent (near the city of Nijmegen); making space for temporary flood water retention by rebuilding farmsteads on mounds in the Overdiepse polder; and 'depoldering' the Noordwaard polder to increase the intertidal wetland area (Yarina, 2024). The Overdiepse polder project, while not without its controversies (Roth & Winnubst, 2014), was especially successful in showcasing Dutch water engineering expertise internationally and received visiting water experts from all over the world.

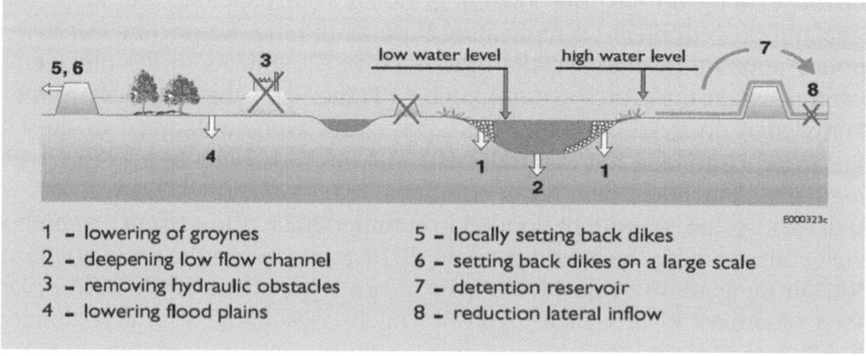

Figure 11.1 Eight different Room for the River measures.

Source: Room for the river | STOWA.

In the 2000s the Netherlands' earlier focus on model-based estimations of peak water levels and related 'design discharges' for embankments gave way to an approach based more on risk and the possibility of embankment 'failure'. Thus, in the post-RftR era,[1] its tenets have continued to support more traditional, 'hard' options – primarily dyke enhancement and the maintenance of other physical infrastructure.

11.2 Challenges and Critiques

'Spatial' approaches to flood-risk governance, such as RftR, are widely celebrated as success stories, and flood security and landscape quality have indeed increased. Nonetheless, the planning and implementation of such complex programmes inevitably encountered obstacles, had unforeseen impacts and were met with resistance on several fronts. In this section we discuss these difficulties with examples from around the world.

11.2.1 Scarce Space and Tensions Between Competing Interests

Projects based on a spatial approach to flood risk mean water literally enters the lifeworlds of citizens and economic activities. This is particularly true for densely populated countries like the Netherlands and Bangladesh. Interventions in the general (national) interest almost inevitably require sacrifice in terms of local interests, potentially involving high economic and other costs and leading to objections and protests. In the Netherlands, plans for controlled flooding in the Ooijpolder sparked friction between layers of government (national and provincial) as well as protests (Roth & Warner 2007). RftR projects often required the resettling of residents and the demolition of houses of cultural-historical value (e.g. Lent, Overdiepse polder and Noordwaard). In France, an NGO (*Levées = danger*) feared the plan to widen the Middle Loire would compromise safety for the sake of ecology. The group scrutinised every detail of the plan, found an error in the hydrological model, and demanded that the modelling be redone. Meanwhile, local mayors opposed a Rhone River restoration project on the grounds that the water entering the floodplain would be polluted and thus threaten organic farming (Guerrin & Barone, 2020). They also pointed to the potential vulnerability of houses in the area to floodwater, despite the extreme rarity of such flooding, and the engineers' not having considered the topography of the plain, which included a motorway and a high-speed rail line. When Thailand's introduction of 'monkey-cheek' retention areas interfered with the fluctuating water requirements of the rice cultivation cycle, it caused crop damage and loss (Trakuldit & Faysse, 2019; Molle et al., 2020). The scope of floodplain spatial plans in Alberta, Canada, was considerably reduced when met with strong opposition from residents facing relocation (Bogdan et al., 2022).

A striking example of the challenges involved in such planning is Bangladesh – a densely populated deltaic country with serious flood threat from both rivers and the sea. In coastal areas, a form of controlled flooding called *ostomashi* – 'tidal river management' – was practised well into the British colonial period. The approach involved making judiciously placed cuts in dykes to let ponded water out while retaining the silt. When over 100 polders were reclaimed in the coastal zone in the 1960s, protected by hard defences, there was widespread ponding ('waterlogging'). To alleviate the resultant muddy landscape, communities reverted to dyke cuts. Since these were now illegal, the government sent enforcers before realising that the technique was actually quite effective. The public works agency BWDB adopted it on a larger scale, albeit with no regard for local variance (Mutahara et al., 2018). The ongoing hydrological, demographic, and socioeconomic delta transformations in Bangladesh are complicated by multiple conflicts between (often contradictory) interests related to land use and water control for (rice) farming, fishing, or settlement, all of which have particular requirements in terms of tidal flows, water levels, and salinity.

Spatial approaches also have distinct limitations in urban environments, not least when planners' imaginations outstrip the physical space available in reality. The megacity of Dhaka, Bangladesh's sprawling capital with aspirations of becoming a 'global city', saw the ambitious Lake Hatirjheel project connect existing urban lakes into one big one, aiming to create a magnificent waterfront while protecting the city from flooding. But slum dwellers were forcibly displaced for a project that ultimately failed since its dimensions proved insufficient to provide flood relief (Nijhum et al., 2019).

In China, rapid urbanisation and industrialisation have dramatically increased downstream flood risk. Persistent lobbying by urban ecologists since the early 2000s, led by landscape architect Kongjian Yu, pressed the government to adopt 'sponge cities' as a green infrastructural alternative to the conventional 'grey' model to manage intense monsoon rainfall. Urban forests, wetlands, agricultural land, public parks, private gardens, ponds, streams and single green elements (street trees, green roofs etc.) are all considered green infrastructure. The goal is to capture water at its source and divert it into winding rivers which slow its flow compared to when it moves in straight lines. A sponge city uses lakes and retention ponds to collect water. These are surrounded by wetland vegetation rather than having large buildings along waterfronts, and coastlines are turned into natural sinks where water can flow out to the ocean. The catastrophic Beijing flood of 2012 and the Guangzhou and Shenzhen floods of 2013 were pivotal moments, after which the sponge city concept was finally adopted in 2014. The approach promotes retention and distribution at the source to prevent rapid runoff as well as the cleaning of polluted runoff. Bioswales, green roofs, and other green and blue spaces were created in major cities. Zeng et al. (2023) argue, however, that the retention capacity and other measures of the sponge city programme will be insufficient to protect China's cities from flooding. This reflects a broader concern that too much rides on the promise of nature-based solutions: it is by no means clear that green infrastructure alone can guarantee the protection of urban centres.

The examples show that spatial approaches may have their disadvantages at the 'receiving end': displaced communities, insufficient space, property damage and local protests. Myopic approaches based on engineering may fail to recognise pre-existing systems that can give room to excess water. In the hunt for spatial solutions, hydrological (and social) realities may become obscured, positive hydrological impacts exaggerated, and an overemphasis put on ideological and financial benefits.

11.2.2 Spatial, Structural, or Both?

Spatial and (infra-)structural approaches should not be seen as a simple dichotomy. 'Spatial' projects often require, or even mainly consist of, large infrastructural interventions like embankments around settlements, river bypasses or retention areas with specifically dimensioned water inlets. As these tend to be costly, authorities may seek to mitigate the expenditure, for example, with riverside property development or selling sand and gravel produced by river widening and deepening works on the construction market, as was seen in the Netherlands' RftR projects. Indeed, where spatial approaches involve significant physical (infrastructure) construction, the distinction between 'spatial' and 'structural' becomes blurred – and the project's priorities unclear to its critics. This happened both in RftR and in more recent interventions under the second Delta programme.

The Netherlands' new Delta programme saw these so-called 'coupled opportunities' (*meekoppelkansen*) strongly promoted within regional planning as a means to address multiple land-use issues, such as urban regeneration and expansion, ecological preservation and infrastructure development, alongside river projects. But alongside the economic and environmental advantages, they can make the project more complicated and even risk overshadowing the goal of flood protection. The proliferation of actors (residents, local government, private developers, NGOs) creates a 'patchwork' of diverse and sometimes conflicting interests that can lack coherence or realism. Plans for the construction of a bypass for the River Waal, for example (post-RftR but a similar intervention), generated serious local opposition and also raised questions as to the necessity and urgency of the water-safety proposal, which was most enthusiastically supported by the province (Gelderland), probably because of the potential co-benefits.

Thus the Netherlands took a relatively broad approach to spatial solutions, developing in line with regional priorities and within a programme of infrastructure. The UK, meanwhile, saw its National Rivers Authority subsumed into the Environment Agency (EA). Created in 1996 with an integrated approach to river management, the EA took a tougher stance on building on flood-prone land. However, its sponsor, the Department of Food and Rural Affairs

(DEFRA), was responsible for drainage, and when it came to funding, it took a much narrower cost-benefit view of green river projects than the Netherlands. The EA's relatively weak regulatory and financial position has prevented an aggressive farmer buy-out scheme as practised by the Dutch authorities.

However, a series of floods in the Midlands (1998, 2000 and 2003) led to the adoption of a RftR policy: in 2005 DEFRA launched the (Labour) government strategy 'Making Space for Water' for flood and coastal erosion risk management. It involved a 25-year plan to restore 75% of the rivers in England to a state 'close to natural'. While the UK's River Restoration Centre reported a 'boost to the feel-good factor within the community' (Mass et al., 2021), projects did not always live up to their promise. The biggest green river project, the Jubilee River bypass on the Thames, was met with anger as it failed to completely prevent flooding and even appeared to increase flood damage at Datchet, where it rejoins the Thames, in 2003, 2014, and 2024 (Warner & Meissner, 2025). This may serve as a reminder that even 'natural' or 'spatial' solutions require 'structural' maintenance. Like all infrastructure, spatial interventions are also technical interventions. Finally, the 'room' created may not always suffice to contain the flood volumes and hence require more 'structural' works (e.g. embankments) to protect adjoining areas against damage.

11.2.3 The Framing and Organisation of Participation

Critical to the success of spatial flood-risk interventions is support from local communities and other interest groups. These projects alter the landscape and living environment, introducing new dynamics and changing land-use patterns. As such, they need inclusive consultation from the start to foster mutual understanding, broad agreement on the need for the intervention, and space for negotiation. Without this basis, projects risk alienating residents, triggering resistance and, ultimately, delaying implementation. In the Netherlands, many RftR projects ran into trouble as early as the planning stage.

Participatory processes have had a faltering introduction into Dutch flood-management projects since around 2000. This unilaterally determined bureaucratic approach often remains limited in scope, as if paying lip service to what is seen as a time-consuming obligation. Focus groups, workshops and similar forums tend to offer little in the way of serious discussion of community input or alternatives. Indeed, local criticism of plans is too easily labelled as NIMBYism. Yet, where concerns are not allayed and there is no other room for dissent, 'participation' is viewed by the local community as misleading or meaningless. In the Ooijpolder, stakeholders realised that the decision had already been made while the participatory process was still under way. Citizens may hesitate to join such official processes for fear of legitimising the very plan they are against. In several RftR cases, eroding trust created a disconnect and fuelled active resistance. An oft-used strategy of citizen groups is presenting counter-expertise, such as retired civil engineers critical of the plans. These power dynamics play a significant role in shaping participation in spatial intervention trajectories. Power frequently resides with developers and government ministries and administrations, whose agendas can dominate decision-making. The concentration of influence raises concerns as to whether the broader public interest and ecological considerations are adequately represented.

The second objective of the Netherlands' RftR programme, 'spatial quality', which carried unexpected weight, exposed the shortcomings of unhelpful power dynamics. Responsibility was handed to so-called 'Q-teams' – Quality Teams of landscape architects. As Hajer et al. (2006) argue, planners are not only tasked with design, but are expected to bring cohesion and concord in a crowded and complex policy field. This approach runs the risk of becoming 'decision by artist's impression' – depoliticised planning in a black box of expert judgement that excludes non-experts and thereby risks derailing sound participatory processes. Moreover, with a growing focus on 'coupled opportunities' (recreational facilities, housing development, nature conservation) there is concern that these are overshadowing or even superseding core flood-safety objectives.

Similar issues with participatory processes are reported from France and Switzerland. In an analysis of floodplain restoration of the River Rhône, participation through consultation committees was, on paper, introduced at an early stage but remained minimal in effect. The idea for the project came from an independent evaluation commissioned by the state prior to the consultative process, but government representatives dominated throughout, leaving no room for participation. Experts and representatives used the forum to convince residents and other stakeholders to support the project, using technical arguments and presenting only its advantages. Critical contributions that highlighted issues of power and inequality, urban and rural, and upstream and downstream, among others were not acknowledged. The 'third correction' of the downstream Rhone in the Swiss canton of Valais ('R3'), the biggest flood-management project in Switzerland to date, was likewise frustrated by narrow and limited participation (Metzger & Véron, 2023). There, however, farmers supported by a right-wing political party managed to break the depoliticised decision-making by successfully demanding a plebiscite, with the upshot of more river deepening than widening and more forest than farmland sacrificed to 'make space'.

Finally, an Asian example of the problematic character of participation concerns the design process of two 'monkey cheek' flood-retention projects in the Chao Phraya river basin in Thailand. Farmers have limited capacity to negotiate their interests and disagreements with the water authority on which they depend for irrigation, and their interests are weakly represented by their official representatives (civil servants who behave like government officials).

11.2.4 Epistemic Capture?

Creating space for rivers is in a discursive field that also comprises nature-based solutions and green infrastructure. Yet, not only is 'nature' in this context far from uncontested, 'nature-based solutions' has been stretched to the point that even large hydropower plants have been given the label.

As Triyanti and Chu (2018) note, however, green infrastructure projects tend to be dominated by ecologists, green engineers, economists and landscape architects; social scientists have barely been involved, let alone non-experts. Likewise, as van Hemert (1999) has it, the design and implementation of RftR projects in the Netherlands make 'room for engineers'. The risk of 'group think' and epistemic capture is real when even key opponents come from the same epistemic circle, with similar ideals of rewilding, which may be at odds with the views of local stakeholders. The Dutch revolving door for water professionals between universities, NGOs, consultants, and policymakers is prone to creating a degree of 'group think' that is hard to challenge. Floodplain restoration projects tend to be presented as economically relevant, paired with the supposed financial benefits of urban renewal and tourism, but end up costing a lot more than planned.

Once concepts such as RftR have been broadly adopted as 'best practice', periodic jamborees such as the World Water Forum and World Water Week can seal their status. They appear globally validated when devoid of the local context and complexities that are decisive in their success. The ubiquitous enthusiasm in the field means they only become controversial once they are ground-truthed at the implementation stage, running into local challenges and even conflict, and seeing significant delays and alternative, compromised designs.

11.3 Conclusion

The spatial turn in flood management has resonated far and wide, enabling bold experiments and impressive results in terms of transformed riverscapes. Yet it has also proved to be full of contradictions. Born of opposition to 'hard' and 'grey' solutions, in practice, the ecological turn (for the Netherlands, see Disco, 2002) in flood policy has manifested more as a complement to such infrastructure. While claiming to be integrative, it has often proved sectoral; while holistic in philosophy, it found itself bogged down in (at times) opportunistic local dealing; and while purportedly participatory, it has become increasingly driven by scientists with their own perspectives on nature and landscape futures. Flood-risk management solutions and regional priorities wedded to green engineering methods can make 'working with nature' seem rather symbolic.

While the 'room for the river' approach was influential in and beyond Europe into the 2010s, the related Sponge Cities concept, which arose in China around the same time, is currently on a global ascent, attracting interest even in the Netherlands and the UK as the 'next big thing' in water management. As an approach, however, it is no less technocratic and thus a missed opportunity in terms of a more integrated take on flood-risk reduction.

Further Readings

Guerrin, J., & Barone, S. (2020). Récits d'action publique et opérations de traductions: la restauration écologique du fleuve Rhône (France) 1. *Politique et Sociétés, 39*(2), 49–79.
Hartmann, T., Slavíková, L., & Wilkinson, M. E. (Eds.). (2022). *Spatial flood risk management: Implementing catchment-based retention and resilience on private land*. Edward Elgar Publishing.
van Staveren, M. (2017). *Controlled flooding in social-ecological delta systems* [PhD dissertation]. Wageningen University.
Warner, J. F., van Buuren, A., & Edelenbos, J. (Eds.). (2012). *Making space for the river*. IWA Publishing.

Note

1 RftR 2.0 was recently announced with objectives of stabilising riverbeds and increasing rivers' discharge and storage capacity during high-water events.

Selected References

Bogdan, E. A., Beckie, M. A., & Caine, K. J. (2022). Making room for nature? Applying the Dutch Room for the River approach to flood risk management in Alberta, Canada. *International Journal of River Basin Management, 20*(2), 153–165.
Disco, C. (2002). Remaking "Nature": The ecological turn in Dutch water management. *Science, Technology & Human Values, 27*, 206–235.
Hajer, M. A., Sijmons, D., & Feddes, F. (2006). *Een plan dat werkt-Ontwerp en politiek in de regionale planvorming*. NAi Uitgevers.
Klijn, F., Asselman, N., & Wagenaar, D. (2018). Room for Rivers: Risk reduction by enhancing the flood conveyance capacity of the Netherlands' large rivers. *Geosciences, 8*(6), 224.
Mass, G., England, J., & Scarr, T. (2021). What's going on in river restoration in England? The blog of the River Restoration Centre, www.therrc.co.uk/blog/whats-going-river-restoration-england
Metzger, A., & Véron, R. (2023). "Sustainable security" through river enlargements: A political ecology of nature-based solutions and flood control in the Rhone Valley, Switzerland. *Développement durable et territoires. Économie, géographie, politique, droit, sociologie, 14*(2), 1–17.
Molle, F., Chompadist, C., & Bremard, T. (2020). *Flood mitigation and the 'monkey cheek' policy in Thailand: Does it hold water?* (Working Paper).
Moreno-Quintero, R., & Selfa, T. (2018). Making space for the Cauca River in Colombia: Inequalities and environmental citizenship. In R. Boelens, T. Perreault, & J. Vos (Eds.), *Water justice* (pp. 134–150). Cambridge University Press.
Mutahara, M., Warner, J. F., Wals, A. E., Khan, M. S. A., & Wester, P. (2018). Social learning for adaptive delta management: Tidal river management in the Bangladesh Delta. *International Journal of Water Resources Development, 34*(6), 923–943.
Nijhum, F. Q., Rahaman, S. T., Hossain, M. J., & Islam, I. (2019). Participatory democracy or state-induced violence? Resettling the displaced people of Hatirjheel in Dhaka. *South Asia Research, 39*(2), 202–217.

NYT. (2017, June 15). The Dutch have solutions to rising seas. The world is watching. *The New York Times*. www.nytimes.com/interactive/2017/06/15/world/europe/climate-change-rotterdam.html

Rijke, J., van Herk, S., Zevenbergen, C., & Ashley, R. (2012). Room for the River: Delivering integrated river basin management in the Netherlands. *International Journal of River Basin Management, 10*(4), 369–382.

Rijkswaterstaat. (2012). *Room for the River. Safety for four million people in the Dutch delta. From higher dykes to river widening. Working together towards a safe and attractive river region* (Room for the River programme).

Roth, D., & Warner, J. (2007). Flood risk, uncertainty and changing river protection policy in The Netherlands. *Tijdschrift voor Economische en Sociale Geografie, 98*(4), 519–525.

Roth, D., & Winnubst, M. (2014). Moving out or living on a mound? Jointly planning a Dutch flood adaptation project. *Land Use Policy, 41*, 233–245.

Trakuldit, T., & Faysse, N. (2019). Difficult encounters around "monkey cheeks": Farmers' interests and the design of flood retention areas in Thailand. *Journal of Flood Risk Management, 12*(Suppl. 2), e12543.

Triyanti, A., & Chu, E. (2018). A survey of governance approaches to ecosystem-based disaster risk reduction: Current gaps and future directions. *International Journal of Disaster Risk Reduction, 32*, 11–21.

van Hemert, M. (1999). Ruimte voor de ingenieur. Rivierbeheer in Nederland eind jaren negentig. *K&M, Tijdschrift Voor Empirische Filosofie, 23*(4), 361–387.

Warner, J., & Meissner, R. (2025). Conflicts over periurban green infrastructure: Unresolved social problems with nature-based urban solutions. *Geoforum, 159*, 104212.

Yarina, L. (2024, February). This river is a model. *Places Journal*. https://doi.org/10.22269/240213. *Accessed 10 March 2026.*

Zeng, C., Aboagye, E. M., Li, H., & Che, S. (2023). Comments and recommendations on Sponge City—China's solutions to prevent flooding risks. *Heliyon, 9*(1), e12745.

Zevenbergen, C., Rijke, J., van Herk, S., Ludy, J., & Ashley, R. (2013). Room for the river: International relevance. *Water Governance, 3*(2), 24–31.

12 *REVISITING*: River Weirs Are Obstacles That Must Be Removed

Régis Barraud and Marie-Anne Germaine

12.1 Introduction

Since the mid-1970s, researchers and environmental activists have popularised the idea of removing dams as a radical solution to environmental degradation caused by intensive river engineering. What was once implausible has become a key part of mainstream ecological restoration programmes. Based on economics and risk management, dam removal aims to restore ecological continuity, particularly for migratory fish (Grant et al., 2017; Perrin, 2018), and the dismantling of major, well-known structures is now commonplace (McCaffery et al., 2024). However, most obstacles that currently fragment watercourses are structures of less than three metres in height (Belletti, 2020) – usually old water mills or irrigation weirs – and their removal can prove a significant challenge for local managers (Barraud, 2017).

The first part of this chapter outlines the theoretical and operational aspects of dam removal as a technical solution, while the second provides a critical analysis of small dam removal as the optimal solution. In line with the book's objectives, we focus here on documenting the pitfalls and unintended consequences of weir removal, providing only selected key references.

12.2 The Trajectory of Dam Removal: From Concept to Implementation

Restoring ecological continuity requires an understanding of the relationship between biological and sedimentary flows and overall ecological quality, including naturalness, biodiversity and heritage value. Projects to restore ecological continuity have benefited from scientific contributions, formalised since the mid-1980s, particularly in river geomorphology and hydrobiology. Among the negative ecological effects associated with the construction of transverse structures, sedimentary and hydrobiological impacts have been particularly highlighted. In the latter case, the impact of weirs and larger dams have attracted most attention, such as the deterioration of water quality, alteration of habitats, homogenisation of flows and fish populations, and significant disruption of fish migration. Indeed, dam removal (DR) is an optimal

DOI: 10.4324/9781003745532-12

technical solution for restoring fish migration; thus, its promotion reflects a failure of previous strategies to assist migration and restock rivers (Thomas & Germaine, 2018), while migratory fish populations continue to decline.

DR was initially conceived in the 1990s as a means of reopening corridors segmented by large dams. The initial steps resulted from three converging dynamics. First, environmental campaigns led by NGOs on both sides of the Atlantic moved on from opposing large dam projects to advocating for the removal of the structures, which were seen as symbols of the ecological damage caused by the 'modern' relationship with nature (Flaminio, 2018). Second, this shift was accompanied by a slowdown in the construction of large dams worldwide and the ageing of existing structures. In the United States, experts alerted the authorities to risks in terms of public safety and financial stability, with production costs and electricity prices ultimately being affected. Thirdly, due to ageing infrastructure, operators found themselves having to consider alternative means of maintaining the dams in their current state. In addition to the cost of repair, there are costs associated with mitigating the ecological impact of these dams, which, in the United States, has provided a 'window of opportunity' to implement an initial series of DRs (Doyle et al., 2003; Drapier et al., 2023).

Since the preservation of wild rivers is a cornerstone of contemporary environmentalism (Palmer et al., 2005), NGOs play a significant part in the global expansion of the movement to remove dams and weirs. In North America, for instance, the NGO American Rivers has made a significant contribution to the growth of DR and maintains a permanent inventory of restored sites. A similar momentum is gaining ground in Europe, notably with organisations such as the WWF and the European River Network. This movement began in 2016 with the establishment of the Dam Removal Europe coalition, which brings together leading international nature conservation organisations. Since 2021 the 'European Open Rivers Programme', funded by the Arcadia Foundation, has lent unconditional support to DR in Europe. Yet, dismantling projects are proving diverse and complex, with their planning phases involving a wide range of institutions, procedures, alliances and opposition.

In the United States, up to five federal agencies may be involved in the decision-making process, together with the highly variable involvement of state administrations (Fostvedt et al., 2020). In Europe, plans remain specific to each country, but instruments such as the European Water Framework Directive (2000), the Eel Regulation (2007) and, more recently, the Nature Restoration Regulation (2024), play a decisive role in shaping public policy. In France, the removal of weirs has been presented as the 'best solution' for the ecological restoration of hydrosystems since the early 2000s. Indeed, a strategic study by the Loire-Bretagne Water Agency (Malavoi & Paris, 2003: 121) asserts that:

> The overall assessment of the advantages and disadvantages of river weirs is largely negative in relation to the objectives pursued by the agency. In

the absence of economic use or major interest, particularly in terms of heritage or landscape, it therefore appears that the best solution is to stop building new weirs and, if possible, to remove or at least significantly reduce the height of existing ones.

Due to the complexity and cost of dismantling large dams, it has proven more convenient to remove small ones (small dam removal, hereafter SDR). Eased by public land ownership in the river valley and either local support for, or a lack of awareness of, the issue, the barriers that are removed tend to be small in size. They belong to an enormous but poorly estimated network of water mills and/or irrigation weirs. In the United States, for example, the national dam inventory lists around 92,000 structures over 7.5 m high but leaves out 2.4 million weirs, most of which are under 1.8 m high.

Europe, meanwhile, is believed to have over a million obstacles fragmenting its rivers, 68% of which are less than two metres high (Belletti, 2020; Zarfl & Lehner, 2020). At least 30% of the obstacles identified are weirs designed to raise the water level and/or power water mills, and around 150,000 are obsolete. In France, the *Référentiel des Obstacles à l'Écoulement* (ROE) reported more than 100,000 in 2024, of which around 15,000 are believed to have been completely or partly destroyed, either deliberately or through neglect. Finally, according to the French Biodiversity Agency (2018), only 10% of barriers would justify total or partial removal due to their impact on ecological continuity.

By 2023, approximately 2,000 dams had been dismantled in the United States, 70% of them in the last 20 years. Highly publicised removals of large dams, such as those carried out in 2024 on the Klamath River (California), mask the fact that most of the operations involve weirs (more than 80%). In California, the Trabuco District Dam Removal Project coordinated by the Cleveland National Forest has led to the removal of 81 weirs in ten years (2014–2024). In partnership with federal agencies, states and Indigenous peoples, American Rivers is currently campaigning to take down 30,000 dams by 2050 (Grabowski et al., 2018). In Europe, 9,003 'barriers' have been removed since 2016, again mainly weirs or culverts. In 2023 only 2% of the 487 obstacles removed were over 5 m high.[1]

12.3 Concerns Over Small Dam Removal

Over the past 20 years small dam removal has emerged as a key mechanism in the ecological restoration of rivers, albeit not without its detractors. The rapid decline in the use of weirs since the second half of the nineteenth century suggests that it is time to reconsider these no longer economically viable structures. The international scientific community has supported, and in some cases backed, the implementation of SDR, highlighting its potential ecological and economic benefits. However, more cautious and critical views on this form of restoration have been present in the fields of environmental and social sciences for over 25 years. Scientific research has covered its hydrological, sedimentary and biological effects (Alp et al., 2024). Doyle et al. (2005) alerted the river management community 30 years ago to the need to accurately assess the potential ecological benefits of weir removal, considering the complexity of time and the nature of the facilities. Indeed, the impact of weirs is still sometimes assessed on the simplistic basis of the results of studies on the effects of large dams. These, of course, have a greater impact, although not proportional to their height (Brown et al., 2024). Other authors have emphasised the need to consider the cumulative impact of weirs in highly anthropised watersheds where rivers are subject to significant pollution and sediment inputs. Conversely, Gangloff (2013: 478) regrets that the 'potentially beneficial effects of small impoundments are often unknown (if not anathema) to conservation practitioners'.

12.3.1 Is There Such a Thing as an 'Ideal River'?

The removal of small dams exacerbates the ambiguity over the reference state used to guide ecological restoration. As with larger dams, the dramatic act of SDR can create the illusion of a return to the past. Yet, clearly, its supporters (scientists and river managers) are not in favour of returning rivers to their pristine, 'wild' state (Dufour et al., 2017). Rather than turn back the clock, the aim is to restore processes that are essential to the health of hydrosystems. At the same time, ecological restoration depends on a certain vision of the ideal river, which can be influenced by imagination and the aesthetics of the wild.

From a technical standpoint, this idealised river is shaped by theoretical models, the interpretation and application of which require scientific updating. Recent studies in the United States and Europe, for example, have shown that morphological forms commonly associated with natural functioning are in fact the product of interactions between physical variables (e.g. climate, hydrology and geomorphology) and human activities (e.g. deforestation, ploughing and hydraulic engineering). Therefore, the form of the single-channel river with regular meanders – used as a reference point for ecological restoration – is the result of these complex interactions. Hydraulic engineering promotes channel rectification and/or stabilises the meanders (Brown et al., 2018).

As Dufour and Piegay (2009) point out, 'a pre-industrial or pre-European settlement state is no longer an achievable reference state'. Nevertheless, examining the long history of river anthropisation is useful in identifying critical processes and clarifying the objectives of a restoration project. For instance, preserving a variety of dynamic wetlands can help to preserve biodiversity, and natural solutions, such as the reintroduction of beavers, can help to achieve this. However, it is also important to recognise that certain inherited landscape features, such as water mills, can be ecologically valuable with a few adjustments. All too often, mill weirs are viewed merely as artefacts. Ecological restoration still struggles to integrate the recent contributions from critical physical geography that encourage us to consider the hybrid nature of long-managed river environments (Dufour et al., 2017) and recognise the value of the long-term interaction of natural processes and human activity (Lespez & Dufour, 2021).

Similarly, models for interpreting the upstream-downstream distribution of fish and benthic communities (organisms living on the bottom of streams) should be used with caution. The discrepancy between these models and the reality of anthropogenic impact on riverbeds over time is so great that rigid application leads to SDR objectives that are overly restrictive and hard to achieve. The historical depth of river engineering, such as water mills and irrigation, has led to a radical transformation of fish communities. This anthropisation has been greatly accentuated by regular restocking and the introduction of non-native species. Indeed, local fishery management persists despite experts calling for a complete overhaul, advocating a more natural approach. The local knowledge and practices associated with these managed rivers – particularly recreational fishing – show the pools formed by the weirs to be refuges for fish during low flow periods. While some sections are less welcoming, others are home to ordinary biodiversity that is often overlooked by public authorities. Similarly, the waterfalls formed by weirs with sufficient flow are identified as a source of oxygenation. Yet, experts are reluctant to accept these nuances or simply see them as anomalies.

Thus, if applied too generally, biophysical models can encourage the concept of an 'ideal river' that is in fact the opposite of the model of a river having developed into a series of stepped pools associated with water mills. The former is based on the free flow of water and the mobility of fluvial features, while the latter suggests the aesthetic value of the waterfall, where water flows over the weir to form a cascade when there is sufficient flow and a pool is formed upstream.

12.3.2 The Hydromorphological and Hydrological Effects of Removing Small Dams

The impact of weirs varies in accordance with hydro-climatic and sedimentary factors (Doyle et al., 2005; Brown et al., 2018). For example, in temperate lowlands, the morphological adjustment of low-energy rivers following weir-removal can be slow to materialise and inconsistent with predictions (Beauchamp et al., 2017). Regressive erosion is often cited as a potential pitfall of SDR under certain hydro-sedimentary conditions (loose and/or mobile sedimentary materials, steep slope, high-energy rivers); however, in most cases, a riffle or hard spot is maintained at the bottom of the channel to prevent this. Contaminants, such as heavy metals, may have accumulated in the sediment, particularly in rivers permanently impacted by industry, and these will be released when the weir is removed (e.g. the Derwent, Derbyshire, UK, Howard et al., 2017).

A limnological approach questions the degree of artificialisation caused by weirs. Donati et al. (2020) challenge the assumption that mill weirs transform free-flowing rivers into stagnant bodies of water. In particular, their thermal effect (warming) does not compare to that of large dams. Moreover, other recent studies have tended to show that weirs and their reservoirs do not increase the potential for denitrification (Hripto et al., 2022).

In hydrological terms, most studies suggest that weirs have little impact on flood control. While this aspect can be easily assessed at the site level, more complex modelling is required when the ecological restoration project involves removing several weirs which spread flood flows laterally to the floodplain. The impact of weirs or their removal on groundwater-river exchange is often poorly assessed. In certain circumstances, the lowering of the water table following the removal of a weir can have negative ecological and/or social impacts. In southern France, five weirs were built in the Gardon d'Anduze in the 1970s and 1980s to compensate for the lowering of the water table caused by gravel extraction. Removing one of these would significantly lower the water table and reduce a nearby wellfield flowrate by 20%–30% (Dewandel et al., 2024).

12.3.3 The Biological Response

The biological response to the removal of a barrier is often rapid in terms of flow, such as the restoration of fish migration and species diversification. However, restoring the fish population structure and the genetic effects of reopening the river corridor take longer. Although migration routes are gradually being re-established, fish flows do not seem to be always recovering. The monitoring of Atlantic shad and salmon migrations in the Loire river basin indicates a severe decline, particularly over the last ten years. In the case of salmon, 'the numbers observed over the last five years have fallen by a factor of eight compared to the estimated reference numbers from the 1970s and 1980s'.[2] The multifactorial sensitivity of migratory fish populations thus contributes to the unpredictability of the ecological restoration process. Low fish stocks may be due to water pollution, parasitism, predators such as martens or cormorants, overfishing in the oceans or the effects of climate change, and may be mistakenly or overly simplistically attributed to dams. On the other hand, the response of living organisms can be interpreted in different ways. For instance, the diversification of fish species caused by the removal of a weir (or a series of weirs) may lead to a reduction in total biomass or the formation of fish populations that are unsuitable for particular recreational fishing. Adverse effects on benthic communities often appear immediately after a structure is removed, but recovery occurs at varying rates thereafter. The health and diversity of these species depend on control variables expressed at the watershed scale. Other studies suggest that particular attention should be given to the presence of pearl mussels downstream of weirs. Populations of this vulnerable mollusc, whose host is brown trout, could benefit from the restoration of flowing water habitats. However, since they are also sensitive to sediment flows generated immediately after a weir is removed, it is sometimes necessary to move them prior to SDR operations.

The effect of SDRs on the spread of invasive species has also been noted. While they initially tend to delay their dissemination, they subsequently create favourable conditions for their development. In all cases, SDR leads to environmental transformation and biological readjustments. It is therefore essential to ensure transparency regarding predicted outcomes (gains, losses and uncertainties) during ecological restoration projects. Furthermore, assessing the impact of SDR on fish populations requires a sufficiently long timeframe to take into account the species' biological cycle and hydrological variability among other features. The spectacular nature of (S)DR can create the illusion of an immediate and complete ecological recovery.

Rigorous monitoring is not always put in place or maintained for long enough to draw robust conclusions. The criteria used to evaluate the results of weir removal are not standardised and reveal strong biases in expertise. For example, in France, the focus is on fish stocks while the riparian vegetation is often neglected. As a result, SDR is designed to benefit specific fish habitats and species but overlooks the adverse effects on wetlands (Dufour et al., 2017).

12.3.4 Patrimonial Values and Attachments

When a particular public policy or project proposal is poorly or inappropriately communicated, it can trigger a 'socioenvironmental' conflict. This is illustrated by the opposition that arise when a dam or weir is framed solely as an engineering structure – essentially no more than a barrier or obstacle (Germaine & Barraud, 2013). In such cases the plan for removal upsets local communities and the users of the sites concerned by failing to acknowledge that, for them, the 'obstacle to flow' is not just a point on a map; a watermill, in particular, will have been a feature of their landscape for generations. Together with the mill race and the mill itself, the artefact of the weir forms a vernacular structure that gives rise to specific practices and attachments. As a landscape motif it lends itself to heritage promotion and has given rise to powerful NGOs or associations in defence of river mills.

In the United States, local opponents to SDR have a similar visceral attachment to their landscape shaped by the triad of mill, weir and pond. This deep connection to place is based on historical, aesthetic and identity values. Indeed, Fox et al. (2016) have recorded respondents as saying 'You kill the dam; you are killing a part of me' and have illustrated the dual process of identification at play in areas affected by SDR: first is the individual's specific commitment to the place and the river as an entity, while the second takes place between the place/river and the community.

It should be noted that scientists have been involved in SDR planning since the early 2000s, identifying certain aspects (e.g. community involvement early on in the process) as critical to the decision-making process. Yet there can still be a significant gap between these recommendations towards caution and inclusion and how SDR is actually carried out.

12.3.5 The Economic and Social Impacts of Small Dam Removal

In France, watermill owner associations argue that SDR threatens the heritage of mills and private property, the economic value of which is tied to the preservation of historic usage rights attached to the property (Drapier et al., 2023). These stakeholders also argue that the financial benefit of producing small-scale hydroelectricity adds to this heritage value. They contend that such production contributes to the energy transition and, therefore, to the global fight against climate change.

An SDR project inevitably raises concerns about water shortages, ecological deterioration and a decline in landscape quality. However, for local users, such transformation may also be taken as a threat to their recreational activities on the river (such as fishing, boating and relaxing), activities which form the foundation of their relationship with the river (Germaine et al., 2022). These recreational uses, historically facilitated by weirs, have shaped an ideal of the river. The local authorities, meanwhile, struggle to justify the cost of removing the weirs in terms of public investment, having hitherto financed their restoration and maintenance, not to mention the fishpasses they may have added to the structure. (The cost of fishpasses on medium-sized streams in France is typically 500,000 to 1 million euros). Removing a structure is generally considered to be less expensive than rebuilding and maintaining it. In Wisconsin, a 2004 financial assessment suggested that rebuilding would cost six times more than demolition. However, this is debatable, given that the cost of river interventions has increased dramatically over the past 20 years, in part because the SDR industry capitalised on this new source of public funding. Most removal or reconstruction operations require funding of several hundred thousand euros.

SDR continues to prove controversial in Europe, with opposition increasingly evident in countries such as France and Spain (Brummer et al., 2017). While the amount of information provided to the public has improved considerably in all the geographical areas studied, it is not always used effectively in the social co-construction of restoration projects. Thus, opposition to SDR cannot be explained by a lack of public awareness but, rather, by a fundamental and intangible difference in how the environment is interpreted and understood (Jorgensen, 2017; Linton & Krueger, 2020).

12.4 Conclusion

For 25 years, SDR has been an accepted 'best practice' to restore the hydro-sedimentary processes hindered by weirs. Despite being hailed as a science-based paradigm shift, it remains controversial in socio-environmental terms and various forms of social resistance have been observed (Sneddon et al., 2017). Critical research in the field of environmental science demonstrates the highly contingent nature of both the positive and negative results of SDR. Indeed, its scientific justifications often fail to take sufficient account of the variation in environmental contexts or the evolving ecology of managed rivers. The practice of SDR is based on an idealised model that conveys the concept of a 'dream river'. It risks selling an illusion to stakeholders that a weir's removal would have the magical effect of restoring a living, clean river that is scientifically and aesthetically consistent with scholarly predictions. In reality, however, the consequences of SDR in terms of the landscape, ecology and local community often prove to be poorly anticipated, with the plans frequently underestimating the interplay of locals' uses of, and relationship with, their river.

This is not intended as a blanket dismissal of small dam removal. Rather, following Boardman and Foster (2023), we offer a critical perspective and issue a warning: the dismantling of weirs, particularly mill weirs, is no trivial operation from an ecological, hydraulic or social standpoint. Thus, the hasty and dogmatic rollout of this policy neglects the complexity of local biophysical factors and the need for inclusive and participatory planning in ecological river restoration projects. The technical solution of SDR must be part of a broader, integrated management plan based at the river basin level that rigorously follows the principle of democratic consultation. Since public funding bodies have adopted it as standard 'best practice', SDR continues to transform not only rivers but the long-standing relationships that local people have with them. It is therefore of critical importance that its implementation is sensitive to a range of scientific and sociological perspectives and allows room for flexibility and compromise (Fox et al., 2016; Barraud & Germaine, 2017).

Further Readings

Barraud, R., & Germaine, M.-A. (Eds.). (2017). *Démanteler les barrages pour restaurer les cours d'eau: Controverses et représentations*. Éditions Quae.

Bravard, J.-P., & Lévêque, C. (Eds.). (2020). *La gestion écologique des rivières françaises: regards de scientifiques sur une controverse*. L'Harmattan.

Palmer, T. (2004). *Lifelines: The case for river conservation*. Rowman & Littlefield Publishers.

Sneddon, C. S., Magilligan, F. J., & Fox, C. A. (2017). Science of the dammed: Expertise and knowledge claims in contested dam removals. *Water Alternatives, 10*(3), 677–696.

Notes

1 See Dam Removal Europe website: https://damremoval.eu/.

2 Authors' translation. Quote taken from the 'migratory species' dashboard for the Loire basin: www.migrateurs-loire.fr/effectif-aux-stations-de-comptage/.

Selected References

Alp, M., Arnaud, F., Barthélémy, C., Bernez, I., Clemens, A., Cottet, M., ..., & Lamouroux, N. (2024). Restaurer la continuité écologique des cours d'eau: que sait-on et comment passer collectivement à l'action? *VertigO, 24*(2). https://doi.org/10.4000/12ppa

Barraud, R. (2017). Removing mill weirs in France: The structure and dynamics of an environmental controversy. *Water Alternatives, 10*(3), 796–818.

Beauchamp, A., Lespez, L., Rollet, A. J., Germain-Vallée, C., & Delahaye, D. (2017). Les transformations anthropiques d'un cours d'eau de faible énergie et leurs conséquences, approche géomorphologique et géoarchéologique dans la moyenne vallée de la Seulles, Normandie. *Géomorphologie: Relief, Processus, Environnement, 23*(2), 121–138. https://doi.org/10.4000/geomorphologie.11702

Belletti, B., Garcia de Leaniz, C., Jones, J., Bizzi, S., Börger, L., Segura, G., ..., & Zalewski, M. (2020). More than one million barriers fragment Europe's rivers. *Nature, 588*, 436–441. https://doi.org/10.1038/s41586-020-3005-2

Boardman, J., & Foster, I. (2023). Are 'free-flowing rivers' a good idea? The challenge of removing barriers from our rivers. *Geography, 108*(3), 121–129.

Brown, A. G., Lespez, L., Sear, D. A., Macaire, J. J., Houben, P., Klimek, J., ..., & Pears, B. (2018). Natural vs anthropogenic streams in Europe: History, ecology and implications for restoration, river-rewilding and riverine ecosystem services. *Earth-Science Reviews, 180*, 185–205. https://doi.org/10.1016/j.earscirev.2018.02.001

Brown, R. L., Charles, D., Horwitz, R. J., Pizutto, J. E., Skalak, K., Velinsky, D. J., & Hart, D. D. (2024). Size-dependent effects of dams on river ecosystems and implications for dam removal outcomes. *Ecological Applications, 34*(6). https://doi.org/10.1002/eap.3016

Brummer, M., Rodríguez-Labajos, B., Thanh Nguyen, T., & Jordà-Capdevila, D. (2017). "They have kidnapped our river": Dam removal conflicts in Catalonia and their relation to ecosystem services perceptions. *Water Alternatives, 10*(3), 744–768.

Dewandel, B., Lanini, S., & Frissant, N. (2024). Steady-state semi-analytical solutions for assessing the two-dimensional groundwater hydraulic head depletion induced by river dam removal. *Journal of Hydrology, 642*, 131801.

Donati, F., Touchart, L., & Bartout, P. (2020). Les seuils en rivière transforment-ils les milieux lotiques en milieux lentiques? *Physio-Géo, 15*, 111–131.

Doyle, M., Harbor, J. M., & Stanley, E. H. (2003). Toward policies and decision-making for dam removal. *Environmental Management, 31*(4), 453–465. https://doi.org/10.1007/s00267-002-2819-z

Doyle, M., Stanley, E., Orr, C. H., Sthi, S. A., & Harbor, J. M. (2005). Stream ecosystem response to small dam removal: Lessons from the Heartland. *Geomorphology, 71*, 227–244.

Drapier, L., Germaine, M.-A., Lespez, L., Magilligan, F. J., & Sneddon, C. (2023). Networks, coalitions and the contestation of dam removal across political and institutional scales in France and New England (USA). *Geographical Review, 113*(2), 246–267. https://doi.org/10.1080/00167428.2021.1953382

Dufour, S., & Piégay, H. (2009). From the myth of a lost paradise to targeted river restoration: Forget natural references and focus on human benefits. *River Research and Applications, 25*(5), 568–581.

Dufour, S., Rollet, A. J., Chapuis, M., Provansal, M., & Capanni, R. (2017). On the political roles of freshwater science in studying dam and weir removal policies: A critical physical geography approach. *Water Alternatives, 10*(3), 853–869.

Flaminio, S. (2018). *(Se) représenter les barrages: (a)ménagement, concessions et controverses* [Doctoral dissertation]. Université de Lyon, Lyon, France.

Fostvedt, M. P., Tullos, D. D., & Tilt, B. (2020). Institutional analysis of small dam removals: A comparison of non-federal dam removals in Washington and Oregon. *Water Alternatives, 13*(2), 24.

Fox, C., Magilligan, F., & Sneddon, C. (2016). "You kill the dam, you are killing a part of me": Dam removal and the environmental politics of river restoration. *Geoforum, 70*, 93–104. https://doi.org/10.1016/j.geoforum.2016.02.013

Gangloff, M. M. (2013). Taxonomic and ecological tradeoffs associated with small dam removals. *Aquatic Conservation, 23*(4), 475–480.

Germaine, M.-A., & Barraud, R. (2013). Restauration écologique et processus de patrimonialisation des rivières dans l'Ouest de la France. *VertigO-la revue électronique en sciences de l'environnement, 16*(HS), 1–19.

Germaine, M.-A., Drapier, L., Lespez, L., & Styler-Barry, B. (2022). How better involve stakeholders in river restoration projects: The case of small dam removals. In B. Morandi, M. Cottet, & H. Piégay (Eds.), *River restoration: Political, social and economic perspectives* (pp. 147–168). Wiley.

Grabowski, Z. J., Chang, H., & Granek, E. F. (2018). Fracturing dams, fractured data: Empirical trends and characteristics of existing and removed dams in the United States. *River Research and Applications, 34*(6), 526–537. https://doi.org/10.1002/rra.3283

Grant, G. E., O'Connor, J., & Safran, E. (2017). Excursions in fluvial (dis)continuity. *Geomorphology, 277*, 145–153. https://doi.org/10.1016/j.geomorph.2016.08.033

Howard A. J., Coulthard T. J., & Knight D. (2017). The potential impact of green agendas on historic river landscapes: Numerical modelling of multiple weir removal in the Derwent Valley Mills world heritage site, UK. *Geomorphology, 293*, 37–52

Hripto, J., Inamdar, S., Sherman, M., Peck, E., Gold, A. J., Bernarsconi, S., ...& Peipoch, M. (2022). Effects of relic low-head dams on stream denitrification potential: Seasonality and biogeochemical controls. *Aquatic Sciences, 84*(4). https://doi.org/10.1007/s00027-022-00894-z

Jørgensen, D. (2017). Competing ideas of 'natural' in a dam removal controversy. *Water Alternatives, 10*(3), 840–852.

Lespez, L., & Dufour, S. (2021). Les hybrides, la géographie de la nature et de l'environnement. *Annales de Géographie, 737*(1), 58–85.

Linton, J., & Krueger, T. (2020). The ontological fallacy of the Water Framework Directive: Implications and alternatives. *Water Alternatives, 13*(3), 513–533.

Malavoi, J.-R., & Paris, P. (2003). *Stratégie d'intervention de l'agence de l'eau sur les seuils en rivière*. Agence de l'eau Loire-Bretagne.

McCaffery, R., Duda, J. J., Soissons, L., & Roussel, J. M. (2024). Large-scale dam removal and ecosystem restoration. *Frontiers in Ecology and Evolution, 12*, 1471146.

Palmer, M. A., Bernhardt, E. S., Allan, J. D., Lake, P. S., Alexander, G., Brooks, S., ..., & Sudduth, E. (2005). Standards for ecologically successful river restoration. *Journal of Applied Ecology, 42*(2), 208–217.

Perrin, J.-A. (2018). *Gouverner les cours d'eau par un concept: Étude critique de la continuité écologique des cours d'eau et de ses traductions* [Doctoral dissertation]. Université de Limoges, France.

Thomas, O., & Germaine, M.-A. (2018). De l'enjeu de conservation au projet de territoire: Le saumon atlantique au cœur des débats. *VertigO, 18*(2). https://doi.org/10.4000/vertigo.22259

Zarfl, C., & Lehner, B. (2020). European rivers are fragmented by many more barriers than had been recorded. *Nature, 588*, 436–441. https://doi.org/10.1038/d41586-020-03440-9

13 *REVISITING*: Environmental Flows Are Necessary to Safeguard Nature

Jason Alexandra

13.1 Introduction

The idea of allocating water to safeguard nature is pragmatically important and intuitively appealing because flows of water are the essence of rivers, quintessential to defining them. Without flowing water (at least some of the time), rivers are not rivers; they are lakes, wadies, drains, or stagnant pools. However, attempts to formally define rivers' needs for water only emerged late in the twentieth century, with the idea of environmental water promoted as an antidote to the excessive exploitation of rivers.

The principle that rivers need environmental flows has become axiomatic within contemporary water governance and, thus, qualifies as one of the 'accepted truths' or 'received ideas' that are the focus of this book.

Regardless of the importance of this idea, or perhaps because of it, there are risks that environmental flows become simplistic panaceas for the problems plaguing humanity's relationships with rivers. Crude solutions risk blinding us to the nuances and complexity of diverse situations.

This chapter explores the origins of environmental flows (or Eflows) and some of the challenges of adopting them as accepted solutions. It draws tentative conclusions about complementing techno-scientific approaches to riverine planning and management with more relational, nuanced, contextual and deliberative approaches.

13.2 History of the Eflow 'Solution' and the Problems It Is Meant to Solve

The concept of needing Eflows to safeguard nature only gained wide acceptance early in the twenty-first century. The idea emerged with great promise, initially fermenting within the unruly offices of environmental NGOs and amongst a small cadre of academics before meandering from these outer margins of policy development to become a central focus of water policy and research. This evolution occurred over about 40 years (Arthington, 2012; Poff et al., 2017).

DOI: 10.4324/9781003745532-13

Eflows offer an attractive approach to the problems arising from the wholesale harnessing of rivers for agricultural and industrial development. During the twentieth century, many of the world's rivers were dammed to secure mass electrification, reliable irrigation, and foreign policy objectives. These nation-building projects became celebrated symbols of modernist technological progress – and its excesses. They harnessed nature's power, reinforced political regimes, and spawned technocratic elites who ruled over modern water's hydraulic mission, promoting its universalising hydro-logics (Boelens et al., 2022). Only late in the twentieth century did the environmental turn challenge this dominance. High-profile protests stopped controversial dam proposals, including the Gordon-Below-Franklin in Tasmania and the Loyettes Dam on the Rhône River. The 'no dams' movement spread internationally, rallying against new dams and promoting the world's remaining free-flowing rivers as iconic symbols of unspoilt nature (Alexandra et al., 2023) to counter the idea of large dams as the modern world's temples (Kaika, 2006). Thus, the origins of the Eflow discourse can be found in the environmentalism of the 1970s and 1980s, when the ideal of protecting and restoring free-flowing rivers became rooted in the body politic (Barraud, 2025).

Initially, most Eflows were specific compensation for rivers whose flows were significantly disrupted by extractions, hydroelectricity generation, and intervalley transfers. These early incarnations focused on the minimum flows needed to sustain specific river reaches, impacted ecosystems, or important fish species, especially salmon in the northern hemisphere (Alexandra et al., 2023). From these early beginnings, Eflow assessment and planning methods evolved, morphed, and mutated, proliferating as they spread internationally (Tharme, 2003), despite Arthington and Zalucki's (1998) warnings about transferring techniques to countries with different ecological and hydrological conditions.

In the 1990s, many local and national non-governmental organisations (NGOs) campaigned for the adoption of Eflows. These efforts were reinforced by big international NGOs, including The Nature Conservancy, Worldwide Fund for Nature (WWF), and the IUCN (Dyson et al., 2003; Pittock et al., 2003). A new global consensus appears to have emerged early in the new millennium. The World Commission on Dams (WCD) reflected the growing crisis of confidence in the hydraulic mission (WCD, 2000), and multilateral banks became increasingly reluctant to fund new dams. In the decades since then, the broad consensus about Eflows' importance survived (at least rhetorically, despite the boom in new dam construction) with the International Water Resources Association (IWRA) promoting Eflows in their brochure 'Environmental Flows: Achieving Harmony between Humans and Nature' (IWRA, 2022).

Defining the epistemic centres of new international movements is notoriously difficult, especially given the rapid movement of ideas. However, Brisbane, Australia, appears to have played a pivotal role in the emergence of this new global consensus by hosting the International River Symposium since the late 1990s. Here, in 2007 and again in 2017, delegates crafted the Brisbane Declaration, which defines Eflows as 'the quantity, timing, and quality of freshwater flows and levels necessary to sustain aquatic ecosystems which, in turn, support human cultures, economies, sustainable livelihoods, and well-being' (Arthington et al., 2018).

13.3 Problems and Questions Raised by Eflows

The previous section outlined the emergence of a consensus on Eflows in the early twenty-first century. Since then, many countries formalised Eflows in laws, policies, and strategies (e.g. EC, 2016), with policy development and the associated research focused on questions like: how much water can be 'sustainably' extracted; how much should be retained in rivers; and how should flows be restored in heavily exploited systems?

The idea that Eflows are needed to safeguard nature is not wrong. However, it has proved challenging to implement (Wineland et al., 2022; Dourado et al., 2023), raising complex political, ontological, and epistemological challenges (Boelens et al., 2024).

This section outlines why, when simplistically applied, the Eflow 'solution' has become deeply problematic.

13.3.1 Implicit Normative Logics and Assumptions

Research and policy on Eflows incorporate and reinforce numerous normative values, logics, and assumptions about rivers and society that are potent and influential but rarely made explicit. These logics and assumptions form part of a new orthodoxy about riverine systems and water resources, guiding how they are valued, understood, and managed. Eflow research and policy contribute to this orthodoxy, perpetuating the dominant ways of knowing and governing water that crowd out different and more diverse ways of relating to water and rivers (Schmidt, 2017; RiverOfLife et al., 2020; Boelens et al., 2024).

A profound concern is that scientifically and technically orientated Eflow methods reproduce and reinforce their normative foundations. For example, they may emphasise that rivers unmodified by humans are of higher value than those more highly modified, or that humans and their needs are separate and different from nature. Therefore, Eflow assessment techniques are not neutral or objective instruments of analysis but culturally constructed methods with embedded logics that reinforce certain accepted assumptions, perspectives, relationships, and norms about society, water, and rivers and their interrelationships (Alexandra et al., 2023).

The ontological impacts and ideological dimensions of scientific research are often silenced or disappear (made invisible or taboo). Therefore, gaining a clear perspective on ideologies and their influence can be difficult. In the case of Eflows, Boelens et al. (2022) argue that they contribute to a dominant power-knowledge regime that strives to establish an overarching river governance worldview, defining 'correct' environmental policies which legitimise authority. This regime attempts to secure one socio-natural order which renders 'one knowledge, ontological and governance frame "natural", as the morally or scientifically "best order", invalidating all others' through dominant river discourses that 'establish concepts, actors, objects, their identity and relations and hierarchy', thereby shaping 'people's feeling, thinking, seeing, talking and behaving in relation to river systems' (Boelens et al., 2022, p. 1134).

13.3.2 Dominant Paradigms Separate Humans from Nature

One of the main problems of Eflows is that dominant paradigms separate humans from nature, perpetuating the dualism that plagues Western thinking in modernity. Eflow research and policy are embedded in and reproduce the two main paradigms of natural flows and ecosystem services. These give expression to and reinforce specific underlying logics about humans and nature and their respective roles in the scheme of things (Boelens et al., 2022; Alexandra et al., 2023).

The basis of the natural-flow paradigm is that mimicking natural flows as closely as feasible will sustain river ecosystems. This is based on the ideal of the intrinsic value of untouched or 'pure' nature and its usefulness in showing us how we should operate in the world (Barraud, 2025). Thus, assessment methods usually compare post-development flows with modelled 'natural' or predevelopment benchmarks. However, restoring flow regimes to approximate 'natural flows' is often challenging, costly or impossible where rivers already have dams, reservoirs and diversion infrastructure producing valuable goods and services like urban water supplies, irrigation and hydroelectricity (Dourado et al., 2023; Molle & Collard, 2024).

The ecosystem-services paradigm focuses on using flows to achieve benefits for humans, often prioritising desirable species (e.g. salmon) or other benefits, like water quality. By emphasising instrumental or economic value, this paradigm embeds the idea that nature's value consists in its delivery of goods and services.

Critics of both paradigms identify their shortcomings. The ecosystem-services approach is critiqued for perpetuating neoliberalism and extending the commodification and financialisation of nature. Additional criticisms revolve around whether accurate valuations of the services provided by riverine ecosystems are feasible and, if so, whether these kinds of valuations support better policy decisions (see Chapter 10 on 'Payments for Environmental Services').

Importantly, both paradigms share some underlying assumptions about balancing modern humans' needs with those of nature, and how they are often fundamentally different and competing. Thus, these Eflow logics focus attention on trade-offs, the scarcity and competition for water and the need for it to be formally allocated between competing users or claimants

(Alexandra et al., 2023). In this conceptualisation, exceptions are sometimes made for Indigenous peoples, leading to the increasingly important claimant category – cultural flows (Leonard et al., 2023).

Some critiques of the natural-flows paradigm focus on the use of modelled, predevelopment benchmarks for river management and restoration (see Chapter 14 on 'River Restoration'), including that these may become increasingly irrelevant due to climate change impacts (Alexandra, 2023). Policies based on naturalistic benchmarks may also marginalise the significance of co-evolved riverscapes in industrialised societies (Linton & Krueger, 2020) and Indigenous cultures (Leonard et al., 2023). The naturalistic ideal, and the benchmarks based on it, implicitly assume that humans are intrinsically damaging to the non-human world (Linton & Krueger, 2020). This underlying logic perpetuates the twenty-first-century sense of angst and foreboding. It also closes off opportunities to explore the potential of co-constituted nature-cultures, which emphasise the intricate interlinkages between humans and rivers (RiverOfLife et al., 2020).

Not only do Eflow logics and paradigms separate humans and nature, but they also rely on increasingly professionalised agencies and their techno-scientific experts to define what is desirable in nature, what nature needs, and what flows must be provided to satisfy these needs. There are risks that a reliance on experts leads to the establishment of technocratic elites, who determine the needs of the modern deity – the idealised and separate River-Nature.

While it can be argued that all nature is culturally co-produced (Boelens et al., 2024), Eflow logics frame nature as separate, manageable, domesticated, and controllable – the managed river. Proponents of Eflows rarely claim that the nature whose capriciousness, violence, and unpredictability deliver damaging floods, devasting droughts, cyclones, and mudslides is the one that needs safeguarding.

13.3.3 Managerialism, Reductionism, and Other River Stressors

Eflows' research and assessment methods were developed to satisfy the growing environmental responsibilities of water agencies when modern water's hydro-developmentalism morphed into Integrated Water Resource Management (IWRM). Despite the proliferation and diversification of Eflow techniques (Tharme, 2003), most planning and assessments use flow-ecology or flow-biota relationships (Gebreegziabher et al., 2023).

Flow-biota methods often rely on key species, like salmon or trout, which serve as a proxy for the river, meaning the latter's needs are equated to those of valuable, 'beloved' or 'well-documented' species. For example, the needs of the recreational angling species trout were the central focus of the Instream Flow Incremental Methodology first developed by the US Fish and Wildlife Service in the 1980s before this method or variations of it spread to many countries.

These technically orientated, natural-science approaches that are acceptable to water agencies appear to be objective and politically neutral. However, Boelens et al. (2022, p. 1134) assert that these 'techno-environmentalist approaches favour "environmental flow" programs that tend to translate river-nature into mathematical biological or physical formulae [that] abstract and disembody rivers'. Nonetheless, in the 1980s, Cavendish and Duncan (1986) explicitly recognised flow-ecology methods as tools for negotiation.

As Eflows became more institutionalised, the social sciences have tended to be relegated to the second tier in Eflow research, despite their usefulness in understanding the social construction of flow goals (Bouleau, 2014) or the political contestations about reallocating water to the environment (Marshall & Alexandra, 2016). Thus, much ecohydrological research provides a relatively safe subject territory that seemingly can avoid the divisive politics of water allocation or the sectors' embedded power relationships, which typically remain taboo subjects.

It is important to recognise that Eflow techniques are rooted in the control of rivers, usually by water-management agencies, using administratively rational, expert-driven planning and management methods. Eflow methods reduce everything into a discharge, which is measurable, manageable, and 'influenceable'. The focus on flow quantities diverts attention from other stressors – temperature, contaminants, pollution, infrastructure, land-use change, deforestation, the grazing of riparian zones, feral species, or climate change – which are often beyond the remit of water agencies. Together, these multiple change drivers and stressors determine the conditions and trends of many rivers, indicating that more integrated or holistic management approaches are needed.

13.3.4 Technical Entrapment and False Precision

One problem, or potential problem, for Eflow praxis is technical entrapment, which may be compounded by the plethora of Eflow planning, assessment, and decision-support tools, developed in diverse basins, climate zones, and cultural and political contexts (Wineland et al., 2022). A possible explanation for this abundance is 'methodolatry' – a neologism which refers to the fetishisation of or preoccupation with methods, data and models, which influences research design, potentially limiting the questions asked to those which the available techniques can answer.

Deferring to technical methods can increase the risk of technical entrapment, making complex problems more difficult to resolve. When Eflow methods constrain policy debates to a few key variables (e.g. flow-biota relationships), they recast complex situations as more neatly defined problems that can yield to simple solutions. If initial quantitative methods fail, more elaborate methods for measurement and assessment can be added, inspiring more detailed models and technical research.

Technical entrapment risks are reinforced by the desire to legitimise decisions scientifically. The belief that science can provide precise Eflow targets can obscure the political nature of water-policy decisions. It is therefore useful to understand Eflows as an assemblage of practices that are inherently technical and political.

An examination of the political economy of research reveals how Eflow researchers became influential actors within the policy networks, promoting Eflows (and their own research) while also legitimising IWRM's vague ambitions of balancing environmental, economic, and social objectives. In the competitive, neoliberal funding environment of the early twenty-first century, self-promotion is a critical skill for aspiring researchers. The emergence of researcher-agency co-dependence on Eflows can be understood in terms of the broader political economy of research, the rise in evidence-based policy aspirations, and the legitimisation (and attempted depoliticisation) of policy choices by using scientifically based justifications or appealing to the idea that science can provide definitive answers to complex problems.

The myth of optimisation is another problem with technically orientated Eflow research. The prospect of optimising extractive and environmental flows is a persistent ideal that inspires many detailed studies with the stated aim of supporting water allocation decisions. However, the precise costs or benefits of altered flows are hard, if not impossible, to determine, partly because river responses are complex, knowledge is imperfect, and social values are diverse and dynamic (Capon & Capon, 2017). Many optimisation studies attempt to locate optimal decisions, despite these contested values, complex variables, uncertain futures, and profound uncertainties about how rivers responded to multiple change factors in the past.

New dam proposals are resurgent in many countries, driven by geopolitical concerns about water, energy, and food insecurity in a changing climate. Analyses of proposed dams and their impacts, like those planned for the Mekong Basin, attempt to understand the social, ecological, and economic impacts of altered flows. However, complex trade-offs are often oversimplified, obscured, or ignored despite these decisions impacting billions of people. Optimisation studies typically struggle to reconcile the uncertainties and complexities of proposed dams and their flow impacts. Thus, big decisions about Eflows, or new hydropower dams, are encounters between contested presents and divergent futures with conflicts that have deep political, economic, and ontological origins (Boelens et al., 2022). Therefore, optimisation is often an impossible ideal when juggling the complex and dynamic variables involved in managing river flows for competing values.

Instead of focusing on more detailed and technical optimisation methods with their false promises of solving complex problems, we need to develop and refine methods for exploring divergent trajectories (e.g. use competing scenarios or worst-case scenarios) and to improve democratic capacity to deal with complex societal choices that inevitably involve winners and losers.

Thus, we need vigilance in how we conceive problems and their solutions. Rather than attempt to reduce complexity and tame wicked problems, we need the determination to probe the complexity and the discomfort, because this work brings new possibilities.

13.3.5 Complexity and the Politics of Eflows

Early Eflow research focused on the specific flows needed to counter the negative impacts of significant flow alterations. Over time, the focus broadened to defining flow strategies to sustain rivers' diverse values, positioning Eflows on the bigger stages of adaptive governance, social-ecological sustainability, and resilience (Richter, 2010; Arthington, 2012). This expansive scope involves societal deliberations about preferred futures and how to achieve them – subjects that are generally the domain of social movements and democratic institutions.

Understanding social change requires different disciplines from those traditionally used to understand rivers and their flows. While social and institutional factors are critical to Eflow planning, most existing Eflow techniques are unlikely to be appropriate for broader societal deliberations about rivers, or for renegotiating how we know, manage, and govern water and water relations (Boelens et al., 2022; Leonard et al., 2023; Boelens et al., 2024).

Therefore, attention to broader socio-political factors is needed when attempting to understand Eflow debates. Many water-policy contestations are not problems to be solved but sustained struggles for justice in people's rights and relationships with rivers (RiverOfLife et al., 2020; Boelens et al., 2022). Social, scientific, and political institutions – with their entangled histories and trajectories – play critically important roles in shaping options (Marshall & Alexandra, 2016).

Within functioning liberal democracies, simplistic solutions to complex problems are rarely viable (or not for long). For example, where livelihoods depend on irrigated agriculture, reducing extractions to ensure environmental flows can be divisive. Political contestations result in intense scrutiny of any scientific methods used to define nature's needs (Molle & Collard, 2024). This includes Eflow methodologies – such as those used in Spain, France, and Australia – that 'reconstitute' in models the predevelopment or 'naturalised flows'. Estimates of extracted flows are added to observed flows, often using incomplete data and a cursory understanding of the impact of land-use change (e.g. deforestation) or surface/groundwater relationships. However, as the science and policy develop, uncertainties tend to fade, and the legitimacy of science-derived numbers are used to justify policies (Molle & Collard, 2024). Alternatively, political imperatives can result in the altering of scientifically derived environmental flow targets to suit the realpolitik, as occurred in the case of Australia's Murray Darling Plan (Marshall & Alexandra, 2016). The tensions between technically orientated Eflow planning approaches and the socially negotiated contestations about governing rivers involve institutions with nuanced histories and procedures for equity and justice. Determining what is desirable, feasible and achievable in terms of Eflows requires an exploration of specific policy options that satisfy various stakeholders and may involve compensation, redefining property rights or augmenting supply. These require an understanding of rivers as complex cultural-social-ecological systems with governing institutions that mediate the relationships between natural resources, riverine ecosystems, and society (Molle & Collard, 2024).

Questions about whether Eflows can safeguard nature sit at the intersections of science and society and invite us to investigate relationships between nature and culture and the ways these are co-constituted (Boelens et al., 2024). Examining how Eflows manifest in practice raises many questions about the evolution of water policy and the role of science in society. Rivers are intrinsically complex, with more detailed scientific studies of their ecohydrology adding new layers to this complexity. Rivers are also large in spatial scale and dynamic, with many context-dependent socio-economic and cultural relationships. Further, understanding the relationship between flow alterations and ecosystem health is a developing science. The uncertainties and limits of the science behind Eflow determinations will be increasingly exposed as we demand more from the concept. In practical terms, many rivers, especially in developing countries, lack hydrological and ecological data to calculate Eflows (Dourado et al., 2023). Thus, imposing the concept for ideological reasons may lead to choosing crude, unrealistic, or unacceptable values.

Planning for Eflows is not just data- or method-dependent, but also often involves conflict over values and the control of natural resources. This indicates a need for better processes and institutions for deliberation about the collective choices involved in allocating water flows to different users and uses.

Employing techniques like citizen juries to enhance democratic deliberations may broaden the scope of Eflow assessments and debates. Indeed, innovative deliberative techniques may help resolve disputes about Eflow allocations. Dealing equitably with competing demands for limited water resources requires more open, discursive, and democratic deliberations about policy choices. However, these will almost certainly make Eflow debates messier and less predictable, which could result in a retreat into simplified or more technical framings of difficult policy decisions. Acknowledging the nuances, changing contexts, and complexity of Eflows in practice is critically important because, as this book emphasises, simple policy prescriptions tend to offer false hope about solving complex water-resource policy problems.

13.4 Conclusions

Disputations about Eflows are conflicts in values and beliefs about power and knowledge, resource access, and preferred futures.

This chapter has outlined the origins and evolution of the concept of Eflow and raised questions about what it means in theory and practice. It has shown that although Eflows are important, problems emerge when they are offered as panaceas. Indeed, solutions based on 'just adding water' or 'giving the river its water' attempt to tame the complexities of humanity's relationships to rivers. Unless we are to accept dominant ideas unquestioningly, we need to probe them with penetrating questions. In this case, these include: what is the nature we seek to safeguard, and who decides nature's needs? Who determines trade-offs and compromises between human needs and those of nature? How should these be debated? Who benefits from defining nature's needs? Moreover, if nature is culturally co-constituted, how can we define its needs separately?

The ethos of caring for rivers, including ensuring enough water to sustain people and biota, is fundamentally sound and widely accepted. Like many intuitively appealing ideas, environmental flows are fraught with challenging complications that persistently bedevilled them in practice. Rather than promoting simplistic solutions, or more technically orientated methods, why not develop bespoke solutions that draw on the best *social, political, cultural, and economic* theory and methods and apply these to the governance of rivers in all their rich and diverse circumstances? We need better ways to explore the options for renegotiating the relationships between rivers, peoples, and governance institutions. We also need critical and reflective research that is attentive to modern water's historical and philosophical foundations and its dominance in framing contemporary water-policy debates.

Eflows provide water for all life forms, including humans, and are part of what is needed to safeguard humans and nature together in an uncertain future. Yet, securing these flows comes at the expense of using water upstream. Thus, we do not need to optimise flows because of the parameters in a model, but rather to respect the precariousness of our condition by broadening the debate on water use and territorial inequities. These issues need critical examination, periodic review and policy reconsideration, as well as governance that enables deliberations to be open, innovative and democratic.

Further Readings

Alexandra, J., Rickards, L., & Pahl-Wostl, C. (2023). The logics and politics of environmental flows—A review. *Water Alternatives, 16*(2), 346–373.

Boelens, R., Escobar, A., Bakker, K., Hommes, L., Swyngedouw, E., Hogenboom, B., ..., & Wantzen, K. M. (2022). Riverhood: Political ecologies of socionature commoning and translocal struggles for water justice. *The Journal of Peasant Studies, 50*(3), 1125–1156. https://doi.org/10.1080/03066150.2022.2120810

Dourado, G. F., Rallings, A. M., & Viers, J. H. (2023). Overcoming persistent challenges in putting environmental flow policy into practice: A systematic review and bibliometric analysis. *Environmental Research Letters, 18*(4), 043002.

Wineland, S. M., Başağaoğlu, H., Fleming, J., Friedman, J., Garza-Diaz, L., Kellogg, W., ..., & Neeson, T. M. (2022). The environmental flows implementation challenge: Insights and recommendations across water-limited systems. *Wiley Interdisciplinary Reviews: Water, 9*(1), e1565.

Selected References

Alexandra, J. (2023). Climate risk assessment in the MDB–A review. *Australasian Journal of Water Resources, 27*(1), 18–30.

Arthington, A. (2012). *Environmental flows: Saving rivers in the third millennium* (Vol. 4). University of California Press.

Arthington, A. H., Bhaduri, A., Bunn, S. E., Jackson, S. E., Tharme, R. E., Tickner, D., & Ward, S. (2018). The Brisbane declaration and global action agenda on environmental flows (2018). *Frontiers in Environmental Science, 6*, 45.

Arthington, A. H., & Zalucki, J. M. (Eds.). (1998). *Comparative evaluation of environmental flow assessment techniques: Review of methods* (LWRRDC Occasional Paper 27/98). LWRRDC.

Barraud, R. (2025). The changing meaning of wild rivers: A review. *Water Alternatives, 18*(2), 181–213.

Bouleau, G. (2014). The co-production of science and waterscapes: The case of the Seine and the Rhône rivers, France. *Geoforum, 57*, 248–257. https://doi.org/10.1016/j.geoforum.2013.01.009

Boelens, R., Hommes, L., & Hoogesteger, J. (2024). Ontological politics in river defence debates: Unpacking fields of contention in eco-centric and non-human turns. *Water Alternatives, 17*(3), 649–668.

Capon, S. J., & Capon, T. R. (2017). An impossible prescription: Why science cannot determine environmental water requirements for a healthy Murray-Darling Basin. *Water Economics and Policy, 3*(03), 1650037.

Cavendish, M. G., & Duncan, I. M. (1986). Use of the instream flow incremental methodology: A tool for negotiation. *Environmental Impact Assessment Review, 6*(4), 347–363.

Dyson, M., Bergkamp, G., & Scanlon, J. (2003). *Flow: The essentials of environmental flows*. IUCN.

European Commission, Directorate-General for Environment. (2016). *Ecological flows in the implementation of the Water Framework Directive* (Guidance Document No. 31). Publications Office. https://data.europa.eu/doi/10.2779/775712

Gebreegziabher, G. A., Degefa, S., Furi, W., & Legesse, G. (2023). Evolution and concept of environmental flows (e-flows): Meta-analysis. *Water Supply, 23*(6), 2466–2490.

IWRA. (2022). *Environmental flows: Achieving harmony between humans and nature*. IWRA.

Kaika, M. (2006). Dams as symbols of modernization: The urbanization of nature between geographical imagination and materiality. *Annals of the Association of American Geographers, 96*(2), 276–301.

Leonard, K., David-Chavez, D., Smiles, D., Jennings, L., Alegado, R. A., Tsinnajinnie, L., & Gomez, A. (2023). Water back: A review centering rematriation and Indigenous Water research sovereignty. *Water Alternatives, 16*(2), 374–428.

Linton, J., & Krueger, T. (2020). The ontological fallacy of the Water Framework Directive: Implications and alternatives. *Water Alternatives, 13*(3), 513.

Marshall, G. R., & Alexandra, J. (2016). Institutional path dependence and environmental water recovery in Australia's Murray-Darling Basin. *Water Alternatives, 9*(3), 679.

Molle, F., & Collard, A.-L. (2024). Sharing water between nature and humans: Environmental flows and the politics of quantification. *Water Alternatives, 17*(2), 510–532.

Pittock, J., Madgwick, F. J., & Tickner, D. (2003). *Establishing river basin organisations for the conservation and sustainable management of rivers* [Paper presentation]. Riversymposium, Brisbane, Australia. http://assets.panda.org/downloads/riversymposiumpaper.pdf (accessed March 11, 2006).
Poff, N. L., Tharme, R. E., & Arthington, A. H. (2017). Evolution of environmental flows assessment science, principles, and methodologies. In A. C. Horne, J. A. Webb, M. J. Stewardson, B. D. Richter, & M. Acreman (Eds.), *Water for the environment* (pp. 203–236). Academic Press.
Richter, B. D. (2010). Re-thinking environmental flows: From allocations and reserves to sustainability boundaries. *River Research and Applications*, *26*(8), 1052–1063.
RiverOfLife, M., Poelina, A., Alexandra, J., & Samnakay, N. (2020). *A conservation and management plan for the National Heritage listed Fitzroy River Catchment Estate* (Nulungu Reports 1). Nulungu Reports. https://doi.org/10.32613/nrp/2020.4
Schmidt, J. J. (2017). *Water: Abundance, scarcity, and security in the age of humanity*. New York University Press.
Tharme, R. E. (2003). A global perspective on environmental flow assessment: Emerging trends in the development and application of environmental flow methodologies for rivers. *River research and applications*, *19*(5–6), 397–441.
World Commission on Dams. (2000). *Dams and development: A new framework for decision-making: The report of the world commission on dams*. Earthscan.

14 *REVISITING*: Rivers Should Be Restored to Their 'Natural' State

Gabrielle Bouleau and Rebecca Lave

14.1 Introduction

Humans have been modifying rivers for thousands of years, but the extent of intervention began to grow dramatically in the mid-nineteenth century. The world's rivers have been subjected to unprecedented levels of gravel extraction, water abstraction or diversion, straightening, embanking, damming, and pollution. Reduced sediment loads in riverbeds and estuaries shrink freshwater habitats and increase risks of submersion and coastal erosion. Habitat destruction, reduced flows, and eutrophication have devastated stream ecology – and all of this is exacerbated by climate change. River modification also has a major impact on humans including increased flood risk and the loss of traditional fisheries and cultural connections. It is no wonder that work to undo this extensive anthropogenic harm to rivers and streams is seen as an obvious good.

Unfortunately, it is not clear whether current river- and stream-restoration practices succeed in repairing the damage. Relatively few projects are evaluated by comparing pre- and post-restoration states (though the number is increasing). Of the projects that are evaluated, many show little to no improvement.

Evaluating the outcome of restoration projects is all the more challenging because there is often disagreement about project goals. Should restoration aim to turn back the clock to recreate the system that existed before human modification or to restore the ecological, chemical, and physical processes best suited to current conditions in the drainage basin? If the goal is to increase biodiversity, should this be prioritised as an intrinsic good or should the focus instead be on species that provide ecosystem services to humans? Should the goal be a system that requires constant human intervention to maintain a high-functioning state or a self-maintaining system that may diverge from human goals, for example, by becoming a home for invasive species? Should we concentrate on expensive actions with an intensive impact over a small area, such as channel reconfiguration, or on low-cost actions at a larger scale but with a more diffuse impact?

DOI: 10.4324/9781003745532-14

In this chapter, we address these questions and other issues that make river restoration more complicated than it initially appears.

14.2 History of Stream and River Restoration

The term 'river restoration' has great semantic elasticity, having been used to justify a wide variety of interests at different periods. For the purposes of this chapter, we define restoration primarily based on the motivation for it: if the goal was to undo anthropogenic harm, whether to improve a river or stream's wellbeing or to increase the ecosystem services it provides, we include it under the restoration umbrella. At the global level, one can broadly group restoration motivations into four sets of concerns: sports fishing (starting in the late nineteenth century), human health and pollution-control (from the late 1950s), environmental flow and sediment load (from the 1970s) (Thompson, 2013), and free flowing rivers (since the late 1980s) (Barraud, 2025). In each period, river scientists debated theories in support of particular restoration approaches, quantification methods, and design solutions.

In the United States (US), Canada, and Europe, it was fishermen – professional and amateur – who first called for rivers to be restored. They were concerned about the loss of gamefish, such as trout, due to pollution from industrial and urban discharges and habitat loss caused by navigation and flood-control projects. Little attention was paid to the concerns, subsistence needs, or even the treaty rights of Indigenous groups. In nineteenth century Europe, fishermen were given financial compensation, but the damaging interventions themselves were rarely challenged (Fressoz, 2013). In the US, wealthy white anglers established private fishing clubs and experimented with instream habitat improvements, such as adding wood (Thompson, 2013). In the early 1900s, fishing clubs sprang up across continental Europe. These private-sector actors developed fish-farming practices that were later encouraged by governments as a way to compensate for fish depopulation without addressing pollution (Bouleau, 2019).

In the US, it was not until the late 1960s that the justifications for stream and river restoration expanded beyond sport fisheries. This was when high levels of pollution and ecological degradation (not least Ohio's Cuyahoga River catching fire over a dozen times) drove a change in national policy towards ecological health and away from solely economic frameworks. Similar concerns about pollution control emerged in Europe in the late 1950s as decolonisation shifted government focus back to domestic water resources. The population growth that followed World War II raised both the prospect of insufficient domestic water supplies and the spectre of health hazards from contaminated water. Many Western countries adopted the 'polluter pays' principle to finance wastewater treatment plants and dams, ensuring a reliable supply of safe tap water without addressing inefficiencies in water use.

While restoration is now widely supported in the US, disputes over environmental flows and endangered aquatic species protection are still common, as are conflicts over appropriate restoration techniques (Lave, 2012). Globally, it was in the 1970s, following the US's first Earth Day, that ecological concerns for rivers and streams arrived on the agenda. Pushed by environmental organisations and by green states coalitions, governments agreed to adopt integrated water resources management (IWRM), which promised to satisfy all uses, while preserving ecosystems, with the help of economic tools. This resulted in the gradual adoption of conventions for the protection of aquatic environments, such as Ramsar (1971) and the Helsinki Water Convention (1992). In the decades since, river restoration has been guided by both modernist promises of control and by ideals of rewilding and relinquishing human control.

Modernist restoration programmes often seek control over nature and society, conceptualising river ecosystems as resources to be maximised or risks to be minimised in quantifiable terms. This depoliticises development and overestimates humans' ability to control ecosystems. This has led to projects aimed at undoing past harm but only while maintaining human dominance over river systems. The OECD has fitted this mitigation approach into the driver/pressure/state/impact/response model (DPSIR), the current dominant framework for environmental policies (Linton & Krueger, 2020). The DPSIR contains a misleading notion of nature that likens it to static physics by presuming a balance between forces of action and reaction (see Figure 14.1).

Natural processes, however, are historical and irreversible, and not like a debt that can be repaid. Modernist dreams of control and fungibility are a poor fit with the messy variability and interconnectedness of ecosystems. Environmental historians and geographers alike have documented how policy paradigms for rivers, purported to be beneficial for all, mostly benefited cities,

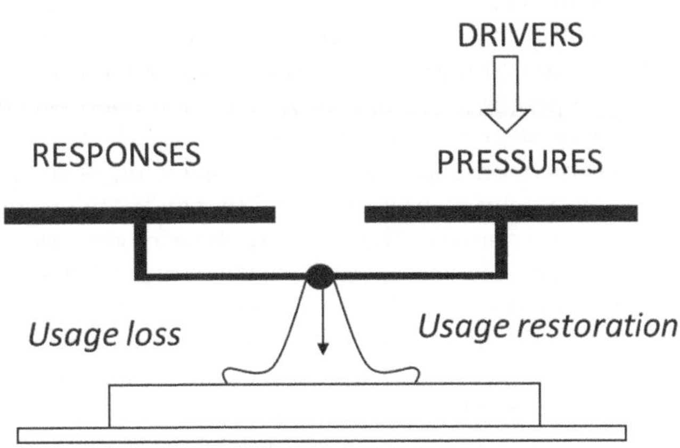

Figure 14.1 Today, environmental policy is generally conceived in DPSIR terms: depoliticised schemes where pressure is mitigated by an adequate response with no irretrievable loss.

capitalists, and global trade, resulting in detrimental environmental impacts for marginalised populations and future generations, not to mention the rivers themselves (e.g. Estes, 2019).

Given that modernist hubris has often led to failures and disappointments, some environmental movements believe humans should relinquish control and allow nature to regenerate itself (Marris, 2013). Advocates for the 'self-determination' of ecosystems have taken a range of positions, including arguing for the legal personhood of rivers, such as the Ganges in India and the Whanganui in Aotearoa/New Zealand (see Chapter 21), that novel ecosystems be retained as sources of resilience in the face of rapid ecological change rather than being targeted for eradication, and that reintroducing keystone species, such as wolves and beavers, can restore lost food webs and reshape entire landscapes and hydroscapes.

14.3 The Promises – and Unexpected Consequences – of River Restoration

Despite the use of financial metaphors such as 'natural capital', river restoration is not as simple in practice as repaying monetary debt, whether motivated by imposing or relinquishing human control. When the 'control' approach falters, as when a dyke is breached, provoking public discontent, a window of opportunity is created to try relinquishing control. Equally, when rewilding has negative effects, such as beaver dams flooding farms or homes, it may lead the public to demand increased control. This dialectic challenges the assumption that actively attempting to restore a river or stream is always the best answer to anthropogenic harm, as we illustrate via three brief examples of flood control, gamefish hatcheries, and rewilding.

Flood defence is a persistent trope in the modernist-control narrative. Starting in the nineteenth century, civil engineers in many parts of the world built dykes to protect large cities from river floods, despite knowing that narrowing channels would raise water levels unless some flood-prone areas remained undeveloped as an escape valve, allowing flood waters out onto floodplains. Yet few, if any, countries have succeeded in protecting rural areas designated for flood expansion, instead allowing development or even extending levee systems to protect them. The predictable result has been rising water levels and damage. The failure of these modernist-control initiatives left many property owners at risk (van Buuren & Warner, 2014) and led to movements to reduce control and restore the natural processes of flood dissipation and infiltration, such as the 'Room for the River' model in the Netherlands (see Chapter 11), 'Engineering with Nature' in the US and 'Natural Flood Management' in the UK. Yet, 'nature-based solutions' themselves require substantial amounts of land and have had mixed results. They can also raise questions of environmental justice, as they are often implemented in disadvantaged areas yet rarely address poverty, unemployment, or neighbourhood regeneration (Cooper et al., 2024).

The attempt to mitigate fish depletion has also demonstrated the dialectic between an initial modernist-control approach and its countering by efforts to restore natural processes, eventually followed by a synthesis between them. In the US, Europe, and the former Soviet Union, conservation agencies used hatcheries to maintain failing gamefish stocks. This reductionist form of environmental managerialism addressed neither the structural causes of the dramatically reduced fish populations (dams, pollution) nor their historical legacies (Secor et al., 2000). Indeed, the approach elicited strong condemnation based on the low survival rates of hatchery-bred fish, concerns over the introduction of non-native species, and fears that hatchery fish would reduce the vitality of native stocks through interbreeding. This led to calls to remove hatcheries and other artificial supports for managed gamefish. Strikingly, there has been a recent reconsideration of the role of hatchery fish, and public support for gamefish can lead to support for beneficial stream policies, such as maintaining environmental flows, improving water quality, and reducing water temperatures (Tadaki et al., 2022).

A final example comes from rewilding practices, which have, in some cases, triggered modernist forms of restoration via control. The reintroduction of wolves to Yellowstone, and the cascading effects of reduced grazing pressure, increased forest cover, decreased erosion, and improved stream health, is often cited as an example of nature's inherent capacity for restoration. However, there are less-successful examples of rewilding, including Oostvaardersplassen in the Netherlands. Here, ungulates such as Heck cattle and Konik horses were reintroduced but struggled to survive, and human observers found it difficult to witness the distressing sight of starving animals. As mentioned above, even ecologically successful rewilding, like beaver reintroduction, can cause human/wildlife conflicts that undermine public support and lead to calls for reasserting control.

14.4 Why Can Restoration Be Problematic?

Undoing anthropogenic harm is a compelling goal. Unfortunately, though, there are many reasons why the current practices of stream and river restoration, whether asserting or relinquishing control, fail to achieve their goals. For example, some river restoration projects are merely greenwashing for big companies or governments hoping to head off public criticism by signalling their environmental commitments without addressing the roots of ecological degradation. Others are explicitly intended as economic development projects in which environmental goals take a back seat (Lee, 2015). Many obstacles can arise even for sincerely motivated restoration projects. These include problems deciding what constitutes a successful restoration project, ignorance of ecological, chemical and physical processes and their complex interrelations, the uncertainty caused by changing conditions in drainage basins exacerbated by anthropogenic climate change, the difficulty of ensuring there is sufficient water in a channel to ensure river health and, given these obstacles, whether restoration is indeed the most effective use of environmental funds.

14.4.1 Defining Success (and Restoration Itself)

Debates over how to define river and stream restoration date back to at least the 1990s (Bradshaw, 1996), and are highly interconnected with the appropriate role of humans in nature (Katz, 1992). If nature is seen as a resource for human use, then projects strengthening ecosystem services would be deemed restoration. Alternatively, if humans are considered part of the community of nature (e.g. Leopold, 1949; Estes, 2024), then only projects with clear environmental benefits would be classed as restoration.

What constitutes success in restoration is a similarly contested question (Lave, 2015). It is surprisingly difficult to declare a restoration project successful, even if it is carefully monitored. In the US, there is no single government standard of success, and relatively few projects have legal performance criteria. The EU's Water Framework Directive (WFD) of 2000 imposed a clear definition and laid out goals for the restoration of European streams. This led to the collection of valuable, standardised data on the ecological status of waterways throughout the EU. Yet, even a clear definition contains ambiguity that can be used to dilute the potential impact of restoration in countries with lower levels of ambition. Thus, even under the WFD, determining whether a restoration project is successful is often more complicated than anticipated.

At least three larger challenges underlie these difficulties in declaring success. First, it is difficult to distinguish and quantify which aspects of a river system's evolution can be attributed to restoration actions versus other influences. Best practice is to use reference sites that are geographically near and biophysically comparable to the restoration site but not subject to restoration. However, the more ambitious the spatial scale of the restoration, the harder it is to find an appropriate reference site. Second, the success of some forms of restoration may only be observable during rare events (such as exceptional floods) or may be historically contingent, as when the order in which species colonise a site has an effect on its composition and future. Lastly, public sentiment towards restoration is dynamic: the view of environmental degradation evolves over generations. Expectations can decline as memories from childhood are idealised as the best achievable state rather than less-damaged conditions further in the past. This shifting baseline also affects the perception of success (Jackson & Hobbs, 2009).

14.4.2 Ignorance

Another reason restoration is not always an appropriate approach to undoing anthropogenic harm is our ignorance, despite the many advances in river science, of key aspects of the ecological, chemical, and physical interrelations in rivers and streams. Some of our ignorance is due to the complexity of river systems, which is only increased by human interactions with biophysical processes. But our ignorance also stems from the fact that science is not an objective mirror of the natural world, but instead a social practice mediated through socially constructed categories that shape our assumptions about the environment (Lave, 2015; Bouleau, 2019). For example, fisheries ecologists have only recently realised that parasites, assumed to be harmful, can in fact protect their fish hosts in highly contaminated waters by accumulating toxins (Molbert et al., 2020). Thus, stream-restoration practices that ignore place-specific conditions and/or try to act on a single driver are unlikely to succeed.

Ignorance about aquatic systems extends to the design and outcomes of restoration projects themselves. Since the complex relations shaping rivers and streams are not fully understood, it is rare that a suitable range of experts is involved in restoration projects (Lave, 2012). Better monitoring of rivers (including those where no restoration has been attempted) would make it easier to learn from mistakes. Monitoring data collected under the EU WFD, for example, is already indicating that some restoration approaches hold more promise than others (Carvalho et al., 2019).

Restoration projects that fail to account for the many uncertainties involved can have unintended and harmful consequences. Indeed, there are cases of channel reconfiguration that have ended up increasing water temperature and nutrient pollution by removing riparian vegetation to give access to construction equipment; while a marsh grass native to the eastern US (*spartina alterniflora*) proved to be invasive when introduced in restoration projects in the western US, choking out native species, clogging flood channels and requiring years of effort to control.[1]

14.4.3 Changing Catchment Conditions

The term 'restoration' implies a return to a prior state. It is often assumed that, as with restoring a historic building, the goal is to make the river look and work exactly as it did at a particular moment in the past. Yet, even if previous river conditions were understood perfectly (which is unlikely, as noted above), and it were physically possible to return a stream or river channel to an exact previous form, there are several key reasons why doing so would be unwise (Dufour & Piégay, 2009). First, it is surprisingly difficult to decide which historical moment to target. In the US, for example, it is common to aim for conditions before European colonisation. But Native Americans actively managed the landscape and hydroscape. Thus, if the goal is to undo human modification, we would need to turn the clock back many thousand years, and our sources of information for how river systems worked then are limited, compounding the problem of ignorance discussed above. In addition, it is not clear that removing human influence is an appropriate goal; if we accept the ethical arguments that humans are part of the community of nature, then there is no intrinsic reason to remove our influence unless it is harmful (Leopold, 1949).

Another big challenge to the idea of restoration as turning back the clock is the enormous and ongoing anthropogenic impacts that have dramatically changed the conditions to which rivers and streams must adapt. These include far more impervious surfaces, reduced sediment loads, and an increase in nutrients and other pollutants (Dufour & Piégay, 2009; Lespez et al., 2015). Because rivers integrate across their drainage basins, they are deeply interconnected to such changes in their surroundings. In many (perhaps even most) cases, restoring previous ecological, chemical, and physical processes would make no sense, since they are not well matched to the current conditions shaping the river system. For example, removing weirs or small dams to restore sediment load will not be effective if gravel extraction upstream has removed the source of the sediment. Finally, the ongoing influence of anthropogenic climate change means that even if a river restoration project is perfectly designed to match current conditions, it is unlikely to be a long-term solution. Predictions vary by region, but in many areas the amount of precipitation is expected to change in both amount and intensity. Increased heat is also a major issue for many of the organisms that live in rivers and streams, which may be unable to survive significantly warmer water temperatures. Taken together, these changing conditions suggest that it would be better to try experimental approaches that are more modest in their alteration of the landscape but ambitious in their monitoring of the results.

14.4.4 Ignoring Root Causes

Most river restoration projects treat the symptoms of degradation rather than their causes because it is quite difficult to tackle broader, systemic issues, such as urbanisation, industrial pollution, agricultural runoff, and deforestation. But without addressing these root causes, even the best restoration efforts are likely to be superficial and temporary. For example, reducing contamination from agricultural runoff is crucial for restoring aquatic biodiversity. Yet, few countries have the political appetite to impose restrictions severely constraining farmers' use of fertilisers and pesticides, meaning any restoration project focused on aquatic biodiversity is unlikely to succeed. Similarly, attempts to restore channel dynamism are unlikely to work as long as the default response to flooding is to build levees.

Perhaps the most fundamental challenge for river restoration is whether humans leave sufficient water in the channel and in aquifers to maintain resilient webs of aquatic life. Particularly in arid and semi-arid areas, enormous amounts of water are appropriated for human use. Famously, the Colorado River no longer reaches the Gulf of Mexico in most years. This drastic overuse is increasingly common, with water syphoned off by competing sectors, including industrial agriculture (irrigation) and energy production (e.g. generating hydropower and/or cleaning solar panels). Rather than prevent the overcommitment of water, however, coalitions of governments and industry often support it for political and economic profit (Molle, 2008).

Unfortunately, ensuring sufficient flow for a river's wellbeing is a zero-sum game. Leaving enough water in the river to protect aquatic organisms means less water for farmers, industry, and city dwellers. Uncertainty and produced ignorance about how much water is needed for a river system to function, as mentioned above, compounds the difficulty of proving that certain (often economically powerful) users must relinquish some of their water rights for ecological purposes (Kroepsch, 2024). But without enough water in the channel, even the best-designed restoration project will fail.

14.4.5 Cost

Despite all these limitations, river restoration is a flourishing market. No-net-loss policies for biodiversity have encouraged mitigation schemes, and consultancy and public-works companies have specialised in this area (Lave & Doyle, 2021). Many of the most common forms of river and stream restoration are very expensive. They involve the purchase of land or a conservation easement, hiring designers, paying for permits, and engaging a construction company to build the project. In the US, the rule of thumb for channel reconfiguration is that it costs $250/linear foot. Removing dams is one of the most effective ways to improve fish habitats and aid many other physical and ecological processes; yet, while far cheaper than repairing them, it is still expensive. In 2023, the United States Geological Service estimated the average cost of removing a dam over 10m in height to be $6.7M (Duda et al., 2023). This only covers putting the restoration project in place. In some countries, it must then be monitored over several years or even maintained in perpetuity with ongoing expenses, such as the removal of invasive plants. The most costly post-construction impact comes from flooding. Already exacerbated by climate change (as described above), flooding can severely damage stream- and river restoration projects, requiring further rounds of design, licensing, and construction. These costs are not trivial.

Given the degree of uncertainty about whether restoration succeeds in undoing anthropogenic harm, it is crucial to consider whether restoration practices that target particular reaches or sections of channel are worth the cost. Would some other form of action to address the root cause of river damage be a better use of limited conservation funds? Might investing in wastewater treatment to improve water quality and address environmental injustice, or reallocating water (potentially requiring the purchase of water rights) to increase flows, be more effective ways to improve the wellbeing of rivers and people?

14.5 Conclusion

The idea of undoing anthropogenic harm to rivers and streams through restoration is very appealing, and yet, there are many reasons to question whether it is always (or even often) the best approach. Overarching motivations for restoration, whether a modernist desire for control or an attempt to expand environmental self-determination, have decidedly mixed records. As described in the previous section, these mixed records are compounded by the difficulties of defining what 'successful restoration' means, overcoming our ignorance of aquatic systems, accommodating rapidly changing conditions, guaranteeing sufficient water flows regardless of economic incentives to use water elsewhere, and crucial questions about the best use of limited funds for environmental priorities.

The strength of restoration lies in acknowledging our ethical responsibility to the landscapes and hydroscapes in which we live. Far from calling that responsibility into question, the issues discussed in this chapter point to the importance of learning how best to fulfil it. We cannot do better until we take seriously the complexity, uncertainty, and potential injustices of river restoration practice, rather than assuming that good intentions and inherited best practices are enough.

Further Readings

Marris, E. (2013). *Rambunctious garden: Saving nature in a post-wild world.* Bloomsbury.

Molle, F. (2008). Why enough is never enough: The societal determinants of river basin closure. *International Journal of Water Resources Development, 24*(2), 217–226. https://doi.org/10.1080/07900620701723646

Morandi, B., Cottet, M., & Piégay, H. (Eds.). (2021). *River restoration: Political, social, and economic perspectives*. Wiley.

Woelfle-Hazard, C. (2022). *Underflows: Queer trans ecologies and river justice.* University of Washington Press.

Note

1 National Geographic website, 'Ecosystem Invaders: Spartina', https://education.nationalgeographic.org/resource/ecosystem-invaders-spartina/

Selected References

Barraud, R. (2025). The changing meaning of wild rivers: A review. *Water Alternatives, 18*(2), 181–213.

Bouleau, G. (2019). *Politicization of ecological issues: From environmental forms to environmental motives*. Wiley.

Bradshaw, A. D. (1996). Underlying principles of restoration. *Canadian Journal of Fisheries and Aquatic Sciences, 53*(S1), 3–9. https://doi.org/10.1139/f95-265

Carvalho, L., Mackay, E. B., Cardoso, A. C., Baattrup-Pedersen, A., Birk, S., Blackstock, K. L., … Solheim, A. L. (2019). Protecting and restoring Europe's waters: An analysis of the future development needs of the Water Framework Directive. *Science of the Total Environment, 658*, 1228–1238.

Cooper, C., Cunningham, N., Bracken, L. J., & Collier, M. (2024). Distribution of nature-based solutions in cities across Europe. *Land Use Policy, 141*, 107160.

Duda, J. J., Jumani, S., Wieferich, D. J., Tullos, D., McKay, S. K., Randle, T. J., ... & Bountry, J. A. (2023). Patterns, drivers, and a predictive model of dam removal cost in the United States. *Frontiers in Ecology and Evolution, 11*, 1215471.

Dufour, S., & Piégay, H. (2009). From the myth of a lost paradise to targeted river restoration. *River Research and Applications, 25*(5), 568–581.

Estes, N. (2019). *Our history is the future*. Verso.

Estes, N. (2024). *Our history is the future: Standing Rock versus the Dakota Access Pipeline, and the long tradition of Indigenous resistance*. Haymarket Books.

Fressoz, J.-B. (2013). Payer pour polluer: L'industrie chimique et la compensation des dommages environnementaux, 1800–1850. *Histoire & Mesure, 1*, 145–186.

Jackson, S. T., & Hobbs, R. J. (2009). Ecological restoration in the light of ecological history. *Science, 325*(5940), 567–569.

Katz, E. (1992). The big lie: Human restoration of nature. *Research in Philosophy and Technology, 12*(1), 231–241.

Kroepsch, A. (2024). Monitored but not metered: How groundwater pumping avoided accounting (and accountability) in the Western United States. *Water Alternatives, 17*(2), 348–368.

Lave, R. (2012). *Fields and streams: Stream restoration, neoliberalism, and the future of environmental science*. University of Georgia Press.

Lave, R. (2015). Stream restoration and the surprisingly social dynamics of science. *WIREs Water, 3*(1), 75–81.

Lave, R., & Doyle, M. (2021). *Streams of revenue: The restoration economy and the ecosystems it creates*. MIT Press.

Lee, S. (2015). Assessing South Korea's green growth strategy. In R. L. Bryant (Ed.), *The international handbook of political ecology* (pp. 345–358). Edward Elgar Publishing.

Leopold, A. (1949). *A Sand County almanac and sketches here and there*. Oxford University Press.

Lespez, L., Viel, V., Rollet, A. J., & Delahaye, D. (2015). The anthropogenic nature of present-day low energy rivers in western France and implications for current restoration projects. *Geomorphology, 251*, 64–76.

Linton, J., & Krueger, T. (2020). The ontological fallacy of the Water Framework Directive: Implications and alternatives. *Water Alternatives, 13*(3), 513.

Molbert, N., Alliot, F., Leroux-Coyau, M., Médoc, V., Biard, C., Meylan, S., ... Goutte, A. (2020). Potential benefits of acanthocephalan parasites for chub hosts in polluted environments. *Environmental Science & Technology, 54*(9), 5540–5549.

Secor, D., Arefjev, V., Nikolaev, A., & Sharov, A. (2000). Restoration of sturgeons: Lessons from the Caspian Sea sturgeon ranching programme. *Fish and Fisheries, 1*(3), 215–230.

Tadaki, M., Holmes, R., Kitson, J., & McFarlane, K. (2022). Understanding divergent perspectives on introduced trout in Aotearoa: A relational values approach. *Kōtuitui: New Zealand Journal of Social Sciences Online, 17*(4), 461–478.

Thompson, D. M. (2013). *The quest for the golden trout: Environmental loss and America's iconic fish*. University Press of New England.

van Buuren, A., & Warner, J. (2014). From bypass to bathtub: Backfiring policy labels in Dutch water governance. *Environment and Planning C: Government and Policy, 32*(6), 1000–1016.

15 *REVISITING*: All Water Uses Can Be Reconciled While Protecting the Environment

Sylvain Barone

The question of whether it is possible to maintain the various anthropogenic uses of water (drinking supply, agriculture, industry, etc.) while preserving the natural environment obviously invokes a series of complexities and nuances. It encompasses, among others, hydro-social configurations, the availability of water, the vulnerability of ecosystems, the extent and type of pressure on the resource, and local hydro-climatic developments. There is also significant uncertainty around anthropogenic usage, such as the difficulty in estimating withdrawals precisely, even in countries best equipped to do so. Issues in quantifying the resource include underestimating climate change, double-counting surface water and groundwater, and lagging hydrological data, as well as how to assess the impact of water uses on ecosystems (PFAS, microplastics, the 'cocktail effect' of certain substances, etc.).

Nonetheless, the idea that existing anthropogenic uses can be reconciled over time – with each other and with the preservation of the ecosystem – has been embraced for several decades in the world of water management. It sees the growing environmental challenges facing our societies, such as those linked to climate change, as manageable *through a number of adjustments and optimisations,* without questioning water use or fundamental economic structures. To understand how we arrived at this point and what its implications are, this chapter analyses the *meta-discourse* of reconciliation, which can take shape in different (sub-)discourses.

15.1 Construction and Dissemination of a Dominant Meta-Discourse

The economic and urban development that characterised the second half of the twentieth century increased competition between water uses, with shifting impacts on ecosystems. Water policies then sought to regulate these uses on varying timelines and in a wide variety of institutional forms: monitoring and control, support for technological innovation, economic incentives, participatory planning, etc. These developed within varied and evolving ideological and normative contexts. Many countries' water management policies came to be justified, and occasionally structured, by a mix of soft doctrines, more

DOI: 10.4324/9781003745532-15

or less vague beliefs, and more precise recommendations, which contributed to the production of a dominant discourse.

The two dominant – and partly intertwined – lineages of this discourse are sustainable development and ecological modernisation. The former seeks to maximise economic and social benefits for populations while preserving ecosystems. Emphasising the need for a holistic and cross-sectoral approach to environmental issues, this concept gained traction in the 1990s – a time of rising ecological concerns, particularly in some Western countries, but also the neoliberal questioning of the role of the state.

Integrated Water Resources Management (IWRM) represents the application of this concept in the water sector. The 1992 International Conference on Water and the Environment in Dublin, in preparation for the Rio Earth Summit held the same year, issued an alarming assessment of the global water situation and articulated several key principles: fresh water is a finite and vulnerable resource essential to life, development and the environment; water management must take a participatory approach, involving users, planners and policymakers at all levels; women play a central role in water provision, management and safeguarding; water has an economic value in all its competing uses and should be recognised as an economic good. In 1993, the World Bank published a policy paper on water resources management, adopting the imperative of integrated management (World Bank, 1993). These principles were adopted and expanded by the Global Water Partnership (GWP) – one of the leading international promoters of IWRM,[1] which then gained influence in water management in many countries (Allouche, 2016), notably in Europe, West Africa, and Latin America. Since then, there has been a proliferation of official declarations and publications advocating more integrated water management, particularly in reports and guidelines from the World Bank, the OECD, the United Nations, and of course, the GWP.[2]

The discourse around IWRM tends to incorporate critiques of the excesses of late-twentieth-century modernisation practices (such as large dams or agricultural intensification in the Global South), asserting that space must be made for all uses. The idea of 'integration' within IWRM is, in fact, multidimensional. Beyond economic efficiency, social wellbeing and the preservation of resources, it must also be applied to all forms of water (surface and groundwater, green, blue or grey), to all actors and public sectors, and, spatially, at the river basin scale. Considerable emphasis is also placed on stakeholder participation, innovation, and the combined use of multiple instruments (regulation, participatory planning, market mechanisms, contracts, etc.) (Mollinga et al., 2006). In Europe, IWRM translations marked a shift from the regulation of water usage to a focus on the environment and participation – often at the expense of equity issues (the 2000 EU Water Framework Directive being a good illustration). Elsewhere, IWRM has been used as a Trojan horse to assign economic value to water (Molle & Berkoff, 2007). Thus, compatibility between uses and with environmental protection is an important dimension of IWRM but far from the only one, and, depending on the case, one dimension often outweighs the other.

The dominant discourse of reconciliation also has origins in ecological modernisation. Strongly associated with environmental sociology and policy, this theory has also been analysed – first in the Netherlands, the United Kingdom, and Germany – as having structured environmental policy since the 1980s (Hajer, 1995). Indeed, ecological modernisation has shaped the way 'environmental problems' are perceived, debated and politically formalised, viewing such issues as 'challenges for social, technical and economic reform rather than as immutable consequences of industrialization' (Mol & Sonnenfeld, 2000, p. 5). It rests on the idea that the preservation of nature is not only compatible with economic growth, but that environmental problems could even represent genuine economic opportunities. Once this condition is met, the problems would stimulate technological progress, encourage efficiency gains, and open up new markets. While all these concepts converge, operational actors tend to refer more to 'green economy' or 'green growth' than to 'ecological modernisation' - just as they also commonly refer to sustainable development and IWRM.

In environmental policy, ecological modernisation oscillates between a weaker and a stronger or 'reflexive' version of the concept. The former refers to the liberal rhetoric of win-win strategies between economy and ecology, while the latter opens pathways for change in public action towards 'ecologisation' (Meadowcroft, 2012). Over time ecological modernisation has tended to hybridise with approaches that are less techno-economistic and more orientated towards inclusion and participation. It aligns with sustainable development and IWRM on the principle of attaining a conciliatory 'optimum'. However, ecological modernisation places much greater emphasis than sustainable development and IWRM on the idea that economic growth and environmental protection mutually reinforce one another. It is also more orientated towards innovation and individual behaviour.

For the UN, OECD, World Bank, and the GWP, good water management is not only indispensable for economic activity but must be at the heart of efforts to build a green economy. As the World Bank puts it:

> Water is essential for jobs, livelihoods, and thriving economies. No country can achieve lasting prosperity without securing its water future. Reliable water supply boosts productivity in agriculture, energy, and industry—sectors that employ millions, especially in low-income countries. [...] On the other hand, smart water management creates opportunities: in the Democratic Republic of Congo, a water access program is expected to generate nearly 30,000 new jobs.[3]

Many of water management's core tools fit the logic of ecological modernisation, such as water rights trading markets (markets serving the environment, often presented as an institutional innovation), precision irrigation, desalination, and the reuse of treated wastewater (technology-driven, with potential markets). By employing digital technology, 'smart' water management aligns particularly well with its prescriptions (Ascençao et al., 2023). Whether in urban networks, irrigation, industry or territorial water management, the use of real-time data collection, automation, smart sensors, big data and artificial intelligence are expected to ensure management is more effective, efficient and sustainable by enabling better leak detection, faster adjustment of consumer behaviour, soil- or weather-based irrigation control, industrial water recycling and real-time modelling and monitoring of basin flows or water quality.

Sustainable development/IWRM and ecological modernisation are mutually permeable approaches. While integrated management embodies a very broad sense of reconciliation, ecological modernisation offers a 'hardened' and more precise interpretation, incorporating methods to achieve its goals. Notably, smart water management is said to advance IWRM and the UN's Sustainable Development Goals. At the same time, other concepts have emerged at the intersection of these different approaches. Promoted since the 2010s by international organisations such as the GWP, the World Economic Forum or the World Water Council,[4] the Water-Energy-Food Nexus is based on the recognition of three interdependent key issues: agriculture accounts for the largest part of freshwater withdrawals worldwide; a significant share of the world's energy production relies on water; and agriculture accounts for a substantial share of the global energy demand. Hence, the nexus seeks to ensure security across these three sectors while fostering sustainable natural resource management through improved cross-sectoral coordination (Benson et al., 2015). While placing less emphasis than IWRM on the river basin, it aligns more closely with ecological modernisation by prioritising optimisation, as well as quantified and model-based approaches.

15.2 A Problematic Approach

The hegemony of these approaches produces – and institutionalises – a standardised vision of problems and potential solutions. It essentially reinforces the idea that by activating the right tools and levers, water management can continue in its current form while preserving ecosystems, downplaying any need to impose restrictions or prioritise uses. As a meta-discourse, it has received strong support in both environmental policy and academic circles. Its critics, meanwhile, mainly come from the environmentalist movement and the social sciences. Rather than discuss these positions, this chapter highlights what becomes problematic when the reconciliation discourse dominates the field of water management.

15.2.1 An Ambiguous Discourse

The strength of the reconciliation discourse lies in its flexibility, as embodied by the concept of 'integrated management'. While this refers to the reconciliation of the ecological, economic, and social dimensions of water, it can equally justify the protection of nature, full cost recovery from consumers, or universal access to the resource. Thus, it is highly consensual – at least in principle, making IWRM appear to be a 'nirvana concept' (Molle, 2008) that can resolve all problems.

In practice, the greening of policies that began in the 1990s was accompanied in many countries by at least a partial renewal of the objectives assigned to water management. Rather than replacing earlier goals these new objectives have overlapped with them, leaving space for anthropogenic uses to persist. In France, for example, the satisfaction of all uses, institutionalised by the 1964 law that created the Water Agencies system, remains one of the major objectives of water management, despite the subsequent implementation of various laws and European water directives orientated toward greening since the 1990s–2000s. Meanwhile, the compromises inherited from the 1960s in terms of instrumenting public action have barely been questioned. Participatory planning still favours major economic users (such as agribusiness), while water policy financing still relies heavily on domestic users and local authorities, far less on industry and even less on farmers.

To prove effective, an approach does not need to be coherent or even prioritise objectives – in fact quite the opposite: its appeal grows when it lends itself to multiple interpretations and manages to accommodate different, even conflicting, interests and perspectives (Palier, 2005). Actors can thus agree on an approach despite differing in their concerns, which is why the discourse of reconciliation is so attractive. The downside, however, is a lack of clarity in action, attributable meaning, accountability, and transparency for users and citizens.

15.2.2 An Easily Instrumentalised Approach

Ambiguity generally benefits the most structured actors. The persistence of trade-offs aimed at maintaining old arrangements and accommodating certain sectoral actors and social groups – despite ongoing greening efforts – can be observed in many countries. Ambiguity also gives more power to decision-makers, who can emphasise at will either the importance of a given use or the need to protect the environment.

But sustainable development or 'green growth' can also conceal political objectives other than environmental protection. Public policy is used to address problems and simultaneously strengthen some political actors. Thus, to varying degrees, an environmental policy will serve to protect the ecosystem, address non-environmental issues (social inequality, economic development, etc.), and also reinforce the power or political legitimacy of its proponents. It is a balancing act that can result in the relegation of ecological objectives to the background.

For example, in the early 2010s, China institutionalised the concept of 'ecological civilisation'. This was intended as a shift from a growth model based on intensive resource exploitation to one where development is 'in harmony with nature'. In the water domain, 'three red lines' were enshrined in the 2012 National Water Resources Management Plan: to limit total water use, improve the economic efficiency of its use, and control pollution. But, as a highly centralised governance instrument, its implementation is uncertain. Indeed, China's water policy represents both a weak version of ecological modernisation and the pursuit of unspoken objectives. The South-North Water Transfer Project illustrates this syndrome with its parallel goals of preventing small farmers from being forced out of agriculture due to high water prices, ensuring food security, reducing conflict among water users, and, ultimately, reinforcing the authority of the Communist Party (Moore, 2014).

15.2.3 Ecological Modernisation Will Not Be Enough

Ecological modernisation is a politically appealing concept as it allows win-win discourses to take precedence over thorny trade-off (compromises on all sides). It promises that even the toughest environmental challenges can be addressed through technology and the market. Yet it offers no indication that this will be sufficient to counterbalance the 'treadmill of production', to borrow a neo-Marxist expression (Gould et al., 2008). The capitalist system requires continuous economic growth for producers to remain competitive. This, in turn, requires the consumption of natural resources such as water. While optimisation may have a positive effect at the local level (factory or agricultural area) or following the adoption of a new law, the *overall* capacity of ecological modernisation to offset this tendency has not been demonstrated (Ewing, 2017; Hickel & Kallis, 2020).

Furthermore, factors that are presented as compatible (water uses for economic production and the protection of resources and ecosystems) also collide with socio-climatic realities. As tensions over water use sharpen, compromise is no longer self-evident. Already palpable today, they are set to intensify. Climate change contributes to droughts, longer low-flow periods, and falling groundwater levels while increasing the variability of these processes. In southern France, the average river flow has decreased by around 30% over the past 50 years. Morocco has seen its renewable resources decline by about 30% over the past 30 years. Mega-droughts in Australia, California and Chile have already led to reductions in irrigated areas (partly offset by groundwater overexploitation, which adds to the problem). Climate change threatens many aquatic species. Water management is thus increasingly challenged – particularly in its ecological objectives – by anthropogenic uses competing for scarcer water.

15.2.4 Counterproductive Effects

Ecological modernisation downplays or ignores a well-identified risk in natural resource management: the 'Jevons paradox'. This occurs when 'rebound effects' generate an increase in resource use that outweighs the efficiency gains achieved by more frugal use, and is well documented in irrigation, for instance (see Chapter 2). Many countries have seen efficiency gains from precision irrigation (better knowledge of plant needs, drip systems, automation, etc.) accompanied by an expansion of the irrigated area, leading to greater overall water consumption (Venot et al., 2017). Thus, after years of supporting Morocco's National Irrigation Water Saving Plan – particularly through drip irrigation – the World Bank eventually acknowledged the existence of this effect in its 2023 Country Climate and Development Report. Not only did the promised water savings fail to materialise, but ecological modernisation often produced the opposite outcome. Side effects are also neglected, including the increased energy needs of pressurised networks or the 'plasticisation' of nature through intensive horticulture (see Chapter 2).

As for water markets, their promoters, who include environmental NGOs in some countries, argue that assigning water its 'true price' will stimulate greater efficiency in production systems while enabling the state, trust funds, or environmental organisations to purchase rights from corporate users and return them to rivers (as in Australia and the United States). Yet, by reducing nature to transferable volumes and market value, certain 'uses' of water are excluded. Not only does this lead to the erasure of their identity or spiritual aspect, but it also ignores the relationship certain social groups, for example, Indigenous peoples, may have with water. Such markets also have their own ecological impact, for example, when water rights are overgenerously allocated to economic use, or when transfers deprive agricultural areas of water in favour of more profitable crops elsewhere, thereby weakening local ecosystems.

15.2.5 A Royal Road to Maladaptation

The reconciliation discourse allows the major economic uses of water to go largely unquestioned, thereby protecting specific private interests, such as intensive agriculture or large polluting industries. These are also often co-producers of water policy, whether through their involvement in water management bodies or through more direct power dynamics and negotiations with political decision-makers, discreetly or otherwise (Barone, 2024). Many studies have documented this trend in Brazil, the United States, South Africa, Morocco, Spain, and France, not to mention the European Union.

The illusion is fostered that superficial reforms and rational optimisation will suffice, without requiring fundamental changes in production or consumption, or addressing issues of inequality or environmental justice. Hence, the narrative of reconciliation may encourage a form of maladaptation to climatic and ecological changes. Critiques of smart water management say it reinforces the eco-modernist approach in which technology 'buys time' by avoiding the question of excessive consumption. This permits a reassuring fantasy that the inexorable march of technology will continually offset the negative impact of growing unsustainability as a form of 'insurance' for individuals and societies (Hartley & Kuecker, 2020).

With ecological modernisation, unlimited growth seems possible and environmental problems surmountable by integrating nature into the economy through technological change and market mechanisms. From this perspective, all that is required is a fine-tuning of the productive system, without breaking with the dominant structures of capitalist production and consumption, including the imperative of accumulation. Some critics of the theory, meanwhile, have focused on its empirical weakness. Others have drawn on the concept of 'treadmill of production', considering that nothing will change unless we tackle the root causes of unfettered productivism. These critics see ecological modernisation as a dangerous form of denial, technological hubris, and a concession to capitalism's near-limitless destructive power (see, for example, York et al., 2010 or Foster, 2012).

15.2.6 A Depoliticising Discourse

By putting efficiency and optimisation centrestage, the dominant discourse of reconciliation leads to an acute form of depoliticisation. Water becomes nothing more than a matter of 'good management' through capital, technology, and calibrating the 'right instruments'. It gives the idea that collective progress could be made towards a single 'best way' by including stakeholders in line with the prescriptions of the 'world polity' on water. Reducing environmental policy or water management to a question of 'common sense' places it 'outside' or 'beyond' politics (Machin, 2019).

It is striking to observe how public water actors at different scales try to depoliticise issues by systematically downplaying controversy and conflict, despite the fact that water management necessarily involves decisions with which not everyone will agree. They also downplay the negative effects of certain uses on the resource and ecosystems. All this, of course, in the name of a peacekeeping imperative: water should unite, not divide. By emphasising discussion and integration rather than tension and discord, they position themselves as mediators, thereby legitimising their central role in the game.

Reconciling the satisfaction of *all* water uses with ecological health means minimising the profound differences in their impacts on water, and claiming instead that they must each play their part in the effort to preserve the resource. Whether corporations, individuals, rich or poor, all uses and users must be accepted, and everyone must change their behaviour. This naturally leads to the promotion of certain solutions, such as water savings for individual users, measures to limit leaks in drinking water networks, optimisation of agricultural irrigation, reuse of treated wastewater (see Chapter 3), desalination of seawater (see Chapter 4) etc. But it also wards off more radical positions and solutions that would discuss the legitimacy of uses, acknowledge the contradictions between economic growth and ecology, or the disproportionate contribution of certain social groups or economic actors to ecosystem degradation. Indeed, the reconciliation framework emphasises the *consequences* of water scarcity and pollution over their collective *causes*. It is more about fixing problems than questioning the structural determinants of the situation.

While politics presupposes dissension – a confrontation of visions – the 'post-political environmental consensus' (Swyngedouw, 2010) negates this, crowding out expressions of alternative socio-environmental futures. By keeping radical transformation off the agenda, this depoliticised approach constantly demands 'more of the same' (Remling, 2018).

15.3 Conclusion

The promise that, under certain conditions, different anthropogenic water uses can be reconciled with one another, as well as with ecosystem preservation, remains very present in the world of water management. It has, of course, partially materialised, with tangible environmental gains and better reconciliation of uses here and there. Yet this discourse is also the subject of numerous critiques based, in particular, on ecological and social justice arguments in a context of increasingly impactful hydro-climatic changes.

Since the 1990s–2000s, a belief in progress towards better environmental protection seemed fairly widespread in the water management world. Yet, current nationalist-conservative or national-populist movements have given rise to a green backlash (Bosetti et al., 2025). The surge of this large-scale dynamic (Brazil yesterday, the United States, Argentina, Europe today) has revived water uses that had recently been called into question due to their environmental impact (e.g. fracking or irrigation programmes). In its own, brutal way, this too has contributed to challenge the myth of reconciliation. This anti-ecologist movement should not, however, make us nostalgic for the deceptive consensus of reconciliation and its devaluation of politics. On the contrary, what is needed is to fully embrace a re-politicisation of water and its management, while ensuring that processes exist to channel disputes, allowing conflict to be managed effectively and collective agreement to be reached.

Further Readings

Foster, J. B. (2012). The planetary rift and the new human exemptionalism: A political-economic critique of ecological modernization theory. *Organization & Environment, 25*(3), 211–237.

Hartley, K., & Kuecker, G. (2020). The moral hazards of smart water management. *Water International, 45*(1), 1–9.

Machin, A. (2019). Changing the story? The discourse of ecological modernisation in the European Union. *Environmental Politics, 28*(2), 208–227.

Molle, F. (2008). Nirvana concepts, narratives and policy models: Insights from the water sector. *Water Alternatives, 1*(1), 131–156.

Notes

1 This network, composed of development agencies, NGOs, universities, local authorities, companies and natural resource managers, was established in 1996 with the support of the World Bank, the United Nations Development Programme and the Swedish International Development Agency. It brings together around 3,000 partners across 170 countries.
2 See, for example, the *IWRM Guidelines at River Basin Level* (2009) or the United Nations' Sustainable Development Goal 6 (Clean Water and Sanitation, 2015), the more recent World Bank guidelines on Integrated Urban Water Management, the *OECD Council Recommendation on Water* (2016) and the accompanying *Toolkit for Water Policies and Governance* (OECD, 2021), etc.
3 World Bank website, 'Water': www.worldbank.org/en/topic/water/overview. The strategy *Toward a Clean, Green, Resilient World for All* (2012–2022) and the report *Reboot Development: The Economics of a Livable Planet* (2025) point in the same direction.
4 The World Water Council is an international organisation composed of NGOs, governments, intergovernmental organisations and private actors seeking to bring the international community together to promote water as a vital political priority. It hosts the World Water Forum every three years.

Selected References

Allouche, J. (2016). The birth and spread of IWRM: A case study of global policy diffusion an translation. *Water Alternatives*, *9*(3), 412–433.

Ascençao, E., Marinangelo, F. M., Almeida, C., Kagan, N., & Dias, E. M. (2023). Applications of smart water management systems: A literature review. *Water*, *15*(19), 3492.

Barone, S. (2024). The state against the environment? Water management and the regulation of tensions between sectoral policies in France. *Environmental Policy and Governance*, *35*(1), 3–12.

Benson, D., Gain, A. K., & Rouillard, J. (2015). Water governance in a comparative perspective: From IWRM to a 'nexus' approach? *Water Alternatives*, *8*(1), 756–773.

Bosetti, V., Colantone, I., De Vries, C. E., & Musto, G. (2025). Green backlash and right-wing populism. *Nature Climate Change*, *15*(8), 822–828.

Ewing, J. A. (2017). Hollow ecology: Ecological modernization theory and the death of nature. *Journal of World-Systems Research*, *23*(1), 126–155.

Gould, K. A., Pellow, D. N., & Schnaiberg, A. (2008). *The treadmill of production*. Paradigm.

Hajer, M. (1995). *The politics of environmental discourse: Ecological modernization and the policy process*. Oxford University Press.

Hickel, J., & Kallis, G. (2020). Is green growth possible? *New Political Economy*, *25*(4), 469–486.

Meadowcroft, J. (2012). Greening the state. In P. Steinberg & S. D. Vandeveer (Eds.), *Comparative environmental politics: Theory, practice, and prospects*. MIT Press.

Mol, A., & Sonnenfeld, D. (2000). Ecological modernization around the world: An introduction. *Environmental Politics*, *9*(1), 1–14.

Molle, F., & Berkoff, J. (Eds.). (2007). *Irrigation water pricing: The gap between theory and practice*. CABI.

Mollinga, P., Dixit, A., & Athukorala, K. (Eds.). (2006). *Integrated water resources management: Global theory, emerging practice and local needs*. SAGE.

Moore, S. M. (2014). Modernisation, authoritarianism, and the environment: The politics of China's South–North Water Transfer Project. *Environmental Politics, 23*(6), 947–964.

Palier, B. (2005). Ambiguous agreement, cumulative change: French social policy in the 1990s. In W. Streeck & K. Thelen (Eds.), *Beyond continuity. Institutional change in advanced political economies* (pp. 127–144). Oxford University Press.

Remling, E. (2018). Depoliticizing adaptation: A critical analysis of EU climate adaptation policy. *Environmental Politics, 27*(3), 477–497.

Swyngedouw, E. (2010). Impossible sustainability and the post-political condition. In M. Cerreta, G. Concilio, & V. Monno (Eds.), *Making strategies in spatial planning. Urban and landscape perspectives* (pp. 185–205). Springer.

Venot, J.-P., Kuper, M., & Zwarteveen, M. (Eds.). (2017). *Drip irrigation for agriculture: Untold stories of efficiency, innovation and development*. Routledge.

World Bank (1993). *Water Resource Management: A World Bank Policy Paper*.

York, R., Rosa, E. A., & Dietz, T. (2010). Ecological modernization theory: Theoretical and empirical challenges. In M. R. Redclift & G. Woodgate (Eds.), *The international handbook of environmental sociology* (pp. 77–90). Edward Elgar.

16 *REVISITING*: A River Basin Begs a River Basin Organisation

François Molle, Doug Kenney, and Bernard Barraqué

16.1 Background to the RBO as a 'Best Practice'

The recognition of river basins as hydrological units began to take shape in Europe and the United States in the nineteenth century, though early Chinese records suggest the concept is much older (Molle, 2009). The notion that water should be governed at the basin level grew from an acknowledgement of the interconnectedness of upstream and downstream users and areas, most visible during droughts and floods, as well as the need to address potential conflict between hydropower and other uses. The origins of the first river basin organisations (RBOs) are disputed. While the concept rose to prominence in the aftermath of World War II, the earliest contenders include the *Emschergenossenschaft* (1899) and the *Ruhrvervand* (1913) in Germany, Spain's first *Confederacion hidrografica* (Ebro, 1926), the Tennessee Valley Authority (TVA) (1933) and the Ontario Conservation Authority (1946).

RBOs were first and foremost boosted by the USA's promotion of the TVA – a federally owned corporation managing flood control, hydropower, and economic development in the Tennessee River Basin. This had ideological underpinnings (for President Truman, the USA had 'demonstrated for all time the efficiency and the humanity of comprehensively planned, multipurpose river basin development') and geopolitical goals (Sneddon, 2019). Marketed alongside a promise of 'grassroots democracy', the TVA turned out to be a major asset in overseas development and diplomacy – 'a new export commodity' in the era of Cold War politics (Ekbladh, 2002). Its model of the 'full development' of river basins epitomised the economic development, state legitimacy, and modernity so coveted by world leaders.

By the 1990s, Integrated Water Resources Management (IWRM) emerged as the dominant framework for global water governance (Conca, 2005). This further elevated the river basin as the preferred unit for water management, promoting integration across sectors and stakeholders at this scale. RBOs became central to the approach, as they were seen as facilitating the implementation of other key IWRM principles: subsidiarity, user participation, clearly defined water rights, and economic tools like water pricing.

DOI: 10.4324/9781003745532-16

The 1992 Rio Earth Summit intensified the global focus on environmental issues, and initiatives like the European Water Framework Directive (WFD) promoted river basin planning. In 1992, France's 'Basin Financial Agencies' transitioned from a purely financial role – collecting fees, financing storage infrastructure, and curbing pollution – to a more holistic approach to river basin management. The WFD required to establish competent territorial authorities and basin planning with user participation, and claimed to incorporate the orthodox 'user pays' and 'polluter pays' principles endorsed by environmentalists on the grounds of reduced use and increased water flows in European rivers. This helped enshrine the principle of planning water at the 'river basin district' level – a vaguer, scale-making concept accommodating geographical and historical specificities.

The institutionalisation of RBOs as a global best practice owes much to transnational policy and epistemic networks (Conca, 2005; Mukhtarov & Gerlak, 2013). These are loosely defined, but their actors, forums, funding, and publications are readily identifiable. Key international promoters of IWRM in general, and RBOs in particular, include the World Bank, the Asian Development Bank, the Interamerican Development Bank and United Nations regional economic commissions, as well as international NGOs, such as the WWF and Green Cross, and networks like the Global Water Partnership (GWP) and the International Network of River Basin Organizations (INBO). The INBO tasked itself with promoting the idea globally, proclaiming in 2011 that 'It is necessary to support the creation and strengthening of Basin Organizations around the world!', since 'integrated and sound water resources management at the level of river basins is obviously essential worldwide!' (INBO, 2011).

At the national level, many countries in the Global South have adopted or adapted RBO models in bilateral cooperation with development agencies from countries including the US, Sweden, Germany, Denmark, France, and the Netherlands. These models are frequently showcased at international events like the World Water Forums and Stockholm Water Weeks, and are taught in IWRM training programmes offered by platforms such as Cap-Net and WaterNet.

At this stage, we should distinguish national RBOs, the focus of this chapter, from transboundary RBOs. The latter respond to the UN Convention on the Law of the Non-navigational Uses of International Watercourses (1997) and the Berlin Rules on Water Resources (2004), which enshrine general cooperation principles and encourage states to set up agreements or RBOs to facilitate transboundary management. The River Basin Organization Design Database, curated by Oregon University, appears to document over 120 transboundary (basin and aquifer) organisations, and these have given rise to a prolific literature, notwithstanding their mixed achievements.

RBOs are one of the 23 'tools' proposed under the 'Institutions and participation' category of the GWP's IWRM toolbox, where they are referred to as 'important institutional bodies responsible for overseeing the development and management of water resources at the watershed level', and thus spawned numerous publications, 'guidelines' and 'keys for success' (UNESCO, 2009). A flurry of indicators soon arrived to benchmark their 'performance' (Hooper, 2006). RBOs gained so much currency in the 1990s and 2000s that most water laws issued in this period, in line with IWRM principles, would include them as a '*universally recognised* objective and principle' of water policy, as one Brazilian water expert put it. Acknowledged to come in different guises, the INBO currently records over 200 organisations worldwide qualifying as major RBOs.

It is worth noting, however, that the reasons for promoting river basin organisations have changed over time (Molle, 2009). The TVA model represented a stage of massive, state-led, post-WWII infrastructural development geared mainly towards hydropower, flood protection, and irrigation, and emulated in countries such as Mexico, India, and Indonesia. The RBO gradually became a home for the integration of environmental management, economic tools, and participation at large. In Europe, it is now the vehicle for monitoring and ensuring the 'good status' of water bodies mandated by the WFD. The popularity of the concept is perhaps reflected in Figure 16.1, which confirms a first hike in usage in the 1960s before another sharp rise in the late 1990s and 2000s, which saw several basin comparisons and studies before attention waned.

Although the 'principles and concepts of IWRM have been widely recognised', UNESCO (2009) admits that

> the implementation of IWRM has not always progressed satisfactorily in many basins perhaps partly because of the disparate perspectives held by the various water sectors concerned, all working with their interests in mind and failing to recognize the need for greater interaction and cooperation.

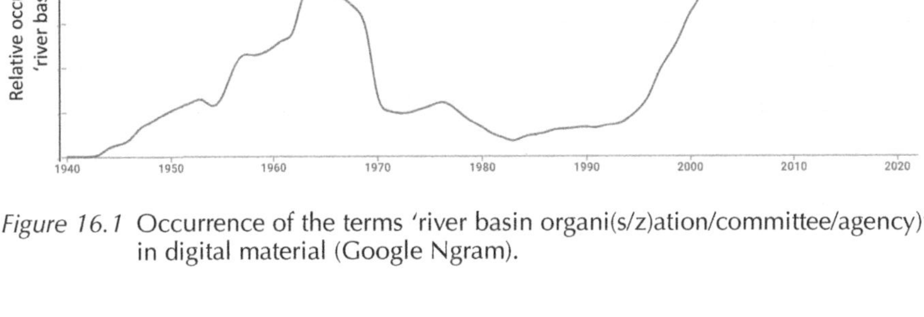

Figure 16.1 Occurrence of the terms 'river basin organi(s/z)ation/committee/agency)' in digital material (Google Ngram).

Most assessments of RBOs tend to be highly prescriptive, with 'to do' lists including sharing knowledge, engaging stakeholders, developing capacity, ensuring political will and self-financing, building institutions, etc. While, on paper, the neat institutional diagrams and the 'integration' rationale and objectives are compelling, many RBO designs and processes appear to have ended up compounding administrative bickering and conflicts. The many barriers and constraints identified in the literature are often minimised to allow the rolling out of the RBO blueprint. Before examining these difficulties in more detail, it is useful to map out the multifarious basin-based organisations that are counted as RBOs.

16.2 The Diversity of RBOs

RBO is a generic term encompassing a large diversity of organisations known variously as authorities, agencies, commissions, committees, boards, councils, partnerships, etc., which differ starkly in objective, remit and organisational structure (GWP & INBO, 2009; Huitema & Meijerink, 2014). Putting aside transboundary organisations, we can conceive of RBOs as including some form of four main constituents that may exist or not and differ in terms of membership and effective power (Figure 16.2).

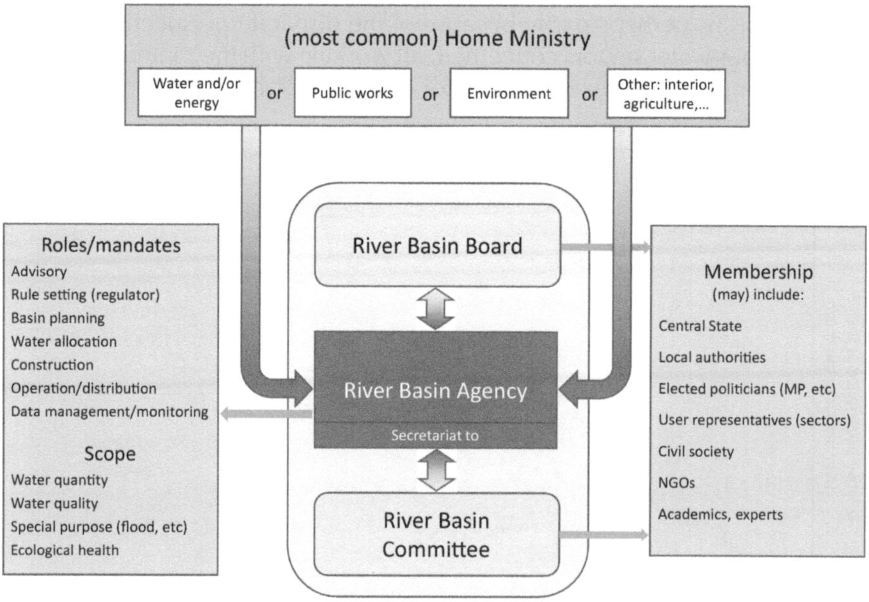

Figure 16.2 Generic structure of an RBO.

RBOs often include a permanent public body (referred to here as the 'Agency'), usually under the direct authority of a parent ministry (or state administration). In some instances, as in Nigeria and Colombia, where the Agency is independent, with broader prerogatives, we refer to it as a River Basin Authority (not shown on Figure 16.2). The nature of the ministry is, of course, a determinant feature of the Agency. (The examples given in Figure 16.2 are indicative only, as actual titles vary widely in practice, and many ministries have hybrid or combined responsibilities.) The Agency is commonly a public body that is deconcentrated and located in the basin concerned (although it may also be in charge of all river basins, as in the state of Ceará, Brazil). Its degree of independence and specific remit can vary greatly but will often include data collection, preparation of basin plans, water allocation (permitting), advising on tariffs, and implementing national water policy (or Water Framework Directive regulations in the EU). The Agency may also manage infrastructure (day-to-day operations), as with the Spanish *confederaciones* and the Genossenschaften in NRW, Germany, and may even oversee construction projects. It may be responsible for quantitative water management only or also for quality, flood-and-drought management, or environmental protection.

The Agency may report directly to its parent ministry and/or be associated with boards and committees. Executive/advisory boards include representatives from various ministries, local government, and main water users. They handle administrative and financial issues and may also establish political priorities. Committees, sometimes known as 'local water parliaments', are consultative bodies that typically advise on priorities and planning. In the EU, the participation requirement for the WFD has led to the rolling out of committees or legitimisation of existing ones. They are associated with Agencies loosely defined as 'the appropriate competent authority', with substantial flexibility in what these may be (Länder water authorities in Germany, *Agences de l'eau* in France, the Environment Agency in the UK, etc.). Membership generally includes (in varying proportions) government representatives, local officials, water users, NGOs and academics. The Agency acts as the secretary to this non-permanent committee.

The Agency will sometimes have a board but no corresponding active committee (e.g. in Morocco). Conversely, the RBO can be limited to this sole committee, which can be merged with the board into an overall governing board, as in South Africa. Some agencies were formed several years after the establishment of the committee, such as those in Mexico (*organismo de cuenca*), which are little more than regional offices of Conagua (National Commission for Water) and are poorly articulated with the committees (*consejos de cuenca*). In many countries, such as Peru, Turkey, and Vietnam, the secretariat/support function is carried out by the dominant water authority (or a weaker one, like the Ministry of the Environment in Thailand) through its regional or basin-level branches.

Sometimes, rather than as a companion to an Agency, the committee is conceived as a cross-jurisdictional coordinating entity for joint policymaking, research, and planning (or management, as in the Paraiba do Sul Basin, Brazil). Typical in federal settings, this committee type is funded by partner organisations, not necessarily involving the general public. Other committees (councils or partnerships) may grow from local actors joining voluntarily to address a sustainability or social issue (such as combatting pollution, providing amenities, or advocating for giving a legal and political voice to a river). The US's 'watershed approach' framework sees partnerships including some or all of these actors: local communities, companies, environmental groups, and local government or federal agencies (Diaz-Kope & Miller-Stevens, 2015; Schlager & Blomquist, 2008). They can be bottom-up or state-driven, as in Indonesia, where watershed conservation efforts involve no fewer than four competing ministries.

These partnerships generally emerge at the local or small-basin level, although exceptions exist, such as the 'Parliament of the Loire' initiative in France. Scale matters greatly: large basins tend to be divided into subbasins, where smaller river basin committees can also be established, as in Thailand and Mexico, the latter distinguishing between nested basins, subbasins, and microbasins. In France, the six main river basins are also partly divided into subbasins, each with a public coordinating body ('river syndicate') and a river (sub)committee. Smaller watershed RBOs may be overseen by regional authorities, or state governments in federal countries like Brazil, Australia, and Mexico, rather than a national ministry.

The diversity in RBO models highlights why a single discussion cannot address all scales and institutional arrangements. Focusing on the most common configuration – an Agency under a ministry, working alongside a committee – we now examine their implementation in greater detail.

16.3 The Devil in the Details

16.3.1 The Implementation of RBOs as Political Process

RBOs always seem like a good idea on paper. They promise decentralisation, improved stakeholder involvement, enhanced coordination, and environmental benefits. Yet the core issue lies in how they reshuffle existing power and incentive structures. Unsurprisingly, the handing out of new powers is often resisted by established actors in direct proportion to how they view the threat to their own authority. Two key factors will determine the RBO's fate: by whom (and why) it is established, and how it is implemented (in terms of process, structure, and power).

It is not uncommon for the idea and design of an RBO to be driven by external actors, such as international development banks or bilateral aid agencies. In such donor-driven cases, pilot RBOs are often introduced as a 'soft' form of conditionality attached to larger loans or as a vehicle for fostering cooperation and financially supporting a dedicated water administration (Lankford & Hepworth, 2010). A good example is the Asian Development Bank's push to introduce pilot RBOs across Southeast Asia in the 2000s. The promotion of this 'best practice' may also align with the interests of certain local officials, who either believe in the approach or benefit indirectly from participating in the process.

Alternatively, the driving force can be internal. RBOs may be viewed as symbols of a 'modern water policy' that appeals to reform-minded officials or as instruments to achieve specific policy objectives. In Brazil, for instance, the federal National Water and Energy Agency promoted the French model in part to secure a dedicated budget for the water sector. RBOs can also be introduced through political windows of opportunity. They may emerge as a response to a historic injustice – as in post-apartheid South Africa – or align with democratic transitions and decentralisation agendas, as in Brazil or Ukraine. In other cases, they are prompted by new legislation, such as the Endangered Species Act in the US or the Water Framework Directive in Europe.

Whether externally or internally driven, however, the introduction of an RBO is usually led by a small circle of actors, who may lack the power to implement the long-term changes needed to disrupt entrenched power structures and vested interests. Indeed, regardless of the benefits they are supposed to deliver, RBOs inevitably disrupt the established bureaucratic/power structure and existing water governance more broadly. While this is not the place to address the full diversity and complexity of possible institutional changes triggered by the introduction of RBOs (see, for example, Streeck and Thelen's [2005] typology of displacement, layering, drift, conversion and exhaustion), a few general illustrations of implementation processes will help make these dynamics explicit.

RBOs that fulfil an unclaimed or limited complementary space (e.g. in water quality or environmental issues) or act as deconcentrated extensions of the main ministry in charge of water typically face less opposition, but their achievements are also more modest. When RBOs are layered over existing governance actors, conflict and turf battles are almost inevitable, irrespective of whether the authority of the latter is reduced or left untouched. Meijerink and Huitema (2017) underscore 'how careful [existing institutions] are to curtail the powers of the new RBOs by arranging for institutional designs that effectively constrain (some might say straightjacket) new RBOs'. This was the case with the Red River Basin Organisation (Vietnam) (Molle & Hoanh, 2011) and Thai RBOs established under the Ministry of Natural Resources and the Environment (MoNRE). RBOs may be tasked with reordering water allocation and thus may also face opposition from powerful constituencies. In South Africa, 27 years after the 1998 National Water Act, only two out of 19 projected catchment management agencies (CMAs) have actually been established. In Chile, RBOs and 'integrated management' have been floated for over 20 years but have been resisted by powerful irrigator associations (*juntas de vigilencia*) that de facto manage (parts of) river basins. Local administrations can also be hostile. Indeed, it often takes a centralised state (e.g. France or Spain as opposed to Italy, Germany or the US) to create strong new institutions whose territories bear no relation to existing administrative divisions (Barraqué, 2001).

Dominant water bureaucracies generally control stakeholder basin committees (where they exist) from their central or regional offices. In Mexico, Conagua supported the management of 25 river basins from its 13 local offices effectively maintaing centralised control over water state commissions seeking more autonomy. Turkey's Basin Management Committees and Peru's river basin boards remain firmly under the centralised control of the ANA (National Water Agency) and DSI (State Hydraulic Works) respectively. The *Agences Hydrauliques de Bassin* in Morocco have no basin committee, and the government has been dragging its feet in introducing one. It is worth noting, however, that committees can be used to legitimise river basin masterplans already drafted by hydraulic bureaucracies, by offering a formal stamp of 'participation' as seen in Turkey, Algeria, and Thailand.

While most RBOs are 'incorporated' in dominant water (or public works) administrations, some are 'excorporated', that is, within other ministries, such as that in charge of the environment. In such cases, the role/power of the committees and their Agencies tends to be commensurate with that of the ministry. In France or Spain, where they are under the Ministry of Ecology, RBOs retain a key role, but in many countries where environmental governance is weak, the opposite is the case.

Introducing a new governance layer at the basin level often triggers unpredictable bureaucratic (see below) and political dynamics. As Hüesker and Moss (2015) show in the Wupper catchment in Germany, the shift toward basin-based governance reshaped the strategies, power relations, and interests of diverse actors, including old water boards, agricultural chambers, green NGOs, and various levels of government. In Peru, the establishment of river basin boards weakened traditional irrigation communities while facilitating the interests of mining and agribusiness sectors (Budds & Hinojosa, 2012). Sometimes, as in France or Spain, river basin committees may be captured, to some extent, by dominant players, such as agribusiness or the main agricultural trade unions.

Meijerink and Huitema (2017) aptly observe that weak RBOs are more likely to survive, while the strong are resisted. Their relative weakness can often be gauged by the limited budgets they receive. Even where there is no deliberate attempt to restrict them financially, ensuring long-term financial sustainability is a persistent challenge. Although some RBOs (e.g. in France or Brazil) are funded from user or polluter fees, most rely heavily – or entirely – on state funding, as in Morocco and across much of Asia. This dependence makes them vulnerable to policy shifts and political change, as seen in Australia, Spain, France, Mexico, and South Africa.

Inadequate staffing is another clear indicator of institutional weakness, especially when RBOs have been imposed from outside. Agencies are often staffed with personnel from hydraulic administrations. In Morocco and Mexico, for example, regional offices of national hydraulic agencies were simply rebranded as basin agencies. In Thailand, the DWR (responsible for overseeing basin committees) was staffed with Ministry of Interior officials with little experience of public participation or environmental concerns. Vietnam's Red River Basin Organisation originated in the Ministry of Agriculture. In short, underfunded, understaffed, and disempowered Agencies are common symptoms of fragile RBOs.

16.3.2 Failing to Keep the Promise of Coordination and Participation

Professed objectives of sectoral integration often remain a dead letter. Notwithstanding the need for, or desirability of, integration, RBOs' mandates frequently clash with those of other ministries and administrations, even when their roles do not explicitly overlap. In Morocco, for example, the regulation of water use is at odds with both the priorities of the Ministry of Agriculture and its Development Offices and the Ministry of the Interior's concern for social stability (Tanouti, 2017). Indeed, RBOs rarely succeed in improving coordination across key sectors, such as domestic supply, agriculture, hydropower, tourism, and the environment. Instead, their introduction may exacerbate bureaucratic competition and sectoral fragmentation. In many countries, RBOs are little more than project agencies, planning infrastructure (e.g. Algeria) and sometimes managing it alongside main irrigation schemes (e.g. the Litani Authority in Lebanon). Lacking regulatory or integrative functions, they are usually deconcentrated rather than decentralised organisations.

Thailand's case vividly illustrates the turf battles that RBOs can provoke. Largely toothless Basin Committees, supervised by the Department of Water Resources (DWR) under the MoNRE, are used to rubber-stamp conventional masterplans. Rather than act as a regulator, the DWR became embroiled in power struggles over planning authority with the Royal Irrigation Department (RID), part of the Ministry of Agriculture. Since planning attracts significant budgets, particularly for infrastructure projects, it remains a hotly contested domain. The rivalry intensified following the historic 2011 floods, when the Yingluck government used the DWR to attempt to take control of flood-control megaprojects and sideline the RID. A similar situation unfolded in Vietnam, where the Ministry of Agriculture and Rural Development (MARD) clashed with MoNRE over infrastructure planning for the Red River Basin (Molle & Hoanh, 2011).

Beyond bureaucratic realities, the complexities of the integration ideal are also linked to the issue of 'spatial (mis)fit'. Improving spatial fit for water can unintentionally create misfits with other policy sectors essential to sustainable water management, such as urban development, agriculture, forestry, transportation, and energy (Moss, 2012). River basin authorities, whose boundaries rarely align with political jurisdictions, often lack democratic legitimacy and struggle to coordinate with these sectors. Furthermore, organising management around ecosystem boundaries has led to an overemphasis on biophysical issues, drawing criticism for being technocentric and elitist despite claims of participation. To be effective, water governance must incorporate the multiple geographies – political, economic, cultural, and physical – involved in resource management rather than focus on ecosystem or jurisdictional scales (Moss, 2012). An alternative to formal river basin institutions is multiagency collaboration that reflects the overlapping realities of governance.

One of the clearest illustrations of the gap between institutional ambition and practical implementation lies in the area of data collection and sharing. While the collective value of developing accessible, centralised water databases is widely acknowledged, efforts in this direction generally fail (e.g. in Morocco), reflecting deeper political and institutional dynamics. For ministries, departments, agencies, and service operators, data can be a powerful resource. This is particularly evident where studies are outsourced to private consultants reliant on access to agency contracts – as in South Africa – creating a disincentive to share data openly.

The promise of participation can be equally unconvincing, especially when (powerless) basin committees are controlled by hydraulic bureaucracies (Morocco, Turkey, Mexico, etc.). In South Africa, it is the Minister of Water and Sanitation who appoints the governing board of CMAs, illustrating that 'it is typically those with the most resources and political clout who have the most influence in such structures' (Meissner et al., 2017). Fatigue generated by the long participatory process of establishing CMAs and the overly technical and technocratic language used discourage small farmers.

In Europe, while the WFD encourages the participation of non-state actors, the proliferation of participatory processes did not stem from any deliberate or voluntary initiative by the traditional powerholders in the water sector. Some analysts have concluded that WFD implementation has included a wide range of actor groups but rarely citizens, rating public consultation as 'insignificant' and suggesting that 'the EU's conviction that participation benefits good water status appears overly naïve' (Rimmert et al., 2020).

16.4 Polycentricity: River Basin Management without Agencies

Agencies may be presented as indispensable, but some countries have shown that river basins can be 'managed' using other types of institutions, often through 'polycentric governance' (Pahl-Wostl et al., 2012; Schlager & Blomquist, 2008), in which the 'lighter' types of RBO described above can have a role. In Germany, with the exception of the 12 Genossenschaften of the Ruhr, the tradition of cities' self-management and the role of the states (Länders) remain strong. They have fostered coordination between Länders, cities, industry, and other users to solve problems on shared rivers. The most powerful actor is not the federal Environment Ministry but the acting president of the LAWA (Länder Arbeitsgemeinshaft Wasser Abfall) – an interstate working party on water and waste.

In the US, meanwhile, small catchment-scale 'watershed initiatives' are thriving, while larger river basin agencies remain scarce, with the obvious exceptions being those formed long ago (such as the TVA) in an era where federally funded water development was a compelling stimulus (Kenney, 2008). Today, the political costs of disrupting established power structures are usually too high to move away from the polycentric governance structures that have evolved incrementally to govern developed basins.

This is particularly evident in the western US, including the Colorado River Basin, where river flows were already formally apportioned among the constituent jurisdictions (states in a federal system) and where basin-scale river development long ago gave way to more localised efforts in water distribution and management (Fleck, 2016). A few systemic management challenges, such as controlling water salinity, make some level of ongoing multijurisdictional coordination desirable, but absent a legal trigger, such as a federally protected salmon run (as seen in the Columbia River Basin), the motivation to coordinate further is lacking. When issues like drought-response necessitate a coordinated approach, technical experts can typically share data and negotiate new rules through informal networks or narrowly conceived RBOs. This approach insulates managers from the obligation to pursue broader (and far more complex) objectives of improved sectoral integration or widened participation, which would presumably be featured in strong RBOs operating within an IWRM framework. Whether that is a governance weakness to be addressed or a strength to be celebrated ultimately depends on one's place within the water community.

IWRM proponents have difficulty identifying RBOs that actually achieve the impossibly lofty outcomes envisioned in theory. Informal and otherwise primitive water networks in many basins have proven sufficient to achieve basic management objectives. This suggests that the argument that all basins need an RBO is probably not only unnecessarily ambitious but also elevates the RBO to a performance standard that these agencies are destined to fall short of in many real-world applications.

16.5 Conclusion

Integrating and coordinating water-related activities at the river basin level holds significant promise, but caution is warranted. Many water challenges extend beyond basin boundaries, involving actors and processes outside the watershed. Moreover, effective water management is possible without a formal RBO, particularly in contexts amenable to polycentric governance. Notwithstanding the variety of organisations labelled as RBOs, their establishment often disrupts existing local government and established constituencies, lobbies, or bureaucratic structures, triggering turf conflict. Strong RBOs may encroach on entrenched authority or be co-opted by dominant actors to consolidate power, especially in key sectors like infrastructure planning. RBOs that emerge from powerful water agencies ('incorporated' RBOs) often reinforce existing hierarchies, while those placed under weaker ministries ('excorporated' RBOs) typically lack authority and capacity. Multistakeholder Committees associated with RBOs frequently prove symbolic or ineffective. Many externally driven RBOs remain underfunded, understaffed, and thus stagnant for years. Whether such entities can eventually catalyse meaningful institutional change in water governance remains uncertain. This is not a call for inaction but, rather, a cautionary warning that establishing RBOs involves far more than a well-intentioned administrative decision.

Further Readings

Abers, R., & Keck, M. E. (2013). *Practical authority: Agency and institutional change in Brazilian water politics*. Oxford University Press.

Huitema, D., & Meijerink, S. (2014). *The politics of river basin organisations*. Edward Elgar Publishing.

Molle, F., & Hoanh, C. T. (2011). Implementing integrated river basin management in the Red River Basin, Vietnam: A solution looking for a problem? *Water Policy, 13*(4), 518–534.

Schlager, E., & Blomquist, W. (2008). *Embracing watershed politics*. University of Colorado Press.

Selected References

Barraqué, B. (2001). La politique de l'eau, le libéralisme étatique et la subsidiarité. In O. Coutard (Ed.), *Le bricolage organisationnel: crise des cadres hiérarchiques et innovations dans la gestion des entreprises et des territoires* (pp. 69–84). Elsevier.

Budds, J., & Hinojosa-Valencia, L. (2012). Restructuring and rescaling water governance in mining contexts: The co-production of waterscapes in Peru. *Water Alternatives, 5*(1), 119–137.

Conca, K. (2005). *Governing water: Contentious transnational politics and global institution building*. The MIT Press.

Diaz-Kope, L., & Miller-Stevens, K. (2015). Rethinking a typology of watershed partnerships: A governance perspective. *Public Works Management & Policy, 20*(1), 29–48.

Ekbladh, D. (2002). "Mr. TVA": Grass-roots development, David Lilienthal, and the rise and fall of the Tennessee Valley Authority as a symbol for US overseas development, 1933–1973. *Diplomatic History, 26*(3), 335–374.

Fleck, J. (2016). *Water is for fighting over*. Island Press.

GWP (Global Water Partnership) & INBO (the International Network of Basin Organizations). (2009). *A handbook for integrated water resources management in basins*. GWP.

Hooper, B. P. (2006). *Key performance indicators of river basin organizations*. US Army Corps of Engineers, Visiting Scholar Program.

Hüesker, F., & Moss, T. (2015). The politics of multi-scalar action in river basin management: Implementing the EU Water Framework Directive (WFD). *Land Use Policy, 42*, 38–47.

INBO. (2011). *Declaration of Mégève*. INBO.

Kenney, D. S. (2008). History of river governance in the United States. In A. Dehnhardt & U. Petschow (Eds.), *Sustainability in river basins: A question of governance* (pp. 109–134). Institute for Ecological Economy Research.

Lankford, B., & Hepworth, N. (2010). The cathedral and the bazaar: Monocentric and polycentric river basin management. *Water Alternatives, 3*(1), 82–101.

Meijerink, S., & Huitema, D. (2017). The institutional design, politics, and effects of a bioregional approach: Observations and lessons from 11 case studies of river basin organizations. *Ecology and Society, 22*(2).

Meissner, R., Funke, N., & Nortje, K. (2017). The politics of establishing catchment management agencies in South Africa: The case of the Breede-Overberg Catchment Management Agency. *Ecology and Society, 21*(3), 26.

Molle, F. (2009). River basin planning and management: The social life of a concept. *Geoforum, 40*(3), 484–494.

Moss, T. (2012). Spatial fit, from panacea to practice: Implementing the EU Water Framework Directive. *Ecology and Society, 17*(3), 2.

Mukhtarov, F., & Gerlak, A. K. (2013). River basin organizations in the global water discourse: An exploration of agency and strategy. *Global Governance, 19*, 307–326.

Pahl-Wostl, C., Lebel, L., Knieper, C., & Nikitina, E. (2012). From applying panaceas to mastering complexity: Toward adaptive water governance in river basins. *Environmental Science & Policy, 23*, 24–34.

Rimmert, M., Baudoin, L., Cotta, B., Kochskämper, E., & Newig, J. (2020). Participation in river basin planning under the Water Framework Directive – Has it benefitted good water status? *Water Alternatives, 13*(3), 484–512.

Sneddon, C. (2019). *Concrete revolution: Large dams, Cold War geopolitics, and the US Bureau of Reclamation*. University of Chicago Press.

Streeck, W., & Thelen, K. (2005). Introduction: Institutional change in advanced political economies. In W. Streeck & K. Thelen (Eds.), *Beyond continuity* (pp. 1–39). Oxford University Press.

Tanouti, O. (2017). *La Gestion Intégrée des Ressources en Eau à l'épreuve du bassin versant – Cas du bassin du Tensift au Maroc* [Doctoral dissertation]. Université Paris.

UNESCO. (2009). *IWRM guidelines at river basin level*. UNESCO.

17 *REVISITING*: Collective Action Needs Water User Associations

Edwin Rap

17.1 Introduction

An influential policy idea that has dominated water governance for four decades already is that collective action requires Water User Associations/Organisations (WUAs/WUOs) to manage shared infrastructure. Since the late 1980s, neoliberal policy has promoted the management transfer of irrigation systems and drinking water facilities from states to various forms of water user organisations (WUAs/WUO/WUG/WUCs).[1] As a result, the organisational figure of the water users' association (WUA) today is ubiquitous in water management, whether we are speaking of participatory irrigation management (PIM), on which we will mostly focus, or of community-based management (CBM) in rural water supply, which we will also engage with Garces-Restrepo et al. (2007). The popularity of this organisational model as policy peaked in the 1990s and early 2000s and has since possibly passed its peak of influence, but its era is certainly not over (World Bank Group, 2019).

This might be an opportune moment to take stock of the origin, promises, and limitations of this policy solution. Despite the apparent attractiveness of entrusting water management to associations of concerned users, its application has often faced diverse obstacles, which can jeopardise the management of water infrastructure. We will first recall the emergence of this 'best practice' and then dwell on its documented shortcomings. In line with the objectives of this book, we have chosen here to focus less on the policy's reported benefits and more on its limitations, while only providing selected key references.

17.2 Background and Origin of Policy Ideas

To understand how WUAs have become so dominant as policy instruments, we need to trace their origins. The roots of this policy idea and its supporting network lie in the recognition that farmers have long managed community-based irrigation systems sustainably around the globe. For millennia, farmers have collectively constructed, operated, and maintained small to larger irrigation systems (Hunt & Hunt, 1976; Coward & Uphoff, 1986). Amongst

DOI: 10.4324/9781003745532-17

the appealing examples of such collective action are water courts around Valencia, Spain; Persian water wheels to lift and distribute water collectively; qanats or underground aqueducts to supply groundwater in the Middle East; and indigenous irrigation systems in Mexico or the Andes. Such collective action was often embedded in larger religious or royal village and state institutions and hierarchies of wealth and power (Aubriot, 2022).

From the late nineteenth century onwards, nation-states have been increasing their role in constructing, interconnecting, and managing irrigation systems as a form of state formation. This was especially true after the Second World War, when national governments made huge capital investments in large-scale public irrigation, increasingly financed through international loans. This led to a huge global expansion of irrigated land surface and its associated public institutions in the following decades. By the 1980s, international experts and donor institutions were progressively disappointed that state bureaucracies were underperforming in the management of large-scale irrigation systems (Garces-Restrepo et al., 2007). Many public irrigation investments in developing countries did not improve irrigation performance (e.g., water use efficiency), long-term infrastructural viability, or productivity according to expectations (Mollinga et al., 2007; Aarnoudse et al., 2018). Donors and governments expected farmers to pay for maintenance, but they did so insufficiently. A 'vicious cycle' in public irrigation thus ensued: The non-payment of farmer fees led to infrastructure deterioration, while low-cost recovery caused poor maintenance and threatened infrastructural sustainability, further decreasing the quality of service and thus farmers' willingness to pay (Garces-Restrepo et al., 2007; Aarnoudse et al., 2018). Then, in the early 1980s, a global economic recession slowed down international lending and forced governments to reconsider their public irrigation expenses. The public nature of irrigation and its typical problems – (over)spending, insufficient cost recovery, poor performance, infrastructural sustainability, and accountability of service providers – increasingly came under question.

More and more experts identified farmer-managed irrigation systems (FMIS) as a solution to poorly performing public irrigation systems. The idea emerged to replicate collective action among water users in public schemes (Hunt & Hunt, 1976; Uphoff et al., 1991). Already since the late 1970s and 1980s, participatory development approaches had promoted the empowerment of 'water users' to improve the performance and sustainability of irrigation systems (Coward & Uphoff, 1986; Chambers, 1988). During the 1980s, user fee payment was increasingly seen as a quasi-market incentive to make public agencies more responsive and improve service delivery in irrigation management (Repetto, 1986; Korten & Siy, 1989; Uphoff et al., 1991). The idea of cost recovery thus became associated with the ideal of user participation. "There is reason to believe that greater farmer involvement in O&M [Operation and Maintenance] activities through user organisations would not only reduce direct government costs but might also produce more appropriate and better-quality O&M results" (Coward & Uphoff, 1986). Under American influence in Asia, early seedlings of an epistemic and policy network emerged, which promoted 'participatory irrigation management (PIM)' and would gradually transplant the idea of WUAs to different corners of the world (Garces-Restrepo et al., 2007; Senanayake et al., 2015). The Ford Foundation and USAID started to build on early transfer experiences and pilot projects with water user participation in the Philippines, Sri Lanka, and the United States. Such ideas became more easily institutionalised and disseminated by science and research with the formation of the International Irrigation Management Institute (IIMI) in 1984, which later became the IWMI (International Water Management Institute).

The idea was that the increased payment of user fees would initiate a 'virtuous circle' that related user satisfaction with performance (Small & Carruthers, 1991). WUAs would consequently be able to recover full O&M costs, improve performance, and reduce public expenditure. The same principles of commodification and participation increasingly underwrote the dominant CBM model in rural water supply (van den Broek & Brown, 2015). Community *participation* in decision-making and maintenance in the 1980s developed into a push for community *management* in the late 1980s and 1990s. Water user committees (WUCs) would subsequently manage collective water supply points to enhance maintenance and sustainability (Harvey & Reed, 2007).

Neo-institutional theory to 'craft institutions for self-governance' inspired both PIM and CBM approaches (Ostrom, 1992). This boosted faith in the social engineering of collective action (Cleaver, 1999), the idea being that WUAs would govern people's rational behaviour to act collectively for the common good. "[T]he core design principles used in long-enduring, self-organised irrigation institutions throughout the world" help to account for "the success of these institutions in sustaining the physical works and gaining the compliance of generations of users to the rules in use" (Ostrom et al., 1993, p. 1907).

The establishment of WUAs as a global best practice is related to broader neoliberal policy paradigms. At the end of the Cold War, the 'Washington Consensus' emerged as a critical juncture of neoliberal policy convergence. International financial institutions imposed structural adjustments and a broad reform package that forced many Southern countries to 'roll back the state', liberalise the economy, and reduce public spending. Two overlapping public policy paradigms in development institutions coalesced: New Public Management (NPM) and Integrated Water Resources Management (IWRM). NPM shifted the focus for accountability from inputs and processes toward outputs and outcomes, as in private business. By leaving public managers 'free to manage' but measuring and rewarding their performance, NPM promised to make public service providers directly accountable to their clients and regulatory authorities. Alongside this, the 1992 Dublin Principles for IWRM considered water an economic good and stakeholder and user participation as important policy principles, expecting WUAs to play a critical role in realising these policy paradigms (Aarnoudse et al., 2018).

17.3 The Policy Network Promotes Solutions

Since the 1990s, international donors increasingly financed loans and projects that decentralised public irrigation management to WUOs/WUAs. This reinforced an epistemic community and policy network that promoted PIM internationally. The Economic Development Institute (EDI), the educational arm of the World Bank, launched the establishment of the International Network on Participatory Irrigation Management (INPIM) (Gorriz & Groenfeldt, 1995). This network of development banks, research institutes, and development organisations further included the Food and Agriculture Organization (FAO), International Water Management Institute (IWMI), and the International Food Policy Research Institute (IFPRI). It played a major role in co-producing the policy models, best practices, and success stories of WUAs and their globalised policy transfer. Expert institutions contributed by organising training and public awareness campaigns, as well as performance measurement and benchmarking projects for promoting and assessing policy ideas (Garces-Restrepo et al., 2007). For example, the FAO developed a programme on irrigation sector reform, organising many events, producing numerous guidelines, and creating a website to orient policy makers and experts on PIM policies (Garces-Restrepo et al., 2007). IWMI launched efforts to measure the performance and success of WUAs (Vermillion, 1991; Johnson et al., 1995).

Strategic training and promotion campaigns were instrumental in forging consensus among international and national stakeholders. Public awareness campaigns to reach water users and a larger public were essential in mobilising popular and policy support. Mexico organised study tours for policy makers and senior officials (Garces-Restrepo et al., 2007). Just like Mexican policymakers had visited advanced user-managed irrigation systems in the United States, around 50 Turkish officials visited relatively privileged pilot projects in northern Mexico, where they could see with their own eyes that WUAs were managing irrigation successfully. A promotional video called 'Seeing Is Believing' contributed to motivating countries to start ambitious Irrigation Management Transfer (IMT) reforms involving WUAs. Mexico, Turkey, Andhra Pradesh, and other regional policy hubs performed models to visiting policymakers from the region, gently steering them towards similar loan-supported reforms and conditionalities. Chambers (1988) argues that such 'islands of salvation' are often unrepresentative but project development myths.

In the neighbouring field of rural drinking water supply, a similar policy network promoted CBM, which also combined commodification and participation. At stake was the same epistemic belief in crafting collective action around a shared water source/point or pump technology and having users pay to ensure their long-term sustainability (van den Broek & Brown, 2015). The motto became, "Now you have a new pump, you have to manage it" (Eguavoen, 2007). About this, a World Bank official jokingly referred to the "unfortunate collusion between structural adjustment and empowerment. 'Elinor Ostrom meets Ronald Reagan', so to speak".[2]

In a wider finance-policy-expert nexus, epistemic beliefs converged around WUAs as the principal solution to the conflated problems of finance and empowerment. Water user organisations (WUO/A/C/Gs) became the international policy recipe for collective action. By organising 'water users' in irrigation or rural water supply, they could assume public tasks and expenses. The initial successes in model countries and the positive language of participation and performance further raised policy ambitions and expectations. A 'positive feedback loop' goes as follows (Aarnoudse et al., 2018): When water users participate and improve performance, the resulting increased productivity will enhance water users' ability and willingness to pay fees and participate in management. With such expectations, World Bank researchers considered WUAs key to achieving 'more efficient water delivery services' (Rusca et al., 2015).

17.4 Limitations

The policy-promoting network raised expectations about performance improvements and therefore overestimated policy impacts. WUAs have accomplished important things and are here to stay. Yet, they did not systematically improve management performance and rural water supply across the board, let alone produce empowerment or increased productivity and infrastructural sustainability. The financial objectives of reducing public expenditure and increasing cost recovery may have been partly and temporarily accomplished, yet all other results have been mixed (Meinzen-Dick, 2007; Ghazouani et al., 2012; Senanayake et al., 2015; Arnoudse et al., 2018). In other words, water user management and cost recovery did not radically improve service provision. More generally, neoliberal decentralisation did not deliver.

We focus here on understanding which limitations have caused WUAs not to function according to policy expectations.

17.4.1 Faltering Financial Sustainability

The policy aim is that WUAs collect sufficient fees to cover all O&M and administrative costs in order to become financially autonomous and sustainably manage the collective infrastructure. Although significant user contributions are reported in the literature, they are often not sustained at that level. The degree to which this occurs varies widely. In total, 61% of worldwide cases reported negative or negligible impact regarding financial viability (Senanayake et al., 2015). In Mexico, full cost recovery was initially reported for WUAs, but self-sufficiency levels have decreased over time, potentially leading to deferred maintenance (Rap, 2006). Similar patterns have been observed in countries as diverse as Tunisia, Kyrgyzstan, Turkey, and the Philippines. In Egypt, the informal cost recovery around tertiary pumping stations could not be replicated in branch-canal WUAs, because the latter had no concrete authority in water management, maintenance, or other functions (Rap et al., 2019). In northern and sub-Saharan Africa, WUAs are also unable to collect sufficient user fees and pay for O&M (Ghazouani et al., 2012; Aarnoudse et al., 2018). Freeman & Lowdermilk (1991) have already explained that "farmers are quick to see that, from an individually rational standpoint, one is foolish to pay water assessments [...] when water service is not substantially affected by making payment".

In rural water supply, the results also do not indicate financial sustainability. For example, a WUC survey in two northern districts of Uganda showed that the majority (77) of 100 water user committees (WUCs) had insufficient funds for minor repairs or no funds at all, indicating that long-term sustainability was at risk (van den Broek & Brown, 2015). In the case of WaterAid, Malawi, full cost recovery was at the expense of the poor and benefited WUA representatives (Rusca et al., 2015). The strong focus on user contributions in CBM allows donors and governments to show a concern for sustainability while distancing themselves from delivering it (Whaley & Cleaver, 2017).

17.4.2 Social and technical engineering produces sub-optimal collective action

For nation-states and international projects, WUAs became instrumental to social engineering and rendering collective action governable. To build water user groups, Ostrom and other neo-institutionalists looked for the right mix of internal rules and incentives, enabling environments, and management tools. However, governments and development projects cannot fully determine the social dynamics, collective action, or human and non-human agency around evolving infrastructures, particularly under time and budget constraints. Building institutions often involves more than the optimal adjustment of institutional nuts and bolts; socially and technically engineering collective action does not mechanically lead to successful outcomes. The donor's belief that 'training' can constitute user groups, independent from politics, community divisions, power imbalances, elite capture, and economic disparities, is rather naïve.

In theory, soft- and hardware interventions develop in parallel, yet in practice they often become disjointed and unequally resourced processes. Despite high expectations, only a fraction of World Bank loans went to the WUA organisation and management 'software'. Infrastructure is usually institutionally prioritised above collective action and interlinking the hard and software of rural water supply systems, therefore, does not necessarily lead to functional collective action (Whaley & Cleaver, 2017). In the Philippines, the National

Irrigation Administration (NIA) used donor funds mainly for system rehabilitation and construction, and only a small portion for 'institutional development' (Oorthuizen, 2003). In the Nile Delta, as in many other regions, top-down infrastructural improvement preceded and became dissociated from WUA organisation, after which many water users refused to organise collectively to assume and pay for unsolicited interventions. The resulting institutional disconnect between technically engineering (collective) infrastructure and socially engineering WUAs limited collective action. Government officials failed to scale up informal collective action, cost recovery, and traditional conflict resolution techniques (known as *orfu*) at the tertiary level to the level of secondary branch-canal WUAs (Rap et al., 2019). Therefore, crafting collective action through blueprint WUA models and standardised hardware solutions is not straightforward (Venot, 2014).

Rehabilitation investments were often insufficient, untimely, and unequally applied, which caused deferred maintenance and jeopardised viability. In Mexico, the policy promise that rehabilitation of irrigation infrastructure would precede IMT did not always materialise. Loan benefits (such as system rehabilitation and maintenance machinery) landed particularly with early-adopting WUAs in irrigation systems that were already more modernised and better maintained. This meant that many water users were left with a deteriorated infrastructure, the maintenance for which they would now have to pay collectively.

Collective action often occurs without WUAs, before and after PIM and CBM. In the Nile Delta, some collective pumping stations are managed, financed, and maintained by a user group without a functional WUA (Ghazouani et al., 2012). In other situations, existing organisations (such as cooperatives, municipalities, or private entities) might be better suited to managing irrigation, so perhaps we need to look differently at water collectives. The life history of a pumping collective in the Egyptian Nile Delta demonstrates the remarkable agency of a pumping collective in renewing and reassembling itself, thanks to – and despite – social and technical engineering by larger institutions. Understanding the transformational capacities of such collectives under water scarcity and abundance may be key to understanding their durability (Rap et al., 2025).

17.4.3 Bureaucratic Dependence, Lack of Empowerment, and Accountability

State commitment to the transfer of resources and power to decentralised authorities is always uncertain, particularly under unstable regimes. In Egypt, the government repeatedly delayed a law which would have strengthened WUAs´ responsibilities and resources at branch-canal level, due to the regime's fear that religious groups and the political opposition would engage. In 2011, Turkey legally reverted WUAs into public entities under state authority, while Albania has now transferred irrigation management to municipalities instead.[3] In practice, thus, states have often not really empowered

WUAs. WUA cost recovery has been imposed as a burden rather than a 'right'. A paradigmatic illustration is the Moroccan WUA law, which reads as a military list of injunctions: only duties, no rights (Ghazouani et al., 2012).

After neoliberal reforms, 'main system management', which is the centralised control and operation of primary infrastructure, remained a state and bureaucratic responsibility (Chambers, 1988). The reasons behind this deserve further exploration and a broader contextualisation elsewhere, but it is probable that nation-states consider the public management of strategic infrastructure as key to their national sovereignty, military security, and political economy. Because the main system management remained a bureaucratic power, which was unaccountable to water users, WUAs that were entrusted with the management of tertiary- and secondary-level infrastructure remained dependent on inadequate bureaucratic performance. This did not put them in good conditions to perform well. In large irrigation systems, the transfer of water management to secondary- or tertiary-level WUAs would imply a contractual agreement between the WUAs and the bureaucracy, defining how much water they would manage. Yet, WUAs hardly ever get such contractual guarantees of a predetermined bulk allocation and just receive what the bureaucratic system can deliver. When supply is variable and uncertain, there is nothing for the WUAs to manage. These structural limitations severely restricted water user empowerment in terms of power, expertise, and financial means.

Nevertheless, neoliberal reforms often also severely weakened water bureaucracies. They responded by making water policy, bureaucratic reforms, and hydraulic projects in order to continue pursuing their hydraulic mission and expert authority (Suhardiman et al., 2014). The *Comisión Nacional del Agua* (CNA) partly restored financial and bureaucratic autonomy in Mexico but did not transform itself into an effective agency to regulate WUA performance, as NPM proposed. Although senior officials in the Philippines pioneered farmer participation in irrigation management, the NIA perceived WUAs as threats to their importance, which weakened user involvement in system management. In Jordan, the Jordan Valley Authority also showed reluctance to meaningfully empower WUAs, despite repeated donor-funded projects. All in all, hydrocracies often escape accountability, preserving a degree of paternalism, corruption, and political meddling, not unlike large companies (Suhardiman et al., 2014).

17.4.4 Political and Elite Capture

Politics penetrates collective action and WUA functioning at different scales. First, it has involved the assertion of a neoliberal doctrine and associated NPM or IWRM ideas. Second, NPM aims to depoliticise irrigation management by focusing on neutral or technical acts of service provision. The ideal of creating lean, clean, efficient, and neutral organisations detached from their social 'thickness' is attractive to donors, experts, and policy makers.

However, this is a political depoliticisation, as politics inevitably permeates the policy process and has impacts (Mosse, 1997). Third, local and regional politics play a major role in WUAs. Ambitious managers in Western Mexico have used WUAs and their resource flows as springboards to secure political positions for their electoral campaigns (Rap, 2006). Political interference, political instrumentalisation of WUAs, or instances of corruption have been widely observed in countries such as Tunisia, Turkey, or Morocco. Fourth, regional politics can jump to higher scales. A national example was Mexican president Vicente Fox (2000–2006), his agricultural minister, and their team, who departed from an influential WUA federation and the governorship of Guanajuato. Fifth, in South Asia and sub-Saharan Africa, elite capture of WUAs reproduced the power and resource control of political elites and vested economic interests, which deteriorated irrigation performance (Aarnoudse et al., 2018). In peri-urban Malawi, WUAs in drinking water provision reproduced local power structures, and elites captured project benefits at the cost of others (Rusca et al., 2015). In Indonesia, the formation of WUA federations reproduced rural elite domination and its involvement in financial management and institutionalised corruption (Suhardiman et al., 2014).

Beyond elite capture, trends that can also weaken WUAs are growing social heterogeneity (traditional farmers versus investors), economic diversification (irrigation-based livelihoods versus hobby farmers), and diversification of water sources and infrastructure. Some users may, for example, turn to groundwater and neglect the collective network (Aubriot, 2022). Sustaining collective action then becomes increasingly problematic.

17.5 Conclusion

Organisational theory shows how the adoption of certain organisational models (WUAs) may improve external legitimacy but not collective action. According to institutional isomorphism, organisations often model themselves on other organisations and mimic earlier 'successes', while international financial and expert institutions reinforce such coercive trends. WUAs' formalised structures further function as powerful myths or ideals that organisations adopt ceremonially. Development interventions frequently formalise community structures, which mirror modern bureaucratic structures, with a hierarchy, a written charter, and perfect bylaws (Cleaver, 1999; Rusca et al., 2015). As a result, WUAs become unquestioned as models of 'good governance' and their mere existence is understood as a sign of success and improved management (Meinzen-Dick, 2007; Aarnoudse et al., 2018). Organisational theory thus partly explains why reforms socially engineer a particular *form* of organisation (WUAs) as opposed to looking at collective action per se. Yet the question of *why* this organisational model is so durable may require further investigation.

Many trumpets have been blown and expectations raised (Feachem, 1980). Neoliberal policy has unrealistically increased expectations about user participation and cost-cutting. Of course, there is nothing wrong with governments trying to balance their expenditures with revenues. But, with hindsight, the ideological tendency to overstate policy expectations of water user participation and empowerment 'worked' for promoting neoliberal donor policy. Behind this 'success' in selling a policy lies a positive epistemic belief in the causal link between user management and performance improvement, which is often disappointed in practice. The underlying philosophy of American pragmatism prioritises the optimistic search for 'what works' above observing processes, effects, and externalities. When applied to WUAs, this belief proved partially effective – but ultimately flawed – and it is problematic when promises of internationally promoted policies are not kept, as this may erode governmental and institutional legitimacy.

Further Readings

Aarnoudse, E., Closas, A., & Lefore, N. (2018). *Water User Associations: A review of approaches and alternative management options for sub-Saharan Africa*. IWMI. www.iwmi.cgiar.org/publications/iwmi-working-papers/iwmi-working-paper-180/

Garces-Restrepo, C., Vermillion, D., & Muňoz, G. (2007). *Irrigation management transfer: Worldwide efforts and results* (FAO Water Reports). FAO. http://agris.fao.org/agris-search/search.do?recordID=XF2009437755

Ghazouani, W., Molle, F., & Rap, E. (2012). *Water users associations in the NEN region: IFAD interventions and overall dynamics* (Report submitted to IFAD). IWMI.

Senanayake, N., Mukherji, A., & Giordano, M. (2015). Re-visiting what we know about Irrigation Management Transfer: A review of the evidence. *Agricultural Water Management, 149*, 175–186.

Notes

1 When using WUA in this chapter, we refer to different organisational forms for water users: associations (WUAs) organisations (WUOs), groups (WUGs), and committees (WUCs).

2 Interview dated 24 June 2024.

3 Interview dated 24 June 2024.

Selected References

Aubriot, O. (2022). The history and politics of communal irrigation. *Water Alternatives, 15*(2), 307–340.

Chambers, R. (1988). *Managing canal irrigation: Practical analysis from South Asia*. Cambridge University Press.

Cleaver, F. (1999). Paradoxes of participation: Questioning participatory approaches to development. *Journal of International Development, 11*(4), 597–612.

Coward, E. W., & Uphoff, N. (1986). Operation and maintenance in Asian irrigation: Reappraising government and farmer responsibilities and rights. *Irrigation and Drainage Systems, 1*(1), 31–44.

Eguavoen, I. (2007). *Now you have a new pump, you have to manage it: Household water management, water rights and institutional change in Northern Ghana* [Unpublished doctoral dissertation]. Center for Development Research, University of Bonn. www.zef.de/fileadmin/template/Glowa/Downloads/Eguavoen_phdThesis.pdf

Feachem, R. G. (1980). Rural water and sanitation: Community participation in appropriate water supply and sanitation technologies: The mythology for the decade. *Proceedings of the Royal Society of London. Series B. Biological Sciences, 209*(1174), 15–29.

Freeman, D. M., & Lowdermilk, M. K. (1991). Middle-level farmer organisations as links between farms and central irrigation systems. In M. M. Cernea (Ed.), *Putting people first: Sociological variables in rural development* (pp. 113–143). World Bank.

Gorriz, C. M., & Groenfeldt, D. (1995). *Proceedings of the International Seminar on Participatory Irrigation Management*, Mexico, February 8–15, 1995 (Washington, DC: World Bank).

Harvey, P. A., & Reed, R. A. (2007). Community-managed water supplies in Africa: Sustainable or dispensable? *Community Development Journal, 42*(3), 365–378.

Hunt, R. C., & Hunt, E. (1976). Canal irrigation and local social organization. *Current Anthropology, 17*(3), 389–411.

Johnson, S. H., Vermillion, D. L., & Sagardoy, J. L. (1995). *Irrigation management transfer. Selected papers from the International Conference on Irrigation Management Transfer, Wuhan, China, 20–24 September 1994* (pp. 1–14). IIMI/FAO.

Korten, F., & Siy, R. (1989). *Transforming a bureaucracy: The experience of Philippine national irrigation administration*. Kumarian Press.

Meinzen-Dick, R. (2007). Beyond panaceas in water institutions. *Proceedings of the National Academy of Sciences of the United States of America, 104*(39), 15200–15205.

Mollinga, P. P., Meinzen-Dick, R. S., & Merrey, D. J. (2007). Politics, plurality and problemsheds: A strategic approach for reform of agricultural water resources management. *Development Policy Review, 25*(6), 699–719.

Mosse, D. (1997). The symbolic making of a common property resource: History, ecology and locality in a tank-irrigated landscape in South India. *Development and Change, 28*(3), 467–504.

Oorthuizen, J. (2003). *Water, works and wages: The everyday politics of irrigation management reform in the Philippines*. Orient Longman.

Ostrom, E. (1992). *Crafting institutions for self-governing irrigation systems*. ICS.

Ostrom, E., Schroeder, L., & Wynne, S. (1993). *Institutional incentives and sustainable development: Infrastructure policies in development*. Westview Press.

Rap, E. (2006). The success of a policy model: Irrigation management transfer in Mexico. *Journal of Development Studies, 32*(2), 1301–1324.

Rap, E., de Bont, C., Molle, F., Bolding, A., & Ismail, A. (2025). Radical reassemblages: The life history of a Nile Delta pumping collective. *Environment and Planning E, Nature & Space, 8*(1), 148–170.

Rap, E., Molle, F., Rap, E., Ezzat El-Agha, D., & Abou El Hassan, W. (2019). The limits to participation: Branch-canal water user associations in the Egyptian Delta. *Water International, 44*(1), 31–50.

Repetto, R. (1986). *Skimming the water: Rent-seeking and the performance of public irrigation systems* (Research Report No. 4). World Resources Institute.

Rusca, M., Schwartz, K., Hadzovic, L., & Ahlers, R. (2015). Adapting generic models through bricolage: Elite capture of water users associations in peri-urban Lilongwe. *European Journal of Development Research Advance, 27*, 777–792. https://doi.org/10.1057/ejdr.2014.58

Small, L. E., & Carruthers, I. (1991). *Farmer-financed irrigation: The economics of reform*. Cambridge University Press.

Suhardiman, D., Giordano, M., Rap, E., & Wegerich, K. (2014). Bureaucratic reform in irrigation: A review of four case studies. *Water Alternatives, 7*(3), 442–463.

Uphoff, N., Ramamurthy, P., & Steiner, R. (1991). *Managing irrigation: Analyzing and improving the performance of bureaucracies*. Sage.

van den Broek, M., & Brown, J. (2015). Blueprint for breakdown? Community based management of rural groundwater in Uganda. *Geoforum, 67*, 51–63. https://doi.org/10.1016/j.geoforum.2015.10.009

Venot, J.-P. (2014). Rethinking commons management in Sub-Saharan West Africa: Public authority and participation in the agricultural water sector. *Water International, 39*(4), 534–548.

Vermillion, D. (1991). *The turnover and self-management of irrigation institutions in developing countries: A discussion paper for the new programme of IIMI*. International Irrigation Management Institute.

Whaley, L., & Cleaver, F. (2017). Can 'functionality' save the community management model of rural water supply? *Water Resources and Rural Development, 9*, 56–66. https://doi.org/10.1016/j.wrr.2017.04.001

World Bank Group. (2019). *Governance in irrigation and drainage: Concepts, cases, and action-oriented approaches—A practitioner's resource*. World Bank Group.

18 *REVISITING*: The Myth of Private Finance and SDG6

David McDonald

After disappointing results with the Millennium Development Goals, the United Nations is trying once again to resolve the global water and sanitation crisis with Sustainable Development Goal (SDG) 6, which seeks to ensure safe drinking water and sanitation for all. Progress has been tragically slow. Although the situation has 'slightly improved' since the year 2000 (WHO/UNICEF, 2019), only 45% of countries are on track to achieve their SDG targets for drinking water, and less than a quarter will meet their sanitation targets (GLAAS, 2022, p. xii). Progress towards SDG 6 'is alarmingly off track' (United Nations, 2023, pp. x–xi), with the achievement of universal access by 2030 requiring a six-fold 'increase in action' on drinking water and a five-fold increase for sanitation. The poorest regions of the world are furthest behind.

And yet, universal access to safe and affordable water supply and sanitation (WSS) is one of the most achievable global challenges we face. The technologies are relatively simple and can be managed in ways that are appropriate to the geographies, cultures, and institutions of different locations, with massive upsides including social, economic, and environmental benefits. Unlike deep-seated tensions around social and technological changes to other essential services – such as energy and education – no one is opposed to the principle of improved water and sanitation services.

What stands in the way? A rash of intractable political factors is partly to blame, including weak legislation, rent-seeking, and cross-border conflicts over (shrinking) clean water supplies (Longo et al., 2022). But the biggest single challenge is arguably that of finance. Estimates vary, but capital expenditures required to reach SDG targets for drinking water and sanitation alone are estimated at US$116–229 billion per year (United Nations, 2023; Joseph et al., 2024), while meeting other SDG 6 targets such as reducing water pollution, implementing integrated water resources management, and protecting water-related ecosystems will cost considerably more, with total global WSS infrastructure development needs estimated at US$6.7 trillion by 2030 and US$22.6 trillion by 2050 (Ajami et al., 2018, p. 5).

DOI: 10.4324/9781003745532-18

The most pressing needs are in the Global South, but high-income countries are also in serious deficit situations (Hutton & Varughese, 2016). Over 75% of countries surveyed by UN-Water reported insufficient funding to implement WSS plans and strategies (GLAAS, 2022, p. xiii). Less than 15% of these countries have sufficient financing to implement their national SDG targets, suggesting a 'need for a radical increase in global financing of water and sanitation investments' (Pickbourn et al., 2022, pp. 1–2), in the range of '2–4 times larger than current practice' (Alaerts, 2019, p. 821). Cost estimates vary widely based on the types of technologies employed – with ongoing debates about 'appropriate' forms of equipment (Swyngedouw, 1997) – but even the most modest proposals will require enormous amounts of financing.

Where will this money come from? Historically, the lion's share of funding has come from public sources (local and national governments as well as donors). The private sector has played a very small role in financing water in all but a handful of regions. In low-income countries, the private sector has been virtually non-existent when it comes to financing WSS infrastructure.

Nevertheless, the World Bank and most other major multilateral development agencies continue to insist that private finance will resolve the world's water and sanitation crises. Since the early 1990s, there have been widespread calls, notably from the World Bank and the OECD, for greater private sector finance investments in water and sanitation infrastructure (World Bank & UNICEF, 2017; OECD, 2019; United Nations, 2023). This pro-private preference is now hegemonic amongst development agencies, mainstream think tanks, and NGOs (McDonald, 2019). Even the World Wildlife Fund insists that 'the only way to achieve the required level of investment in the water sector is by significantly leveraging the private sector' (WWF et al., 2018, p. 8). This 'myth' persists despite the realities on the ground, driven by little more than an ideological commitment to privatisation and a perpetual faith that new and 'innovative' forms of private finance will come to the rescue.

This chapter provides a brief overview of the history of contemporary WSS financing from public sources and then outlines the failures of private finance to live up to its multilateral directives. The result, sadly, is a shrinking pool of public money for WSS as governments spend less in the false hope of massive infusions of private capital. But all is not lost. There are alternative forms of public finance that can be tapped into, such as public banks, which are outlined briefly at the end of the chapter.

18.1 Predominance and Challenges of Public Finance

WSS is a diverse sector with multiple agencies working across different jurisdictions, making it difficult to fully assess sources of financing (Joseph et al., 2024). Nevertheless, most funding comes from public sources and end users, as outlined below.

18.1.1 Water Operators and Cost Recovery

For most water and sanitation operators, the largest source of income is revenue generated from their own operations. Globally, an average of 90% of water budgets comes from a combination of cost recovery from consumers and government spending (GLAAS, 2017), and 'although governments have an important role in [WSS] finance, households are the largest source of funding' (GLAAS, 2022, p. xiii). According to Bartram et al. (2018, p. 3), it remains common practice for water and sanitation to be 'largely funded from domestic resources, principally tariff payments by users and public finance derived from taxation'.

But it is equally true that cost recovery efforts are woefully inadequate. For most water operators, tariffs do not even cover operation and maintenance (O&M) costs, and there are no known examples anywhere in the world where tariffs cover all capital expenditures (GLAAS, 2022, p. xiii). This funding shortfall is endemic in low-income countries and rural areas, with only 15% of such water utilities matching operational expenditures or generating a cash surplus (World Bank & UNICEF, 2017, p. 14). A survey of 30 public utilities in Africa found that only 36% met their full O&M costs and only 9% met O&M costs plus a portion of their capital costs (UNICEF, 2019, p. 25). These revenue shortfalls are particularly acute in sanitation, where the challenges of applying and collecting fees is more problematic, resulting in deep infrastructure spending shortfalls and a greater reliance on higher tiers of government spending (Hall & Lobina, 2008).

The realities of poverty have meant that low-income households struggle to pay for basic levels of water and sanitation, with tariff structures in many parts of the world disproportionately burdening the poor. But even if tariffs could be increased to help fill the financing gap, it is foolhardy to rely on full cost recovery in one sector, with knock-on effects in other key areas such as health, education, and energy, potentially undermining the viability of these other services in an effort to finance water and sanitation.

18.1.2 Local Governments

Most local governments also lack the necessary resources, particularly in the Global South. The majority of towns and cities in Asia, Africa, and Latin America have a low (or no) municipal tax base to draw on, while downloading and rapid urbanisation over the past few decades have made it even more difficult to finance infrastructural challenges on their own.

Options for raising external capital are also constrained, limited by weak (or no) credit ratings as well as a lack of capacity to manage the funds necessary to address WSS spending. Municipal bonds – a form of borrowing which offers the 'most affordable and accessible financing source' for WSS in many countries in the Global North (AMWA & NACWA, 2016) – are notably lacking in most countries in the Global South, restricted by laws that impede issuance at local levels, perceived risk by investors, lack of economies of scale, and political problems associated with pooling risks and costs between municipalities (Bagchi & Kundu, 2003; Pearce-Oroz, 2006). There is potential for municipal bonds to play a larger role in financing WSS, but it will take time to develop these bond markets, and they are unlikely to assist smaller and rural municipalities in the foreseeable future (Singh & Dhanda, 2021). They also run the risk of financialising WSS funding, a topic we return to below.

18.1.3 National Governments

For many water operators in the Global South, funding for WSS comes from national governments, due in part to the highly centralised and politicised nature of water services. But once again, monies are insufficient.

In some cases, it is a lack of political will. Water and sanitation are relatively easy to ignore when it comes to low-income households with little collective bargaining power, particularly when infrastructure is buried underground and lacks the eye-grabbing attention of other infrastructure investments such as roads, airports, and electricity. Corruption and rent-seeking are also a concern, with WSS spending often used to garner political support or punish political opponents, further entrenching political power with central governments (Colon & Guerin-Schneider, 2015).

Even if the political will does exist – with many national governments claiming to prioritise water and sanitation in their official development plans – few countries have the resources required to meet their SDG 6 targets. Combined with the structural features of a global economy that have left many Asian and Latin American countries – and most African countries – heavily indebted and unable to generate sustainable economic growth, the likelihood of dramatic increases in national-level spending on WSS is extremely low (Ndung'u et al., 2021).

This is not to absolve national governments of the moral, economic, health, and environmental necessities of investing more money in water and sanitation services. Public spending on WSS represents only 1.2% of national budgets globally (compared to 5% in energy), and even these meagre budgets often go underspent, with an annual 'budget execution rate' of only 72% on average (60% in the case of Tanzania in recent years) (Joseph et al., 2024, pp. xviii, xxi). Public spending on WSS also tends to benefit wealthier households and urban areas (Bakker et al., 2008).

Nevertheless, it is unrealistic to expect the majority of national governments in the Global South to cover the costs of meeting SDG 6 on their own. A significant portion of WSS funding will need to come from other sources.

18.1.4 Bilateral Donors and MDBs

For many low- and middle-income countries, multilateral development banks (MDBs) such as the World Bank, the Asian Development Bank, and the

African Development Bank are critical players in financing WSS, accounting for about 15% of total investments in the sector (Heidler et al., 2023, p. 2). MDBs provided over US$15 billion for water and sanitation infrastructure investments between 2010 and 2020 (OECD, 2022), and because of their simultaneous role as 'knowledge banks', MDBs also act as information brokers and gatekeepers, shaping the institutional and financial nature of WSS (Kramarz & Momani, 2013).

Despite this oversized role for donors and multilateral banks, WSS forms only a small portion of official development assistance (ODA), with less than 5% of ODA allocated to the sector over the 2016–2020 period on average (UNESCO, 2023, p. 164; Joseph et al., 2024, p. xix). And even these low figures are shrinking. ODA commitments to WSS declined by about 20% over the 2012–2015 period (WHO/UNICEF, 2019), and a further 12% from 2015 to 2021 (United Nations, 2023, p. xi), while the gap between ODA commitments and actual disbursements to the WSS sector grew from US$100 million in 2016 to over US$2.6 billion in 2019 (Pickbourn et al., 2022). These gaps are geographically uneven as well. In sub-Saharan Africa – the region most desperately in need of WSS investments – the proportion of aid to the sector dropped from 32% to 23%, while in Central and Southern Asia it increased from 12% to 20% (GLAAS, 2022, p. xiii).

18.2 From Private to Blended Finance

Despite several decades of pushing for private investment in WSS, it has not materialised outside of a handful of countries (notably France, the United Kingdom, and Germany). On average, 'private-sector finance for water has remained minor' (Alaerts, 2019, p. 8). This is particularly true in the Global South, where private-sector financing of WSS accounts for only 7% of the total; in sub-Saharan Africa the figure is less than 1% (Leigland et al., 2016, p. 4).

The World Bank's own research confirms these conclusions, noting that globally, private sector spending in WSS accounts for a mere 1.7% of investments (Joseph et al., 2024, p. xvii) and a tiny portion of total private-sector infrastructure financing (Kolker et al., 2016, p. 3; cf. Wu et al., 2016). The United Nations Inter-Agency Task Force on Financing for Sustainable Development (the body tasked with addressing the financing needs of the 2030 SDGs) notes, moreover, that private-sector investment appears to be decreasing and is 'well below the peak reached in 2012' (IATF, 2019, p. 61). The World Bank draws the sobering conclusion that, for most public water operators, 'private finance is almost non-existent' (Kolker et al., 2016, p. 1).

Why has the expected level of private finance failed to materialise? The reasons are various and speak to the structural barriers to increased private-sector investment in water services. Much of the aversion stems from risk–return considerations. Early expectations of large profits and trillions of dollars of potential revenues quickly evaporated in all but the wealthiest countries or in locations where guaranteed rates of return had been negotiated (Bakker,

2013). Today, low levels of cost recovery and the political challenges of collecting payments from low-income households have meant that many private water companies have scaled back activities in risky locations, changed tactics to focus on more value-added niche markets and services (such as desalination), or simply withdrawn altogether from money-losing contracts (Bauby, 2014). From the perspective of private investors, this makes sense; their remit is not to maximise public goods for the community but to maintain and increase private returns for their shareholders (cf. Lazonick & Shin, 2020).

Political backlash against water privatisation may also help to explain why private water companies and financiers have reduced their financial exposure in the sector (see Chapter 19). Widespread and often violent protests have served to erode expected returns (and sully the reputations) of many private water companies (Barlow & Clarke, 2017). The growing trend towards remunicipalising water services (that is, making them public after a period of private-sector delivery) has created additional concerns in the private water market, particularly after high-profile cities such as Paris made their water services public once again (with French multinationals Veolia and Suez losing contracts in their own backyards) (Le Strat, 2014). Private water firms have been relatively silent on the matter, but there are signs that they are increasingly worried about what remunicipalisation means for their future (Ulmer & Gerlak, 2019).

With a growing political backlash and a difficult risk–return outlook, it is little wonder that private water companies are unwilling to invest more in a sector with shrinking opportunities for recovering costs and political threats to their very existence. These realities speak to the 'serious structural constraints' of relying on private capital to fund WSS needs (Leigland et al., 2016, p. 2). Most private investors see 'water as a financial risk' and for that reason have not invested in the sector (Jägerskog et al., 2016). As the United Nations (2023, p. xii) bluntly commented in its *Synthesis Report on Water and Sanitation*, the water sector is 'unattractive for investment'.

Having recognised this reality, international organisations such as the World Bank have been promoting policies of blended finance – i.e., 'the strategic use of public taxes, development grants and concessional loans to mobilise private capital flows' (World Bank & UNICEF, 2017, p. vii). Blended finance is now the centrepiece of the World Bank's 'billions to trillions' SDG agenda and its related 'Maximising Finance for Development' strategy. Although the emphasis has been on emerging and frontier markets where private investments are lowest, public incentives for private finance can be found in virtually every part of the world.

The UN Inter-agency Task Force on Financing for Development writes that, '[b]y shifting some of the risk or cost of a project from the private to the public sector, blended finance can enhance risk–return profiles for private creditors or investors'. The aims are to 'leverage additional funds for the sector and reduce borrowing costs compared to a fully commercial arrangement' (IATF, 2019, p. 86). In the water sector, it is argued that blended finance will 'help overcome affordability and/or political constraints to borrowing [...] [and] can create new relationships and opportunities between the water and financial sectors, which can promote the long-term goal of increasing commercial financing' (World Bank & UNICEF, 2017, p. vii). As UNESCO (2023, p. 166) insists, blended finance 'can help transform "nearly bankable" projects into viable projects'.

There is nothing particularly original about this blending model, although there are new actors and new financing vehicles such as web-based 'fintech', as well as expanded social mandates such as green financing. There is also an increased emphasis on microlending, particularly in low-income countries (World Bank & UNICEF, 2017).

Novelty aside, evidence suggests that blended finance is having little impact on WSS, particularly in the countries it aims to target most. To date, blended finance in developing economies has not been widely used at scale in the water sector. A few isolated experiences have been supported by international donors, but these have mostly been in middle-income countries, and they have so far failed to be replicated at scale (Leigland et al., 2016, p. 4). Very few of these often generously subsidised efforts have met with success:

> Despite mechanisms that promised to leverage private financing at the national or regional level, or within sub-sectors (e.g. sanitation), there is not one internationally recognised financing instrument that has facilitated significant, sustainable private financing into the WASH sector in the last decade.
>
> (Kolker, 2022)

As Bernards (2024) has observed, private finance is generally not interested in blended finance.

The blended finance that does exist is concentrated in highly profitable sectors such as financial services and energy, and flows largely to a small cluster of emerging economies such as Turkey, Nigeria, India, Brazil, and China. Only 2% of this amount is allocated to WSS (IFC, 2017). Data from the Inter-agency Task Force on Financing for Development reinforces these claims, demonstrating that flows to middle-income countries are nearly ten times those to lower-income and least-developed countries, while only 7% of blended finance in water is aligned with SDG 6 as 'most blended deals are concentrated in sectors with significant potential for economic returns' (IATF, 2019, pp. 87–88).

As Summers and Singh (2024) note – the former a previous World Bank President – blended finance has 'not raised capital [or] substantially changed financing practices'. In fact,

> 2023 was a disaster in terms of support for the developing world [...]. [R]ising interest rates and bond and loan repayments meant that nearly

> $200 billion flowed out of developing countries to private creditors in 2023, completely dwarfing the increased financing from the international financial institutions. "Billions to trillions" […] has become "millions in, billions out".

As a result, the vast majority of WSS funding continues to come from public finance. The public sector 'dominates spending on water', providing over 90% of infrastructure investment (Joseph et al., 2024, p. xvii). According to the High Level Panel on Water (convened by the World Water Council and the OECD), 'private money can rarely fully substitute for public finance in major water infrastructure – it can only be a junior partner in most cases, and even then will need comforts of various kinds' (WWC & OECD, 2015, p. 58).

Despite this, the majority of policy-making related to financing WSS continues to focus on the private sector. Although aware of the dominance of public funding, the World Bank insists that it is 'essential to mobilise up-front financing from private or commercial sources' (Leigland et al., 2016, p. 1). The United Nations (2023, p. xii) also continues to maintain that 'new investments [in WSS] must be attracted by better enabling environments'.

This emphasis on private financial solutions is also reflected in the academic literature. A comprehensive survey of research on the topic found that over 75% of publications written between 1989 and 2015 focused on private-sector financing for WSS, while only 5% explored public investment (Kumari & Sharma, 2017). Clearly, more research is needed on the potential for public sources of financing to meet WSS requirements.

This ideological push for private investment also appears to have undermined commitments to public spending. As Heidler et al. (2023, p. 6) note in their review of 60 years of MDB involvement in the water and sanitation sector, calls for greater private finance investments have 'delegitimise[ed] the role of public finance', arguing that 'if this discourse has had any effect at all it has been to reduce public investment in water supply and sanitation'.

18.3 Public Banks to the Rescue?

There is one set of public financial institutions that has been largely ignored in this debate about financing WSS: public banks. Broadly defined as financial institutions majority owned by the state or another public entity, public banks have been involved in financing WSS for decades (McDonald et al., 2021). There has also been a resurgence of interest in public banks, with a growth in their size and numbers, as more and more governments see their potential in tackling some of the world's biggest challenges (Marois, 2022). Public banks have also begun to assert their own influence in these financing debates, most notably with the creation of the Finance in Common (FiC) network in 2020, which sees itself as 'the global movement for all public development banks', managing more than US$23 trillion in assets and US$2.5 trillion in annual investments (FiC, 2023).

Public banks cannot resolve the water and sanitation financing crisis on their own. They simply do not have the capacity or the resources to meet SDG 6 funding needs – not to mention other SDG priorities. Resolving the WSS crisis will require injections of capital from all levels of government. Nor are there any guarantees that public banks will provide lending for public water services, let alone in ways that are sustainable, equitable, and transparent. Public banks, like all public institutions, are diverse and complicated entities, with no predetermined trajectory. Public banks are neither inherently good nor bad, with significant differences within and across jurisdictions, making any simple or singular assessment of their potential to play a meaningful role in financing WSS misleading and inappropriate (Marois, 2021). In short, there are no simple answers when it comes to the role of public banks in financing WSS.

Nevertheless, public banks do offer a unique and potentially transformative vehicle for addressing short- and long-term financial challenges in water and sanitation services around the world. A growing body of research has begun to demonstrate this potential, critically examining promising examples by highlighting the strengths and weaknesses of different models and suggesting ways in which these public banking systems might be improved and potentially replicated elsewhere (Marois and McDonald, 2023; Marois et al., 2025). The results of research to date are as mixed as the countries and institutions investigated, but they do point to one common outcome: There is significant potential (and considerable appetite) for progressive forms of public bank financing of water services.

Further Readings

Joseph, G., Hoo, Y. R., Qiao, W., Aroha, B., & Luis Alberto, A. (2024). *Funding a water-secure future: An assessment of global public spending*. World Bank Group.

Marois, T. (2022). A dynamic theory of public banks (and why it matters). *Review of Political Economy, 34*(2), 356–371.

Marois, T., McDonald, D., & Spronk, S. (2025). *Public banks and public water in the Global South financing options for sustainable development.* Routledge.

McDonald, D., Marois, T., & Spronk, S. (2021). Public banks + public water = SDG 6? *Water Alternatives, 14*(1), 117–134.

Selected References

Ajami, N., Habicht, H., Fewell, B., Lattimer, T., & Ng, T. (2018). *Water finance: The imperative for water security and economic growth*. Stanford University. https://waterinthewest.stanford.edu/sites/default/files/Water_Finance_Water_Security_Economic_Growth.pdf

Alaerts, G. J. (2019). Financing for water—water for financing: A global review of policy and practice. *Sustainability, 11*(3), 821–835.

AMWA & NACWA. (2016). *Affordable water infrastructure relies on municipal bonds.* Retrieved March 25, 2022, from: www.nacwa.org/docs/default-source/resources---public/tax-exempt-muni-bond-hub/2017-02-16muni-bonds.pdf?sfvrsn=dd86eb61_4

Bagchi, S., & Kundu, A. (2003). Development of municipal bond market in India: Issues concerning financing of urban infrastructure. *Economic & Political Weekly, 38*, 789–790.

Bakker, K. (2013). *Privatizing water: Governance failure and the world's urban water crisis*. Cornell University Press.

Bakker, K., Kooy, M., Shofiani, N. E., & Martijn, E.-J. (2008). Governance failure: Rethinking the institutional dimensions of urban water supply to poor households. *World Development, 36*(10), 1891–1915.

Barlow, M., & Clarke, T. (2017). *Blue gold: The battle against corporate theft of the world's water*. Routledge.

Bartram, J., Brocklehurst, C., Bradley, D., Muller, M., & Evans, B. (2018). Policy review of the means of implementation targets and indicators for the sustainable development goal for water and sanitation. *NPJ Clean Water, 1*(1), 3.

Bauby, P. (2014). Water globalization: The strategies of the two French "Majors". In G. Schneier-Madanes (Ed.), *Globalized water: A question of governance* (pp. 45–61). Springer Netherlands.

Bernards, N. (2024). Where is finance in the financialization of development? *Globalizations, 21*(1), 88–102.

Colon, M., & Guérin-Schneider L. (2015). The reform of new public management and the creation of public values: Compatible processes? An empirical analysis of water utilities. *International Review of Administrative Sciences, 81*(2), 264–281.

Finance in Common (FiC). (2023). *Final communique: 2023 Finance in Common summit*. Retrieved October 10, 2023, from: https://financeincommon.org/sites/default/files/2023-09/FICS%202023%20Final%20Communiqu%C3%A9_2_0.pdf

GLAAS (UN-Water Global Analysis and Assessment of Sanitation and Drinking-Water). (2017). *UN Water-GLAAS 2017: Financing universal water, sanitation and hygiene under the Sustainable Development Goals*. United Nations.

GLAAS. (2022). *Strong systems and sound investments: Evidence on and key insights into accelerating progress on sanitation, drinking water and hygiene: GLAAS report 2022*. UN-Water.

Hall, D., & Lobina, E. (2008). *Sewerage works: Public investment in sewers saves lives.* Public Services International Research Unit. Retrieved June 29, 2020, from: www.ircwash.org/sites/default/files/Hall-2008-Sewerage.pdf

Heidler, A., Nesi, M., Nikiema, J., & Lüthi, C. (2023). Multilateral development banks investment behaviour in water and sanitation: Findings and lessons from 60 years of investment projects in Africa and Asia. *Journal of Water, Sanitation and Hygiene for Development, 13*(5), 362–374.

Hutton, G., & Varughese, M. (2016). *The costs of meeting the 2030 Sustainable Development Goal targets on drinking water, sanitation, and hygiene*. World Bank. http://hdl.handle.net/10986/23681

Inter-Agency Task Force on Financing for Development (IATF). (2019). *Financing for development report 2019*. United Nations.

International Finance Corporation (IFC). (2017). *DFI Working Group on blended concessional finance for private sector projects*. World Bank Group.

Jägerskog, A., Lexén, K., Clausen, T. J., & Engstrand-Neacsu, V. (2016). *The water report 2016*. Stockholm International Water Institute.

Kolker, J. (2022, June 1). How to improve water-sector financing in developing countries. *World Bank Blog*. Retrieved March 15, 2023, from: https://blogs.worldbank.org/water/how-improve-water-sector-financing-developing-countries

Kolker, J., Kingdom, B., Trémolet, S., Winpenny, J., & Cardone, R. (2016). *Financing options for the 2030 water agenda* (Knowledge Brief). World Bank Group.

Kramarz, T., & Momani, B. (2013). The World Bank as knowledge bank: Analyzing the limits of a legitimate global knowledge actor. *Review of Policy Research, 30*(4), 409–431.

Kumari, A., & Sharma, A. K. (2017). Infrastructure financing and development: A bibliometric review. *International Journal of Critical Infrastructure Protection, 16*, 49–65.

Lazonick, W., & Shin, J.-S. (2020). *Predatory value extraction: How the looting of the business corporation became the US norm and how sustainable prosperity can be restored*. Oxford University Press.

Le Strat, A. (2014). Discussion: The remunicipalization of Paris's water supply service: A successful reform. *Water Policy, 16*(1), 197.

Leigland, J., Trémolet, S., & Ikeda, J. (2016). *Achieving universal access to water and sanitation by 2030: The role of blended finance*. World Bank.

Longo, P. J., Schutz, A., & Scott, J. M. (2022). Borders and water conflicts: Mitigating conflicts with love and cooperation. *Natural Resources Journal, 62*(1), 125–141.

Marois, T. (2021). *Public banks: Decarbonisation, definancialisation, and democratisation*. Cambridge University Press.

Marois, T., & McDonald, D. A. (eds.). (2023). *Public banks, public water: Exploring the links in Europe*. Routledge.

McDonald, D. A. (2019). Will the empire strike back? Powerbrokers and remunicipalisation in the water sector. *Water Alternatives, 12*(2), 348–359.

Ndung'u, N., Shimeles, A., & Manda, D. K. (2021). Growing with debt in African economies: Options, challenges and pitfalls. *Journal of African Economies, 30*(Supplement 1), i3–i13.

Organisation for Economic Co-operation and Development (OECD). (2019). *Making blended finance work for water and sanitation: Unlocking commercial finance for SDG 6* (OECD Studies on Water). OECD Publishing.

Organisation for Economic Co-operation and Development (OECD). (2022). *Detailed aid statistics: ODA Official development assistance: Disbursements [Data set]*. OECD International Development Statistics. https://doi.org/10.1787/data-00069-en

Pearce-Oroz, G. (2006). The viability of decentralized water and sanitation provision in developing countries: The case of Honduras. *Water Policy, 8*(1), 31–50.

Pickbourn, L., Caraher, R., & Ndikumana, L. (2022). Does project-level foreign aid increase access to improved water sources? Evidence from household panel data in Uganda. *World Development, 158*, 105994.

Singh, S., & Dhanda, S. (2021). Municipal bonds: Financing urbanization in India. *South Asian Journal of Management, 28*(4), 145–164.

Summers, L. H., & Singh, N. K. (2024, April 15). The world is still on fire. *Project Syndicate*. Retrieved June 4, 2024, from: www.project-syndicate.org/commentary/imf-world-bank-spring-meetings-need-to-get-four-things-right-by-lawrence-h-summers-and-n-k-singh-2024-04

Swyngedouw, E. (1997). Power, nature, and the city. The conquest of water and the political ecology of urbanization in Guayaquil, Ecuador: 1880–1990. *Environment and Planning A*, *29*(2), 311–332.

Ulmer, R., & Gerlak, A. K. (2019). The remunicipalization of water services in the United States. *Environment: Science and Policy for Sustainable Development*, *61*(4), 18–27.

UNESCO. (2023). *The United Nations world water development report 2023: Partnerships and cooperation for water*. UNESCO.

UNICEF. (2019). *The State of WASH financing in eastern and southern Africa: Regional level assessment*. UNICEF. Retrieved May 5, 2023, from: www.unicef.org/esa/sites/unicef.org.esa/files/2019-10/UNICEF-ESARO-2019-WASH-Financing-Regional-Assessment.pdf

United Nations. (2023). *Blueprint for acceleration: Sustainable Development Goal synthesis report on water and sanitation 2023*. United Nations.

WHO/UNICEF Joint Monitoring Programme. (2019). *Progress on household drinking water, sanitation and hygiene 2000-2017*. WHO/UNICEF Joint Monitoring Programme.

World Bank & UNICEF. (2017, March). *Water and sanitation for all: How can the financing gap be filled?* [Discussion paper]. World Bank & UNICEF.

World Water Council (WWC) & Organisation for Economic Co-operation and Development (OECD). (2015). *Water: Fit to finance? Catalyzing national growth through investment in water security*. WWC and OECD. Retrieved April 10, 2019, from: www.worldwatercouncil.org/en/publications/water-fit-finance

World Wildlife Fund (WWF), Boston Consulting Group, & ING Bank. (2018). *Seizing the water opportunity: How private and public capital can join forces to secure sustainable water supplies and improve freshwater ecosystems*. Retrieved May 12, 2023, from: https://wwfint.awsassets.panda.org/downloads/seizing_the_water_opportunity_report_pages_2_2.pdf

Wu, X., Schuyler House, R., & Peri, R. (2016). Public-private partnerships (PPPs) in water and sanitation in India: Lessons from China. *Water Policy*, *18*(S1), 153–176.

19 *REVISITING*: Private Water Services Are More Efficient and Reliable

Kate Bayliss

19.1 Introduction

This chapter revisits the once-hegemonic claim that the private sector can make water services more efficient and reliable. For decades, there has been a widely held perception that the private sector brings a range of benefits to public water systems, such as being more effective at managing infrastructure and better able to raise funding. According to the World Bank, for example, the private sector can 'help governments fund much-needed investment and bring technology and efficiency that can improve the performance and the financial sustainability of the water sector'.[1] In policy circles, then, the private sector has become associated with increased efficiency and an ability to get things done in ways that often elude the state agencies, but, as this chapter shows, this view has weak theoretical foundations and is not borne out by empirical analysis.

'Privatisation' is often ill-defined and has attracted a host of policy objectives in the past three decades, during which time the modalities of private sector involvement in water have shifted. Rarely does water privatisation refer to the private sector taking full ownership of the water system. This has been the case in England and Wales (hereafter EW) and in parts of Chile, but these are extreme examples. Elsewhere, privatisation refers to the issuing of long-term concession contracts for the management and operation of water infrastructure for a set time period (often decades). Other forms of privatisation may take the form of variations on the build-operate-transfer (BOT) model, where the private sector builds and operates infrastructure such as a desalination wastewater treatment plant before transferring it to the public sector. Privatisation may also take the form of joint ventures between the state and private companies. More recently, the emphasis has shifted to innovative financing interventions (see Chapter 18). The language has also changed, as 'privatisation' has been replaced by 'public–private *partnership*' (PPP) to suggest a cooperative, mutually beneficial enterprise, rather than something of the contested, contractual nature that is usually the reality of these arrangements.

DOI: 10.4324/9781003745532-19

This chapter deals with the broad themes that lie behind these different forms of privatisation.

19.2 How the Privatisation Myth Evolved

19.2.1 How It All Began

Water privatisation emerged in the 1980s as part of the wider neoliberal project adopted in many countries: free markets, deregulation, and a reduction of the state. The privatisation of state-owned enterprises was seen as key to addressing the perceived problems with public sector provisioning. Public choice and principal-agent theories were used to justify the political stance, but the drive came more from shifting politics and the expansion of global capital. In addition, the end of the Cold War led to an acceleration of privatisation, with little regard for the underlying institutional prerequisites. This approach was described as 'privatise now and regulate later', which, at the time, according to Stiglitz, 'seemed a reasonable gamble' as 'no-one knew how long the reform window would remain open' (Stiglitz, 1998). In EW, privatisation was part of a wider programme to reverse 'the corrosive and corrupting effects of socialism' (Thatcher, 1993, cited in Parker, 2004). Privatisation thus had a strong ideological underpinning. The public sector was strongly denigrated, depicted as bureaucratic, inefficient, and corrupt, while the private sector was perceived as efficient and innovative in responding to changing circumstances. For example, according to a 1994 World Bank report, *Adjustment in Africa* (World Bank, 1994, p. 99), 'the public sector lies at the core of the stagnation and decline in growth in Africa'.

The privatisation project was, then, driven by a political shift, which in turn led to the development of an analytical framework whereby privatisation accumulated a collection of objectives. In a 1992 World Bank Report (Kikeri et al., 1992, p. 49), even though there were acknowledged risks from privatising monopolies, it was considered that

> the potential economic and financial benefits may be worth the risks. Privatizing badly managed firms that provide critical upstream goods and services ... helps accelerate modernization and growth and removes constraints on private sector development. Privatizing a few large loss-makers can have an enormous budgetary impact.

But privatisation was also associated with more profound neoliberal transitions. Even if not privatised, utilities were encouraged to adopt market-like measures such as cost-recovery pricing. The potential risks of privatising a monopoly providing an essential service were supposedly overcome by competitive tendering, such that the absence of competition *in* the market was compensated for by competition *for* the market. Water was acknowledged as a special case due to its essential nature and monopolistic structure. But even here, private was expected to be better than public. The World Bank's 1995 policy report *Bureaucrats in Business* considered that there were potential pitfalls from both public and private water systems but declared that 'the evidence more often tilts towards privatisation' (World Bank, 1995, p. 37).

At the beginning of the 1980s, private management – let alone ownership – of water systems was rare. France was an outlier with its history of private water concessions, and some private water companies existed elsewhere, such as Aguas de Barcelona in Spain. But in general, water was provided by public, local, municipal, regional, or national public bodies. Water privatisation began to be adopted at scale, with utilities transferred to private shareholders, in the 1980s in Chile and EW. Concession contracts were adopted elsewhere, such as in Spain, Portugal, Germany, and other European countries.

Privatisation of water also became a core development policy, featuring in the Washington Consensus set of structural adjustment policies. A study of 276 World Bank loans for water supply from 1990 to 2002 showed that privatisation was a requirement for 30% of these. Hopes for privatisation were so high that donor spending on infrastructure fell in the expectation that the private sector would take up the slack (report cited in Molina & Chowla, 2008; see Bayliss, 2020).

The expansion in the global South provided new markets for private water companies emerging from privatisation movements in the North. Water privatisation across Latin America and in many countries of sub-Saharan Africa (SSA) and Asia allowed French, British, Spanish, and Portuguese water companies to diversify internationally (Bakker, 2013). By 2002, privatisation of water was carried out to some degree in at least 14 countries in SSA (Bayliss, 2003).

19.2.2 But Privatisation Stalled

Despite strong policy support, results were disappointing. Water consistently attracted the least interest from investors of any infrastructure sector, and contracts were difficult to achieve and to sustain. Some countries struggled to get privatisation off the ground. In Cameroon, for example, only a single bid was received for contracts tendered (Bayliss, 2003). Once signed, contracts were often quickly renegotiated. One study found that 76% of contracts in the water sector were renegotiated shortly after being awarded (Guasch & Straub, 2009), suggesting opportunistic behaviour on the part of investors, strategically underbidding and seeking better terms once they had secured the contract (Guasch et al., 2008).

There were also high rates of contract cancellations. This was much higher in Africa, where 29% of contracts were prematurely terminated, than in other developing regions. Just a handful of contracts remained active in the region by the end of the 2000s (Banerjee & Morella, 2011). Many concession contracts awarded in Latin America in the 1990s were also terminated early, with some widespread protests and cancellations such as the Guerra del Agua in Cochabamba (Castro, 2016; Bakker, 2013) or Tucuman and Buenos Aires in Argentina (Castro, 2016). Meanwhile, shorter-term management contracts such as those in Uganda and South Africa were not renewed.

Water privatisation has now largely stagnated in Latin America (outside Brazil) but re-emerged in urban Southeast Asia and Southeast Europe. With the shift to PPPs, which incorporates a range of private sector involvement, China has witnessed particularly rapid expansion, accounting for just over 50% of all water PPP projects since 1990 and around 24% of total investment over the period (Qian et al., 2019).

In a number of locations, privatised water has reverted to public control. Kishimoto and Petitjean (2017) document 267 remunicipalisations in the water sector, 106 of which are in France, including in Paris. Water has also been taken into public hands in cities such as Barcelona and Berlin, and some governments have taken legal steps to prevent the future privatisation of water: for example, the Netherlands, Uruguay, and the city of Baltimore in the USA.

19.2.3 Privatisation Rebooted

In the wake of the disappointing results from concession contracts, the nature of private engagement has shifted to lower-risk contracts that are smaller in value and less complex. Stand-alone projects such as water treatment or desalination plants under BOT contracts provide predictable incomes for investors and attract less attention than traditional privatisation contracts (Pierce, 2015; see also Chapter 4 on desalination).

More recently, there has been a shift towards the private sector as important for filling the financing gap. Calls for 'blended finance' in particular stipulate that public funds be used to attract private sector finance. This is to engage with a different set of global actors in order to access the trillions of dollars that rest in the hands of global capital in the form of financial investment funds and assets (see Chapter 18 on the financing of the WSS sector).

No longer does the World Bank call for water to be privatised. Rather, the aim is to 'optimise' the contributions of the public and private actors and to 'facilitate greater engagement of the private sector in the provision of capital, innovation and expertise' (Khemka et al., 2023). In essence, the approach is for the public sector to construct water systems in a manner that will make them more attractive to international capital.

19.3 Why Is the Solution Problematic?

An important starting point in this discussion is the challenge of objectively measuring the effects of privatisation. It is not possible to have a counterfactual, as each case is contextually specific, and the public and private are intertwined. Privatisation is often tied up with other policies, such as pricing, and the effects are difficult to disentangle. The operations of the private sector depend fundamentally on the state for regulation and matters of operation, such as pricing. Regardless of ownership, water remains a political matter.

19.3.1 Analytical Contradictions

Privatisation, however, was based on shaky theoretical foundations; the analytical arguments to support the private sector in water raise some inherent contradictions. First, efficiency is supposed to come from the profit motive, but what is important for driving efficiency is competition, and that factor is absent in water systems, as these tend to be monopolistic. If a monopoly makes gains in productive efficiency, there is no motive for it to share them with the wider society. Moreover, when it comes to the provision of water, productive efficiency is likely to put downward pressure on costs and upward pressure on prices and sales volumes. Environmental issues are unlikely to be a priority for the profit-maximising private company.

Second, the entry of the private sector is typically intended to address the limitations of government bureaucracy. But the successful implementation of privatisation relies on a strong state. A weak state is thus a reason to bring in the private sector – but also to *not* do so (see below). Third, the entry of the private sector does not immunise the water supply from political machinations. The essential nature of water means that its supply will always be political, and the allocation of procurement contracts can be a site of political interference, lobbying, and corruption.

Fourth, water is not necessarily an attractive investment. In much of the world, water privatisation has failed simply because the private sector was not interested. In some locations, the inability to attract investors led to some water projects being taken back in-house (for example, in Hamilton, Canada; see McDonald, 2018), and lack of profitability led to a withdrawal of investors from Latin America (Lobina & Hall, 2007). Water has attracted the least amount of investment in LMIC (low–medium income countries) infrastructure, and very little of it has gone to SSA, where it is most needed.

Finally, water still has to be paid for, either from user fees or from taxes, and private finance is not a substitute for public funds. In EW, one of the reasons for privatising was to raise investment finance in a way that would not require an increase in government borrowing or taxation. Companies were privatised with zero debt in 1989. Some 36 years later, the total net debt in the sector comes to nearly £70bn. These liabilities of the privatised water utilities are paid for and serviced by user fees; the entry of the private sector simply transfers investment costs from taxpayers to water consumers, who are likely to be mostly the same people. But taxation is a progressive way of raising revenue, while water bills take a larger share of household income from poorer consumers.

Privatisation was perceived as a way to replace a lumbering bureaucracy with a slick, efficient operator, but the detailed pathways by which this would be achieved were never made clear, so it is no surprise that overall, the policy outcomes have been disappointing.

19.3.2 Little Empirical Evidence of Superior Performance

Despite the widely held perception of private sector efficiency, there has been little empirical analysis to support this. There are different types of efficiency, and the choice of measure is not necessarily clear cut. For example, a focus on productive efficiency may exclude financing costs, which in EW are much higher since privatisation. In addition, there is the challenge of attribution: For example, investment may increase following privatisation, but only because of government guarantees of reimbursement or the release of donor funding. Upfront private investment is eventually funded by bills or taxes. Finance that comes from the private sector eventually returns to the private sector, with a profit: In the long run, the outflows are as important as the inflows. What might be efficient for the private sector might not be in the best interests of society; when water was remunicipalised in Paris, prices were reduced in part because surplus funds were reinvested rather than being diverted to private shareholders (Lobina et al., 2021).

There is not much strong evidence of efficiency gains from water privatisation. There was some success from short-term management contracts in areas such as revenue collection and service continuity, but little impact was felt on such intractable issues as reducing unaccounted-for water and expanding access (Banerjee & Morella, 2011). Marin (2009), following a study of more than 65 large water PPPs, notes that private operators brought about improvements in bill collection because they had a direct financial incentive to do so where their remuneration was based on revenue collected. PPPs were also considered successful in terms of boosting labour productivity – because staff were laid off. The study found that layoffs ranged from 20% to 65% of the initial labour force.

Privatisation has typically failed to increase access. Where connections did expand, this was usually due to public funds. One study which found that privatisation in Buenos Aires was associated with a fall in child mortality failed to account for the source of finance, which seems to have been from donor and government funds (Bayliss, 2011).

Water PPPs often fail to be effective in addressing weak state capacity. Privatisation has failed to transform poorly performing utilities but proved profitable where utilities were already running well. Some privatisations went badly wrong. Following the collapse of the contract with UK firm Biwater for management of the water utility in Dar es Salaam, a review by the World

Bank concluded: 'The primary assumption on the part of almost all involved, certainly from the donor side, was that it would be very hard if not impossible for the private operator to perform worse than DAWASA. But that is what happened' (Christen et al., 2005, p. 28, cited in Bayliss, 2020).

In EW, for the first 15 years or so after privatisation in 1989 following years of public underinvestment in the sector, the privatisation of water was considered a success. By the end of 2003/04, companies had invested around £50bn in new assets and maintenance, and the number of significant pollution incidents fell substantially (from 374 to 87) between 1995 and 2000 (Ofwat, 2006). But now, 35 years since privatisation, the sector is facing both financial and ecological crises (see Section 19.3.3 for more on EW). In the French system, where there has been a long history of municipalities delegating water services to the private sector, there has been a growing trend to revert to public management of water resources (Chiu, 2025).

There is no clear evidence that the private water sector in England is more efficient than other public water systems. The regulator in neighbouring Scotland, where water was not privatised, claims that the customer service and operational efficiency of publicly owned Scottish Water are at least as good as the leading privatised water companies in England and Wales.[2]

19.3.3 From Privatisation to Financialisation

In line with wider developments in capitalism, investors have ramped up their approach to extracting value for shareholders from the water system, following a shift to financialisation. This entails the expansion of financial investors into water, as well as the extraction of shareholder returns via financial rather than productive means. In short, the supply of water has become a financial asset for global capital.

Financialisation has been particularly prevalent in EW. Initial investors were standard private utility companies such as Veolia and RWE, but in the 2000s, private equity investors bought up some of the water utilities, and they revolutionised the way the sector was financed. Private equity-owned water company investors used financial engineering to increase the debts of the water utilities while transferring surplus funds to shareholders. The finances, in other words, were organised to suit the interests of investors rather than consumers. This shift from privatisation to financialisation means that revenues from future water bills have been securitised and converted into debt in ways that have facilitated payouts to shareholders and that risk the future sustainability of the water utilities. Thames Water, the largest utility serving London and the surrounding area, accumulated high levels of debt and is now under significant financial strain following an increase in interest rates and inflation, which have pushed up their borrowing costs. In addition, Thames is one of the worst performers in the country when it comes to pollution and leakage. This is far from the vision of an efficient privatised company envisaged at the time of privatisation!

Similar financialisation practices have been observed in other case studies such as large desalination plants in California and Chile (see Pryke & Allen, 2019 and 2022), and global finance has become increasingly involved in privatised water in Brazil (Britto, 2025). In Sao Paulo, critics pointed to the amounts paid to shareholders in dividends rather than investing that money in infrastructure as a contributor to water shortages in 2015 (e.g., *Brasil Wire*, 2015). In another example, in Portugal, some contracts guaranteed compensation to the private provider if there was a reduction in water use or in number of consumers, and in one case assumed a minimum monthly consumption level that was double the actual level. Meeting the terms of these guarantees to investors meant that the cost of water effectively doubled (Teles, 2015). Rather than improving performance for users, some privatised water companies have been manipulated to favour the short-term interests of shareholders – and the costs of meeting investors' needs can be high.

The private sector is depicted as a cooperative partner, as a benign force looking for partnerships with states. Mostly, the literature promoting water privatisation ignores the predatory nature of capital, which does not always behave as policy makers anticipate. The entry of financial investors has led to an expansion of more aggressive financial restructurings to boost returns at the expense of the water user. For investors, the rationale for putting money into water investments stems from its potential to generate shareholder returns; there is little concern for social or environmental goals such as achieving SDG6.

19.3.4 Privatisation Needs a Strong State

While the private sector is intended to make up for the failings of the state, the success of the policy relies on a strong public sector. Baer (2014) points to this inherent contradiction, with weak state capacity cited as both an incentive for privatisation and an explanatory factor for privatisation failure.

Private sector involvement leads not to a withdrawal but merely a reconfiguring of relations between the state and capital. What may appear to be successful private sector engagement may in practice be down to backstopping by the government: State subsidies for water consumption can make sure that investments are profitable. In Chile, a government-funded subsidy programme helps poor households pay their water bills. This allows water companies to ensure that tariffs can be high enough to finance investment (Baer, 2014). Risk transfer is seen from an investor perspective, but PPPs raise risks for governments, including contributing to debt distress. As Alston (2018, p. 7) states, 'Corporations take the profits but Governments will bear much of the losses if they are significant'. As a case in point in EW, the government now faces the risk that Thames Water will collapse, in which case the failing company will be taken into state hands via a Special Administration process.

Regulation is central to the private water model, but there are numerous intractable challenges to this. Information asymmetries are profound. These go beyond practical details such as knowledge of input costs to more far-reaching aspects of private sector operations such as corporate strategy. The crises in the water system in EW, in terms of the risk of financial collapse of Thames Water and the lack of investment in sewerage, have only come to light now that they can be externally observed. But these crises were a long time in the making, and the information was only available to those inside the firm.

Regulation is demanding. Part of the issue with raw sewage leaks in EW was that the environmental regulator saw its budget cut by two thirds during the 2010s and so was unable to carry out audits of water companies. Regulation has been slow to respond to the practices of private companies. In part this is because the state has to follow government processes. Thus, while the private sector may be more agile, the state bureaucracy still has a role to play, and this is even slower when regulating, rather than providing, water.

Regulation is depicted as an external rule-setting function, but in practice it is a process of mediating across sets of agents with different interests and different capacities to influence regulatory outcomes. Regulation is tasked with the conflicting goals of attracting the private sector and protecting social and environmental interests. In EW, there is a revolving door between the staff of the regulator and water companies, which means that the financial culture of the water system is perpetuated. The private sector lobbies the regulator and has far more influence than other stakeholders such as consumers. Regulatory stability is seen as crucial to maintaining investor interest, which lessens the state's capacity to discipline private capital. Regulation, then, is not simply about policies and processes; it is a question of cultures and capabilities, as well as politics (Bayliss et al., 2022). If the challenges are huge, if not insurmountable, in England, where regulatory skills are not in short supply, what hope is there for water regulation in less wealthy parts of the world?

Of course, public water providers are not without their limitations, but these are not necessarily resolved through privatisation, which can create a different set of issues, although attribution is difficult. Both supporters and detractors are selective in picking out positive and negative outcomes. In Manila, for example, the IFC claims that privatisation has increased access, reduced diarrhoea cases, and lowered prices, while critics say that the process led to continual price hikes, legal challenges, and unfair returns to shareholders.

19.3.5 Water Is a Public Good

The nature of water management is such that service provision encompasses far more than productive efficiency. There are wider issues that need to be taken into account, such as democratic oversight and the need to operate in the long-term interests of society and the environment. The supply of water for profit often does not sit well with the public, and there has been substantial social opposition, which has limited the extent of implementation.

Protests were widespread in some locations, particularly in Latin America when the policy failed to deliver on promises of improved water access (Baer, 2014). Conflicts and protests were often due to high price increases from private contractors. In Spain, in Alhaurín de la Torre in 2004, after a series of incidents and mutual accusations, the local government seized the bank accounts of the private water managers, and a public company was created in 2005 to manage the water service. Deekshit (2019) provides a comprehensive review of reasons for the stagnation of water PPPs in India. The wave of contract failures around this time was a deterrent to investors, and global water operators began to retreat from developing countries (Castro, 2016, p. 51).

19.4 Conclusion

The privatisation of water services and associated interventions, such as through PPPs, were promoted in large part on the idea that a dynamic private sector would bring innovation and efficiency. This enthusiasm overlooked a number of mundane issues. Water is a monopoly and a vital good that requires strong state intervention even (or maybe especially) when privately managed. The context is important for outcomes. A handful of cases have been considered successful, such as in Senegal, Gabon, Cote d'Ivoire, and Manila, not least because they continue to run. But on the whole, privatisation in water has not lived up to the optimism of the 1990s. It has been difficult to achieve and, where implemented, the evidence of greater efficiency or increased access is far from compelling.

Rather than leading to a reduction of state involvement, privatisation demands that the public sector create an attractive investment opportunity and regulate private investors. When the broader picture is considered, there are even wider potential concerns with regard to equity and social justice. Everyone needs water, and where payment of water bills generates returns for wealthy shareholders, questions need to be asked about the way the resources are managed. In LMICs, this effectively means that the world's poorest are funding the returns on shares and pensions in the richest nations. Using private finance for water investment will perpetuate and exacerbate pervasive inequality, boosting returns for investors from low-income households seeking to have their basic needs met. These important outcomes tend to be overlooked in analyses of relative efficiency.

Further Readings

Bayliss, K. (2020). Moving the goalposts: Reconfiguring the role of the private sector in the provision of water. In J. Gideon & E. Unterhalter (Eds.), *Critical reflections on public private partnerships* (pp. 79–97). Routledge. https://doi.org/10.4324/9780429280450

Bayliss, K., van Waeyenberge, E., & Bowles, B. O. L. (2022). Private equity and the regulation of financialised infrastructure: The case of Macquarie in Britain's water and energy networks. *New Political Economy*. https://doi.org/10.1080/13563467.2022.2084521

Lobina, E., Weghmann, V., & Nicke, K. (2021). *Water remunicipalisation in Paris, France and Berlin, Germany: A PSIRU report commissioned by the City of Barcelona, Spain and Aqua Publica Europea, the European Association of Public Water Operators*. PSIRU. https://shorturl.at/56Q7s

Prykc, M., & Allen, J. (2022). The everyday construction of value: A Canadian investment fund, Chilean water infrastructure, and financial subordination. *Finance and Society*, *8*(2), 189–208.

Notes

1 See the World Bank website, 'Water & Sanitation PPPs', https://ppp.worldbank.org/public-private-partnership/water-and-sanitation/water-sanitation-ppps, accessed 12 June 2025.

2 WICS website, 'Encouraging better performance from Scottish Water' https://wics.scot/what-we-do/encouraging-better-performance-scottish-water#:~:text=Scottish%20Water's%20service%20quality%20performance,companies%20in%20England%20and%20Wales.

Selected References

Alston, P. (2018). *Report of the Special Rapporteur on extreme poverty and human rights* (A/73/396). United Nations General Assembly.

Baer, M. (2014). Private water, public good: Water privatization and state capacity in Chile. *Studies in Comparative International Development*, *49*, 141–167. https://doi.org/10.1007/s12116-014-9154-2

Bakker, K. (2013). Neoliberal versus postneoliberal water: Geographies of privatization and resistance. *Annals of the Association of American Geographers*, *103*(2), 253–260.

Banerjee, S., & Morella, E. (2011). *Africa's water and sanitation infrastructure: Access, affordability, and alternatives*. The World Bank.

Bayliss, K. (2003). Utility privatisation in sub-Saharan Africa: A case study of water. *Journal of Modern African Studies*, *41*(4), 507–531.

Bayliss, K. (2011). A cup half full: The World Bank's assessment of water privatisation. In K. Bayliss, B. Fine, & E. van Waeyenberge (Eds.), *The political economy of development: The world bank, neoliberalism and development research* (pp. 73–98). Pluto Press.

Bayliss, K. (2020). Moving the goalposts: Reconfiguring the role of the private sector in the provision of water. In J. Gideon & E. Unterhalter (Eds.), *Critical reflections on public private partnerships* (pp. 79–97). Routledge.

Brasil Wire. (2015, January 28). Sao Paulo water crisis: Solutions? Retrieved from www.brasilwire.com/sao-paulo-water-crisis-solutions/

Britto, A. (2025). Neoliberal changes and perspectives for financialization in the management of Brazil's water and sanitation services. *Geoforum*, *160*. https://doi.org/10.1016/j.geoforum.2025.104220

Castro, J. E. (2016). *An examination of the politics of privatization of water and sanitation services in Africa, Europe, and Latin America (1990–2004). Comparative analyses*. WATERLAT-GOBACIT Working Papers. https://doi.org/10.13140/RG.2.1.2321.3682

Chiu, V. (2025). The public water services in France: Between public and private management. *Journal of Agricultural and Environmental Law, 20*(38), 195–221. https://doi.org/10.21029/JAEL.2025.38.195

Christen, R., Gökgür, N., Nellis, J, & Rühl, O. (2005). United Republic of Tanzania: Privatization Impact Assessment – Infrastructure, Report No, 68869, https://documents1.worldbank.org/curated/en/625791468172138286/pdf/688690ESW0Whit0privatization0Report.pdf

Deekshit, P. (2019). *Private sector participation in urban water: Literature review focused on India*. Working Paper EQUIPPPS Network.

Guasch, J., Laffont, J.-J., & Straub, S. (2008). Renegotiation of concession contracts in Latin America: Evidence from the water and transport sectors. *International Journal of Industrial Organization, 26*, 421–442.

Guasch, L., & Straub, S. (2009). Corruption and concession renegotiations: Evidence from the water and transport sectors in Latin America. *Utilities Policy, 17*, 185–190.

Khemka, R., Lopez, P., & Jensen, O. (2023). *Scaling up finance for water: A WBG strategic framework and roadmap for action*. World Bank Group.

Kikeri, S., Nellis, J., & Shirley, M. (1992). *Privatization: The lessons of experience*. IBRD.

Kishimoto, S., & Petitjean, O. (2017). *Reclaiming public services: How cities and citizens are turning back privatisation*. Transnational Institute.

Lobina, E., & Hall, D. (2007). *Water privatisation and restructuring in Latin America*. Public Services International Research Unit, University of Greenwich.

Marin, P. (2009). *Public-Private Partnerships for urban water utilities: A review of experiences in developing countries*. World Bank.

McDonald, D. (2018). Remunicipalization: The future of water services? *Geoforum, 91*, 47–56.

McDonald, D. & Swyngedouw, E. (Eds.) (2019). Special themed section: Remunicipalization: The future of water services? *Water Alternatives*, 12(2).

Molina, N., & Chowla, P. (2008). The World Bank and water privatisation: Public money down the drain. *The Bretton Woods Update*, (62). www.brettonwoodsproject.org/publications/62-2/

Ofwat. (2006). *The development of the water industry in England and Wales*. Water Services Regulation Authority. Retrieved from www.ofwat.gov.uk/wp-content/uploads/2015/11/rpt_com_devwatindust270106.pdf

Parker, D. (2004). *The UK's privatisation experiment: The passage of time permits a sober assessment* (CESifo Working Paper No. 1126).

Pierce, G. (2015). Beyond the strategic retreat? Explaining urban water privatization's shallow expansion in low- and middle-income countries. *Journal of Planning Literature, 30*(2), 119–131.

Pryke, M., & Allen, J. (2019). Financialising urban water infrastructure: Extracting local value, distribution value globally. *Urban Studies, 56*(7), 1326–1346.

Pryke, M., & Allen, J. (2022). The everyday construction of value: A Canadian investment fund, Chilean water infrastructure, and financial subordination. *Finance and Society, 8*(2), 189–208.

Qian, N., House, S., Wu, A., & Wu, X. (2019). Public–private partnerships in the water sector in China: A comparative analysis. *International Journal of Water Resources Development, 36*(4), 631–650. https://doi.org/10.1080/07900627.2019.1685951

Stiglitz, J. (1998). *More instruments and broader goals: Moving toward the post-Washington consensus*. WIDER Annual Lectures.

Teles, N. (2015). *Financialisation and neoliberalism: The case of water provision in Portugal* (FESSUD Working Paper No. 102).
Thatcher, M. (1993). *The downing street years*. HarperCollins.
World Bank. (1994). *Adjustment in Africa: Reforms, results and the road ahead*. Oxford University Press.
World Bank. (1995). *Bureaucrats in business: The economics and politics of government ownership*. Oxford University Press.

20 *REVISITING*: Off-grid, the New Water Supply Solution

Alexandre Gaudry, Marine Colon, and Catherine Baron

20.1 Introduction

'Off-grid solutions' to provide drinking water have emerged recently as a way to reach Sustainable Development Goal (SDG) 6.1: universal access to safe drinking water supply. Historically, the gold standard in the supply of drinking water was a centralised, piped network (a 'grid'). Pipes connect key components: abstraction (freshwater withdrawal from rivers, lakes, and aquifers), production (water treatment), storage in reservoirs, and distribution (from tanks to households or any other users). Introduced in Europe in the nineteenth century, the model not only shaped infrastructure design (and urban development), but it served as the structural backbone around which the governance of the sector has been organised. However, in 2019, a World Bank report advocated considering 'off-grid' solutions as an adjunct to centralised systems (Misra & Kingdom, 2019) and not only a transitory solution.

The 'off-grid' moniker, borrowed from the energy sector, includes a wide range of alternatives to the conventional, centralised, piped water system. International donors see these options as representing a response to the widespread failure of authorities and water-utility operators to reach universal provision. Off-grid 'solutions' encompass various sociotechnical and governance configurations: from formal to informal actors, from devices plugged into main networks to fully autonomous systems disconnected from the network. They include small networks operated by community-based organisations (CBOs), bottling treated water and selling it at water kiosks, reselling tap water in containers, digging wells and boreholes, and harvesting rainwater. What they all have in common is private investment, whether by NGOs, subsidiaries of big companies, or local entrepreneurs, mostly with little or no regulation.

The term 'off-grid' was absent from Frenoux's (2016) bibliometric analysis of 1,366 publications on 'non-conventional services' in the water sector. The label gained visibility with Misra and Kingdom (2019), who used it to argue that non-utility provision can legitimately complement networks, while also documenting its limitations and avoiding universal prescriptions.

DOI: 10.4324/9781003745532-20

The terminology used to describe non-network arrangements has remained inconsistent. In 2022, Misra and Kingdom referred to 'supplementary urban water service providers', but there has not, so far, been a wide uptake of this label in the literature. This chapter keeps to 'off-grid' as a shorthand for alternative configurations, many of which interact with centralised infrastructure. It should not be read as a precise analytical category.

In 2019, the World Bank raised the question of why the international community should keep funding the development of centralised networks with no respect for the efforts made by private actors providing alternative access to water. Should a development bank invest public money in such schemes? Should these alternative approaches be supported or expanded? This chapter seeks to make sense of the recent shift towards recognising longstanding alternative approaches. It then looks at the many unintended consequences and glossed-over issues that hamper the rollout of off-grid solutions in the water sector. It focuses exclusively on urban and peri-urban areas, where networked water infrastructure remains the dominant model and therefore serves as a relevant point of reference. In rural contexts, by contrast, the extension of centralised networks cannot be assumed and, in many cases, may not represent a technically viable option. This chapter highlights the problems with treating off-grid as a universal 'best practice'.

20.2 The Emergence of a Discourse Framing Off-Grid Alternatives as Solutions

Since many countries, particularly in the Global South, have yet to achieve universal access to a piped network, alternative solutions are not new. Indeed, they have been studied since the 1980s (Whittington et al., 1989). This recognition, however, has not substantially altered the direction of national water policies or the priorities of major international donors. In the 2000s, public-private partnerships (PPPs) were trialled in many Southern countries, with mixed results (Marin, 2009). In some cases, they ended in the failure of the concession, as in the infamous case of Buenos Aires (Botton, 2004). Meanwhile, Marin (2009) highlighted the contribution of local, informal operators to the PPP in Maputo, Mozambique. These operators invested in small networks in peri-urban areas that were not supplied by formal operators through the central network. However, these were intended as temporary solutions until such time as the central grid expanded to reach all users. More recently, the term off-grid has made its way in the water sector and has profoundly changed how alternatives to conventional access are seen.

World Bank experts Misra and Kingdom (2019, 2022) called for a paradigm shift (a 'mindset shift') from the traditional logic of states and donors that systematically favours centralised networks managed mainly by public or private monopolies. They point out that large segments of the Global South's populations are not connected to these networks and therefore rely on off-grid water access. These seminal reports invite recognition of alternative solutions not as temporary or inefficient stopgaps but as legitimate and complementary components of water provision in these regions. By adopting an approach focused on access rather than the nature of the service provider, the term off-grid evolved into a generic concept encompassing a wide range of mechanisms, contexts, and actors (in terms of status, size, and activity). Since 2019, complementary models of water provision have been embraced and advocated by researchers, NGOs, and entrepreneurs (such as Sachdeva et al., 2023), so why this shift?

Taken from the electricity sector, the term 'off-grid' originally referred to an independent energy supply disconnected from the grid. Such solutions are based on renewable energy (such as solar, wind, or hydraulic), although hybrid systems with fossil fuels exist. Thus, off-grid solutions in the energy sector have two main advantages: they bring electricity where the grid cannot reach and provide a decarbonised supply. These devices tend to be seen as more affordable, easier to use, and more environmentally friendly.

The idea of an off-grid water supply has been propelled to the status of a 'fashionable' solution. These alternative arrangements are no longer solely the remit of small, local entrepreneurs investing in spaghetti networks in peri-urban areas. Indeed, transnational entrepreneurs and NGOs have been promoting off-grid drinking water solutions through market-shaping, standard-setting, financing, and advocacy, while outcomes remain context-dependent. Examples of the various actors operating internationally include 1001fontaines (a donor-backed NGO), Aqua for All (a blended-finance foundation), and Danone Communities (a global social business investor). They are fuelling an argument branding off-grid options as robust solutions rather than makeshift bricolages. In other words, off-grid tends to be seen as 'cool'.

Providers of off-grid devices and services interviewed at the Global Forums in Dakar (2022) and Bali (2024) asserted that 'People don't want to rely on the grid anymore. It is up to people to take back control'. A company proposing on-demand home delivery via mobile phones prided itself as 'the Uber of Water', while an NGO promoting water-kiosk microenterprises in rural and peri-urban areas announced, 'We want to send the following message: we are the ones that will stay. We are, forever, your public drinking water infrastructure'. Thus, off-grid actors are not limited to informal local entrepreneurs; they include NGOs and commercial enterprises with marketing narratives to circulate at World Water Forums, who are thereby contributing to a broader transformation of the sector's representations.

A further reason for the growing recognition of off-grid solutions is the United Nations' adoption of the SDGs in 2015. Aiming to "ensure availability and sustainable management of water and sanitation for all", SDG 6 simultaneously addresses resource protection and the provision of safe drinking water and sanitation services. Target 6.1 is to achieve universal and equitable access to safe and affordable drinking water for all by 2030, as monitored by the proportion of the population using 'safely managed' drinking water services (United Nations General Assembly, 2015). Table 20.1 shows drinking water

Table 20.1 Service ladder of drinking-water categories

Source type	*JMP service ladder*	*Definition*
Improved: • Piped supplies (from tap water in the dwelling, yard, or plot, or public stand pipes) • Nonpiped supplies (from boreholes/tube wells; protected wells and springs; rainwater; packaged water, including bottled and sachet water; and delivered water, including from tanker trucks and small carts)	**Safely managed**	Drinking water from an improved water source that is located on the premises, available when needed, and free from faecal and priority chemical contamination
	Basic	Drinking water from an improved source, provided collection time is not more than 30 minutes for a round trip, including queuing
	Limited	Drinking water from an improved source for which collection time exceeds 30 minutes for a round trip, including queuing
Unimproved	**Unimproved**	Drinking water from an unprotected dug well or unprotected spring
Surface water	**Surface water**	Drinking water directly from a river, dam, lake, pond, stream, canal, or irrigation canal

Source: Mitlin et al. (2019: 12), adapted from WHO and UNICEF (2017: 8, 52).

provision ranked by source, from direct, then unimproved surface water, to limited, basic, and finally safely managed 'improved' water.

Improved drinking water sources are defined very broadly as those with 'the potential to provide safe water by virtue of their design and construction' (WHO and UNICEF, 2017). Since these may include 'nonpiped supplies', such as: 'boreholes/tube wells, protected wells and springs, rainwater, packaged water including bottled and sachet water, including from tanker trucks and small carts', SDG 6.1 acknowledges that access to safe drinking water can be achieved through independent, alternative solutions. It does not mandate a unique service model. What should be universalised is access to water of a certain quality, rather than a service defined by a particular governance structure, distribution system, or standardised infrastructure. The goal, therefore, could be achieved by any appropriate means, for example, in terms of service level, access modality, governance structure or pricing scheme, and regardless of the nature of the provider-user relationship (operator/user versus water vendor/customer).

The 2030 deadline prioritises solutions that can be implemented quickly and yield measurable short-term outcomes. Off-grid models, such as bottled water and decentralised water kiosks, are valued for their lower capital requirements and fast installation process. This pace can rarely be matched by the deployment of centralised networks, which typically involves long timelines and high investment costs (Larsen et al., 2016). Uncertainty about the possibility of achieving the SDGs by 2030 has led to a broadening of scope and adaptations that in turn serve to support arguments for off-grid. Some academics (Eggimann et al., 2018) have called for 'modular' solutions, prompting a reassessment of centralised networks and revisiting key messages from recent research. Indeed, the centralised network is seen as suffering two major limits: it is highly capital-intensive and has long development timelines. Rapid population growth in the Global South demands a pace of investment in infrastructure that is beyond the reach of most governments. Relying on external funding to expand networks, they are unlikely ever to ensure universal access. Moreover, informal population centres often lack the necessary conditions to implement conventional networked infrastructure. Such factors have prompted a shift in perspective among actors who had previously promoted centralised systems while criticising off-grid solutions as inefficient. This is reinforced by a recognition that, in many cities of the Global South, financial and political barriers limit the feasibility of expanding centralised networks.

International frameworks such as the SDGs have thus further fuelled the shift by legitimising alternative service-delivery models. As a result, even the dominant paradigm favouring centralised networks is being reconsidered, not least because, once laid, the centralised networks lack flexibility and are intended to last decades. This rigidity is increasingly problematic given the unpredictable impact of climate change, where an extreme event could cause irreparable damage. Combined with the increasingly erratic patterns of urbanisation, large infrastructure is facing a challenge in terms of feasibility and even desirability. Indeed, given its cost, it may well lose relevance within a decade. Hence, less capital-intensive, off-grid solutions may be seen as more readily adaptable to changes in local needs. As such, off-grid solutions are accepted as a key tool in reaching SDG 6.1, helping to build resilience in cities threatened by climate change, being environmentally friendly, and fostering social enterprise.

20.3 The Multifaceted Challenges of Off-Grid Water Solutions

While off-grid water solutions have gained traction as innovative alternatives, a critical analysis of the literature reveals a number of problems with their use in urban settings. These range from technical, economic, and environmental issues to social and political challenges. There are five major issues involved.

20.3.1 Off-Grid Solutions May Depend on or Become Entangled with Existing Infrastructure in Urban Areas

Despite their label, off-grid solutions are rarely autonomous. In practice, they are often embedded in, or dependent on, elements of centralised infrastructure. For example, water kiosks and automatic dispensers in Dakar, while operated by private companies, simply sell tap water processed by reverse osmosis. At the same time, 98% of Dakar's population has access to the central network. Those with wells use the water for domestic use but avoid drinking it. Thus, the off-grid solutions provided by kiosks, automatic dispensers, and bottled water represent a supplementary, rather than alternative, service. Likewise, the NGO 1001fontaines, a big promotor of off-grid, derives approximately 30% of its water supply from main networks in Cambodia, its primary area of operation.[1]

In Bandung, Indonesia, only 65% of the population has access to water from a central network,[2] the rest relying mostly on private or communal wells (Suryani et al., 2019) and water kiosks for drinking purposes. Kiosk entrepreneurs operate within a complex value chain, as exemplified by the case of the Ujung Berung district (Nastiti et al., 2017) located in a mountainous area where surface water is both accessible and abundant. Local entrepreneurs have built infrastructure to collect and store water, which is metered and taxed by the municipality. Tanker trucks then transport the raw water to kiosks. Transport costs vary based on factors such as distance, destination, and water quality (e.g. pH level). Operated by small businesses or franchises, these kiosks are equipped with treatment systems (UV or reverse osmosis) and distribute water through various channels, including home delivery, walk-in refills, mobile vendors, and informal resale. This model highlights the complexity of integrated off-grid systems that encompass collection, transport, treatment, and distribution (Baron et al., 2025).

Creating further complexity in terms of administration, governance, and regulation is unhelpful in places already marked by multifaceted challenges. Moreover, the fact that many arrangements optimise a single function (such as potable quality) means gaps often remain in continuity, pressure, volume, or suitability for non-drinking uses. Thus, in most urban settings, complementary sources are still required.

20.3.2 Few Off-Grid Solutions Offer a Complete Alternative

Most off-grid solutions cannot fully replace conventional access to water for domestic use, from drinking, cooking, hygiene, and washing to watering gardens or animals, or commercial use. For instance, market-oriented alternatives, such as kiosks, dispensers, or bottled water, are optimised primarily for drinking purposes. Their design (small containers for the individual customer) and economic model (high price per litre) prevent them from serving other uses. Likewise, with the notable exceptions of wells and mini-networks, most off-grid solutions target drinking-related use as they are too expensive to cater to broader water needs. The 2019 World Bank report (Misra & Kingdom, 2019) highlights that, while off-grid solutions are crucial for providing basic services, they face significant limitations in terms of quality, quantity, and affordability. As a result, most off-grid systems only partly meet needs and must be supplemented. While praised for their adaptability to local conditions, off-grid solutions are often highly specialised. Each system requires specific technical configurations – pumping, filtration, bottling – making it difficult to pool them for multiple uses. This represents context-specific adaptation rather than a universal or integrated response.

20.3.3 **Off-Grid Solutions May Slow Network Expansion**

A persistent challenge identified in the literature is the erosion of trust in public services and centralised water utilities, fuelled by performance issues. Beard and Mitlin (2021) observed that in 12 of the 15 cities in their comparative study across Southern countries, including Sub-Saharan Africa, Latin America, and South Asia, households connected to the central network experienced frequent service interruptions, potentially affecting water quality and leading to widespread user dissatisfaction. Off-grid entrepreneurs capitalise on this to promote their solutions. In cities where a large proportion of the population is connected to the network, users who can afford it often turn to off-grid options, such as bottled water (Cohen & Ray, 2018) or automatic water dispensers, as in Dakar.

In areas not covered by a centralised water network, off-grid options can help meet local demand, thereby reducing pressure on the public authorities responsible for network expansion. Where the centralised operator is perceived as unreliable, residents are even less inclined to push for connection to the formal network. This dynamic makes it more difficult for decision-makers to justify the high capital investment needed to extend and upgrade the network, especially given the uncertainty surrounding future demand. Households may opt to maintain alternative solutions when they prove more reliable or accessible. Allen et al.'s (2017) study of 'non-networked infrastructure' takes the cases of Cochabamba, Dar es Salaam, and Calcutta to demonstrate that, while these solutions address immediate need, they can become permanent, reducing the incentive to expand the centralised network. Moreover, the diversity of local infrastructure creates a range of standards, complicating future integration into a unified system.

This issue is of concern since households without access to centralised networks often rely on fragmented and unregulated alternatives for their water supply. Playing into existing social disparities, it hinders equitable access to essential services, thereby reinforcing existing divides in urban water-provision systems. Finally, donor policies can play a decisive role in these dynamics, depending on whether they continue to prioritise investment in centralised systems or shift their support away from them towards off-grid actors.

20.3.4 Off-Grid Solutions: Not That Eco-Friendly

Water is considered a non-substitutable good: the resource can regenerate but cannot be manufactured. This intrinsic characteristic makes it imperative that it is protected and the associated externalities managed effectively. Historically, these factors have positioned water governance as a core responsibility of the state, leading to the predominance of centralised-network governance models – whether public or private monopolies – as the primary means of ensuring (urban) potable water access. The emergence of off-grid water solutions warrants a critical examination of their environmental implications. For instance, although informal recycling practices can be observed, plastic sachets and bottles frequently contribute to urban pollution, as observed in cities like Dakar (Valentin, 2010). Similarly, in Bandung, the aforementioned complex supply chain for conditioned water involves diesel-powered transportation and home deliveries, leading to increased fossil-fuel consumption and heightened extraction pressures on natural water sources.

Even solutions marketed as eco-friendly (including those powered by solar energy featured at the 2022 World Water Forum) raise significant concerns. Primarily designed for deployment in remote areas, where they may offer practical benefits, empirical studies have found environmental problems associated with certain off-grid options. Aberilla et al. (2020) conducted an environmental assessment of domestic water supply in remote communities and revealed that bottled water, which is widely used, is among the least sustainable off-grid alternatives, exhibiting a higher environmental impact than diesel-powered desalination systems. While the integration of photovoltaic energy with reverse osmosis desalination systems can mitigate their environmental impact, such as greenhouse gas emissions, it does not eliminate concern related to energy consumption, chemical usage, or the generation of waste (Razman et al., 2023). Meanwhile, Tashtoush and Alshoubaki (2024) found solar-powered, atmospheric water-harvesting systems demonstrate promising results, although their efficiency is dependent on factors such as variations in climate, the type of installation, and the cost of maintenance.[3] They concluded that additional research would be needed to assess the system's long-term sustainability and potential for use in a variety of settings.

20.3.5 The Challenge of Regulation

Regulating the operators of centralised water networks has never been straightforward. It involves the state's intervention into the market to enforce compliance with the societal values of fair access to the resource and respect for the environment, consumers, workers, etc. In theory, it aims for the best service at the best price for users, with a long-term perspective. Without proper regulation, investment may slow down, user protection may be ignored, the resource may be depleted, and coordination with other utilities may fail.

The primary argument for regulation is that, in principle, users cannot choose their water provider, since they operate as local monopolies. Nonetheless, it poses complex challenges, involving human and financial resources, political will, and administrative capacity. While this is true with regard to central-network operators in most countries in the South, regulating the numerous, informal, and formal off-grid providers present additional challenges. Although these providers help fill gaps in formal water provision, they also strain existing governance frameworks by introducing the need to manage decentralised systems.

Regulation is essential, however, to meet environmental, health, social, and economic standards.

- Price regulation: where prices are regulated by the market, a policy is needed to ensure drinking water is affordable for all. The risk of predatory practices is an enduring challenge (Hailu et al., 2011), as illustrated by the rise of water-tanker mafias in cities such as Karachi and Mexico City, where the existence of monopolies or near-monopolies in some areas permits excessive pricing. With centralised networks, cross subsidies can be used, where the lowest-income consumers enjoy a subsidised tariff funded by a higher rate, for example, for business customers. A pricing structure such as this is impossible with off-grid systems.
- Water quality/sanitary regulation: water quality testing to ensure raw and treated water meet national/international standards. The quality of the filters used in small treatment works should also be monitored.
- Environmental regulation: monitoring of water volumes extracted by off-grid operators and controlling their access to avoid resource depletion or pollution.
- Continuity of service: ensuring that entrepreneurs/associations/communities deliver a continuous supply (flow, pressure, quality). Without regulation, there is no guarantee that the off-grid provider will ensure quality service or not discontinue business unexpectedly.

Off-grid entrepreneurs should be protected to enable them to access finance and keep developing their activities, to be able to create good working conditions for the workers, and to increase their capacity to develop a high-standard service. However, although this does not necessarily require the presence of a centralised operator, it nonetheless calls for robust and precise oversight. This becomes particularly challenging in environments marked by a proliferation of heterogeneous, non-standardised off-grid solutions, which characterise many contexts across Southern countries (Caprotti et al., 2022; Furlong & Kooy, 2017). These indeed involve a wide array of actors who differi significantly in terms of scale and legal status

(Frenoux, 2016), as well as in the technologies deployed, the modalities through which users access services (Baron et al., 2025), and the quality and pricing of the services delivered.

20.4 Conclusion

The growing recognition of off-grid solutions in the drinking water sector marks a profound transformation in the governance of water services. Once viewed as provisional or second best, these decentralised approaches have gained legitimacy in policy, practice, and discourse. They have been boosted by the urgency of SDG 6.1, the failure of centralised systems to reach universal coverage, and the push for more adaptive, modular responses to climate and urban transitions. Far from offering an independent alternative, many off-grid solutions remain embedded (technically, financially, or institutionally) within larger, networked systems. Their operational limits in terms of scale, affordability, and water use reveal that they rarely represent a complete substitute, instead offering partial and fragmented add-ons to complement the primary system. They may even jeopardise that system, since their widespread adoption can, intentionally or not, delay or hinder investment in public infrastructure, as well as reinforcing unequal service provision and dependence on expensive, sometimes unregulated alternatives. More fundamentally, the regulatory vacuum in which many of these operators thrive is a cause for concern. The proliferation of actors, devices and delivery models challenges the very notion of water as a public service governed by principles of quality, equity, and accountability, and since it is often hard to establish consistent and effective regulation of pricing, environmental protection, business and worker protection, water quality and continuity of service, the presence of off-grid solutions can end up exacerbating inequality of access and environmental degradation.

Rather than framing off-grid and centralised systems as binary opposites, the critical task for public institutions is to govern them in combination. This requires effective coordination and comprehensive regulation, as well as the establishment of long-term objectives extending beyond the 2030 SDG deadline. The future of urban water provision will hinge less on technological innovation and more on the political commitment and institutional capacity required to ensure that all modes of access are firmly anchored in core principles: equality of access, environmental and socioeconomic resilience, continuity of service, quality of service, and the preservation of the resource.

Further Readings

Amankwaa, G., Heeks, R., & Browne, A. L. (2022). Water ATMs and access to water: Digitalisation of off-grid water infrastructure in peri-urban Ghana. *Water Alternatives, 15*(3), 733–753.

Binz, C., Sedlak, D., O'Callaghan, P., Truffer, B., Nesi, M., Morgenroth, E., Lesch, D., Miörner, J., Maurer, M., Narayan, A., Schelbert, V., Lüthi, C., Aalbu, S., & Wellauer, S. (2024). *Mainstreaming decentralized urban water management solutions for sustainable cities*. Synthesis report on an international expert workshop held at Eawag in June 2023. Eawag, Duebendorf, Switzerland. https://doi.org/10.55408/eawag:33039

Furlong, K. (2014). STS beyond the "modern infrastructure ideal": Extending theory by engaging with infrastructure challenges in the South. *Technology in Society, 38*, 139–147.

Nastiti, A. (2017). *Beyond access. The multifaceted water supply in urban and peri-urban areas of Bandung and Jakarta, Indonesia* [Doctoral dissertation]. https://repository.ubn.ru.nl/bitstream/handle/2066/175602/175602.pdf

Notes

1 Source: interview on the sidelines of the 2024 Global Water Forum with two representatives from the NGO, 18 May 2024 (11 am–12 pm).

2 This does not mean that 65% of the population has access to water through pipes; we estimate the connection rate to be around 20%.

3 Problems similar to those posed by the centralised network.

Selected References

Aberilla, J. M., Gallego-Schmid, A., Stamford, L., & Azapagic, A. (2020). Environmental assessment of domestic water supply options for remote communities. *Water Research, 175*, 115687. https://doi.org/10.1016/j.watres.2020.115687

Allen, A., Hofmann, P., Mukherjee, J., & Walnycki, A. (2017). Water trajectories through non-networked infrastructure: Insights from peri-urban Dar es Salaam, Cochabamba and Kolkata. *Urban Research & Practice, 10*(1), 22–42. https://doi.org/10.1080/17535069.2016.1197306

Baron, C., Gaudry, A., & Nastiti, A. (2025). « Solutions » hors réseau et arrangements marchands: Vers une nouvelle conception du service d'eau potable en Asie du Sud-Est? In G. Facal & J. Samuel (Eds.), *L'Asie du Sud-Est 2025: Bilan, enjeux et perspectives* (pp. 59–78). Institut de recherche sur l'Asie du Sud-Est contemporaine. https://doi.org/10.4000/13g83

Beard, V. A., & Mitlin, D. (2021). Water access in global South cities: The challenges of intermittency and affordability. *World Development, 147*, 105625. https://doi.org/10.1016/j.worlddev.2021.105625

Botton, S. (2004). Les « débranchés » des réseaux urbains d'eau et d'électricité à Buenos Aires: Opportunité commerciale ou risque pour les opérateurs? *Flux, 56–57*, 27–43.

Caprotti, F., de Groot, J., Bobbins, K., Mathebula, N., Butler, C., Moorlach, M., Schloemann, H., Densmore, A., & Finlay, K. (2022). Rethinking the off-grid city. *Urban Geography, 43*(8), 1217–1230. https://doi.org/10.1080/02723638.2022.2036928

Cohen, A., & Ray, I. (2018). The global risks of increasing reliance on bottled water. *Nature Sustainability, 1*(7), 327–329. https://doi.org/10.1038/s41893-018-0098-9

Eggimann, S., Truffer, B., Feldmann, U., & Maurer, M. (2018). Screening European market potentials for small modular wastewater treatment systems – An inroad to sustainability transitions in urban water management? *Land Use Policy, 78*, 711–725. https://doi.org/10.1016/j.landusepol.2018.07.031

Frenoux, C. (2016). *Institution et transactions: Déterminants et performances des services non conventionnels d'approvisionnement en eau dans les villes en développement. Le cas des entrepreneurs privés locaux dans les petits centres urbains du Cambodge* [Unpublished doctoral dissertation]. Université Toulouse 1 Capitole.

Furlong, K., & Kooy, M. (2017). Worlding water supply: Thinking beyond the network in Jakarta. *International Journal of Urban and Regional Research, 41*(6), 888–903. https://doi.org/10.1111/1468-2427.12582

Hailu, D., Rendtorff-Smith, S., & Tsukada, R. (2011). *Small-scale water providers in Kenya: Pioneers or predators?* United Nations Development Programme.

Larsen, T. A., Hoffmann, S., Luthi, C., Truffer, B., & Maurer, M. (2016). Emerging solutions to the water challenges of an urbanizing world. *Science, 352*(6288), 928–933. https://doi.org/10.1126/science.aad8641

Marin, P. (2009). *Public Private Partnerships for urban water utilities: A review of experiences in developing countries*. World Bank Publications.

Misra, S., & Kingdom, B. (2019). *Citywide inclusive water supply: Adopting off-grid solutions to achieve the SDGs*. World Bank. https://doi.org/10.1596/32046

Misra, S., & Kingdom, B. (2022). *Citywide inclusive water supply: Role of supplementary urban water service providers*. World Bank.

Mitlin, D., Beard, V. A., Satterthwaite, D., & Du, J. (2019). *Unaffordable and undrinkable: Rethinking urban water access in the global south* (Working Paper). World Resource Institute.

Nastiti, A., Meijerink, S. V., Oelmann, M., Smits, A. J. M., Muntalif, B. S., Sudradjat, A., & Roosmini, D. (2017). Cultivating innovation and equity in co-production of commercialized spring water in peri-urban Bandung, Indonesia. *Water Alternatives, 10*(1), 160–180.

Razman, K. K., Hanafiah, M. M., Mohammad, A. W., Agashichev, S., Sgouridis, S., & AlMarzooqi, F. (2023). Environmental performance of a photovoltaic brackish water reverse osmosis for a cleaner desalination process: A case study. *Science of the Total Environment, 896*, 165244. https://doi.org/10.1016/j.scitotenv.2023.165244

Sachdeva, P. K., Biswas, A. K., Tortajada, C., Arora, O., Leneveu, E., Adamjee, R., & Sharma, A. (2023). *Off-grid water supply: Should it be mainstreamed?* Springer Nature. https://doi.org/10.1007/978-981-99-4060-8

Suryani, N., Ichiki, A., Shimizu, T., & Maryati, S. (2019). Investigation of the water supply system and water usage in urban kampung of Bandung City, Indonesia. *Journal of Water and Environment Technology, 17*(6), 375–385. https://doi.org/10.2965/jwet.18-068

Tashtoush, B., & Alshoubaki, A. Y. (2024). Solar-off-grid atmospheric water harvesting system: Performance analysis and evaluation in diverse climate conditions. *Science of the Total Environment, 906*, 167804. https://doi.org/10.1016/j.scitotenv.2023.167804

United Nations General Assembly (2015). *Transforming our world: The 2030 Agenda for Sustainable Development*, A/RES/70/1.

Valentin, M. (2010). Bouteilles et sachets en plastique: Pratiques et impacts des modes de consommation d'eau à boire au Sénégal. *Autrepart, 55*(3), 57. https://doi.org/10.3917/autr.055.0057

Whittington, D., Lauria, D., Okun, D., & Mu, X. (1989). Water vending activities in developing countries: A case study of Ukunda, Kenya. *International Journal of Water Resources Development, 5*, 158–168. https://doi.org/10.1080/07900628908722429

WHO and UNICEF. (2017). *Progress on drinking water, sanitation and hygiene: 2017 update and SDG baselines*. World Health Organization (WHO) and the United Nations Children's Fund (UNICEF).

21 *REVISITING*: Rivers Should Have Legal Rights

Erin O'Donnell[1] *and Julián Suárez*

21.1 The Rise of 'River Persons'

Rivers are in trouble. Freshwater ecosystems are in poor health around the world, with human-induced climate change accelerating this decline, as droughts and catastrophic floods intensify. This unfolding natural disaster poses a real problem for people: we all depend on freshwater ecosystems for drinking water and sanitation, livelihoods, recreation, amenity, and cultural connection to place. There is a growing sense of urgency to halt this decline, and a growing concern that existing water and environment laws and regulations are not up to the challenge. Law and policy makers, community advocates, and Indigenous communities around the world are reaching for a radical idea: what if rivers had rights in law? What if rivers and freshwater ecosystems could protect themselves?

This is no longer a theoretical question. This big idea is now embedded within water governance and water-law scholarship in many parts of the world and is starting to shape the way we regulate rivers and waterways. It is time to take a critical look at how well it is working. In line with the objectives of this book, this chapter focuses on documented shortcomings, providing selected key references.

21.1.1 River Rights and Rights of Nature

Rights of nature is one of the fastest growing legal movements of the twenty-first century, with almost 500 initiatives in 40 countries, over 100 of which focus on the rights of rivers (Putzer et al., 2025). The non-binding *Universal Declaration of the Rights of Rivers*, developed by International Rivers and the Earth Laws Centre in 2020 and now endorsed by over 150 organisations, declares rivers to be living entities with legal standing and substantive fundamental rights, including the right to flow, to be free from pollution, and to be fed by and feed aquifers.

The rights of nature has a long history. Over 50 years ago, Christopher Stone described the need for nature to have standing: to be able to go to

DOI: 10.4324/9781003745532-21

court and protect its own interests, rather than relying on the courts to protect human interests, which may then have tangential protections for the environment (Stone, 1972). For example, people need water of reasonable quality to use for crop irrigation. This may have the result of ensuring that water in rivers meets certain quality standards, but on its own will not necessarily protect water quality to the standard required by river ecosystems. In the 1970s, when environmental law was in its infancy, Stone considered legal standing essential for nature's rights to be protected. In 2008, 36 years later, Ecuador amended its constitution to create legal rights for all of nature, including the ability for any person in Ecuador to represent nature in court. The first case to rely on these new rights was in 2010 and involved the Río Vilcabamba – a river that had been heavily impacted by the construction of a road. The court eventually held that the river's rights had been impacted and required the road developer to remediate the area.

In 2017, rivers emerged strongly as a focus for rights of nature laws. In Aotearoa, New Zealand, the Whanganui River was recognised as both a living and indivisible whole and the holder of all the rights, powers, duties, and liabilities of a legal person. In India, the Ganga and Yamuna Rivers (as well as their glaciers) were recognised as legal and living persons. The Colombian Constitutional Court declared the Atrato River to be a legal subject, whose own rights interacted with community biocultural rights. In Australia, the Birrarung (Yarra River) in Victoria was given the status of a living entity in law. Since then, rivers, lakes, and wetlands around the world have continued to be recognised in law as legal and living entities, thanks to ongoing local advocacy, as well as international advocates such as International Rivers and the Global Alliance of Rights of Nature.

21.1.2 What Is a River Person?

The idea of river rights and river personhood assumes that we can identify the holder of these rights. However, rivers are messy subjects with complex hydro-social relations. In hydrological terms, a river could encompass the surface water flow, any flow into and out of aquifers, and rainfall across the river catchment. River sources are often hard to identify precisely, as they can emerge from multiple points in their upper catchment. The river is both the water and the land, and potentially the connection to the sea and the sky, as well as the human interface of water extraction and run-off (Linton & Budds, 2014). The phenomenon of personhood is not limited to rivers, either; lakes, wetlands, and estuarine lagoons have all gained rights. This chapter uses 'river person' as a shorthand for all waterway persons.

River rights are extremely diverse in their creation and implementation. From a legal perspective, it is necessary to distinguish between the rights and powers created by the different forms of river person (Table 21.1). Legal persons and legal subjects both have legal standing, which is the ability to go to

Table 21.1 A legal typology of river person

River person	*Rights and powers*	*Example of river person*
Legal person	An assumed bundle of rights, which includes the right to own property and the power to enter into and enforce contracts	Whanganui River
Legal subject	Specific rights that have been articulated in the law, such as the rights to protection, conservation, maintenance and restoration	Atrato River
Living person/ moral person	Similar to a natural person, such as a human being	Ganga and Yamuna Rivers
Living entity/ being	No legal rights or powers	Birrarung/Yarra River

Source: Based on O'Donnell (2020).

court, to sue to protect the river's own interests (and potentially to be sued). A living person, which is an idea that emerged in India, appears to overlap with the rights and powers of human beings, although this has not been comprehensively tested in court. Lastly, a living entity or living being does not have any specific legal rights or powers, but it is a significant shift in the way that rivers are constructed in water law, which treats them as resources for human domination and exploitation. These categories of river person, while distinct, are not exclusive. A river may be assigned more than one, such as the Whanganui River, which is both a living entity and a legal person. The remainder of the chapter uses 'river persons' to encompass all of these options.

River persons are endowed with rights, which translate into novel environmental protection duties in their favour and may be a higher standard than existing laws (see Kauffman & Martin, 2023). Duties and liabilities, on the other hand, remain largely uncertain: it is still unclear whether the river could be liable for damage it causes. This was one of the reasons for the appeal against the decision to recognise the Ganga and Yamuna Rivers as rights-holders (final decision still pending in 2025).

Given that river persons are positioned as a response to the problem of inadequate environmental law, it is also helpful to consider the way that different forms of legal creation of river persons fit within typologies of rights of nature norms (Table 21.2). The practical effect of these legal mechanisms varies significantly, with resolutions and declarations signifying intent but no legally binding requirements, whereas rules codified in statute and court rulings will have legal force and effect.

The level at which the river rights laws are made will also affect their ability to drive change. Local laws, for instance, enable a community to embed

Table 21.2 A typology of the mechanisms of creation for river rights

Typology of rights of nature norms	*Legal mechanisms*	*Legal content*	*Examples of river person*
Legal Declarations of Political Intent	Local resolutions Soft law declarations	Recognition of a river's intrinsic value and need for rights-based approach	River Ouse, UK
Programmatic Environmental Standards	Constitution Statute Local ordinances Court rulings	Principles about river value and the relationships between humans and rivers	Ecuador Constitution arts 71–74
Public Interest Environmental Protection Rules	Statute Court rulings	Substantive and procedural rules and institutional framework	Mar Menor Act 19/2022, Spain
Local Participative Environmental Management Rules	Statute	Rules about creation of rights, establishment of management boards, interaction with other legislation	Te Awa Tupua Act 2017 (NZ)

Source: Based on Suárez (2023).

river rights relatively quickly, but they will not protect against a development enabled under a state or federal law. Constitutional change provides one of the most robust mechanisms for river rights creation, but this is extremely hard to achieve. In 2022, voters in Chile overwhelmingly rejected a new constitution that would have included rights of nature protections.

21.1.3 What Is the Problem That River Persons Are Trying to Solve?

River rights emerge in response to multipolar problems stemming from complex human-nature relations with waterways. They materialise a political and institutional window of opportunity for substantive action where different perspectives on river development, political control of resources, and 'rights' converge (Immovilli et al., 2022). In Bangladesh, the devastation caused by the overdevelopment of riverbanks (river grabbing) and river pollution led to the recognition of all rivers as legal persons and living entities. In India, the court connected the existential threat faced by the Ganga and Yamuna Rivers to their status in the Hindu religion to support their recognition of the rivers as legal and living persons, which in turn has been heavily criticised in the context of Hindu nationalist politics. The Mar Menor, in Spain, was recognised as a legal person in response to a public outcry about the decline in the health of the lagoon. In the Whanganui, although river health is heavily impacted by water extraction and land clearing, the problem being addressed was a long-standing failure by the New Zealand government to uphold Māori Treaty rights. For Australia's Birrarung, the problem was two-fold: a failure by the settler state of Victoria to recognise Indigenous interests in the river, as well as a fragmented and poorly coordinated system of river governance.

What is clear is that river rights and river persons are presented as solutions to an array of different problems (and can sometimes even appear as a solution seeking a problem). A river person can:

- level the playing field with human rights holders by overcoming legal-standing problems,
- be a pathway towards genuinely ecological jurisprudence, in which the river's needs and interests are fundamental features of water law, and/or
- reflect the desired relationship between people and rivers.

What is less clear is that river persons are an adequate mechanism to address the specific problem confronting a particular river. Is water law failing to protect rivers *because* rivers lack rights? And what new problems might river persons create along the way?

21.2 River Persons: Pitfalls and Performance Issues

This section identifies a range of problems that the creation and implementation of river persons has generated for effective riverine protection. We note that river persons are sometimes created for strategic reasons that their advocates may consider outweigh the difficulties that these problems can cause. But, for this big idea to become established as best practice in water management, the problems will need to be recognised and addressed.

21.2.1 What Is Success?

It is extremely difficult to point to clear indicators of success that flow from recognising river persons in law, and this stems from a lack of clarity. Measuring the success of river persons requires them to have both a clear objective (what is the problem they are trying to solve?) and a clear mechanism for intervention (how will river personhood solve that problem?).

Sometimes the objective of creating river persons has very little to do with river protection, such as the Whanganui River, where the new river person was the result of a treaty dispute settlement. Attempting to measure the success of this legislation solely in terms of improved river health would be a mistake.

In other cases, while there is clear reasoning from the courts that they are creating river rights as a response of last resort to the intractable problem of the ongoing decline in river health, it is not always clear how they expect the new river person to solve this problem. When there is a sense that something must be done, courts can sometimes create river persons without a clear pathway to success. This creates problems for implementation (and ultimately in defining whether the recognition of the river person was successful). For instance, in Bangladesh, the court required the National River Conservation Commission to be the guardian of the rivers, but this organisation does not have the legal powers it needs to intervene (Agnew & Jahan, 2021). In the United States, the local ordinance that created legal rights for Lake Erie did not survive its first legal challenge, which was lodged the very next day, rendering the exercise a waste of time and resources.

21.2.2 A Lack of Integration with Existing Law

River rights often lack coherent integration with existing water-law frameworks. In almost every case, even when rivers ostensibly hold the right to flow, river persons do not hold rights to the water flowing between their banks (O'Donnell, 2020). River persons remain reliant on existing environmental flows (if any) and limits on water extraction within water law. The sole exception to this is the Aquepi River judgement, in which the Ecuadorian Constitutional Court found that the river's constitutional rights included a right to flow and that this could be understood as a requirement to maintain an ecologically meaningful minimum flow in the river (Kauffman and Martin, 2023) similar to environmental flow protections already present in many jurisdictions. However, even in Ecuador this right to flow has not been consistently applied. A later case involving the Monjas river, where its ecological water flow had been severely affected by urban discharge and rainwater drainage, was not accorded the same protection (Tănăsescu et al., 2024). As a result, river persons may not substantively improve environmental flows or river health beyond the protections already established in water laws.

Where substantive rights, such as the rights to exist, to be protected, and to be restored, have been defined by legislation, this does not always result in clear, unconditional, or automatic duties towards the environment or a particular ecosystem (Bétaille, 2019). For instance, the Mar Menor lagoon and its basin hold the right to be restored, entailing redress of its natural dynamics, resilience, and ecosystem services. But the Spanish Mar Menor Legal Personhood Act 19/2022 is silent regarding baselines, goals, and recovery activities and standards in EU and national nature-restoration law (Suárez, 2024). Some of these may be developed in the future by the Scientific Committee and the Representatives' Committee.

River rights may also be superfluous to, or overlap with existing environmental law, adding complexity without improving on-ground outcomes. For example, the Mar Menor lagoon already holds Ramsar Convention and Barcelona Convention status, is a Special Area of Conservation and a Special Protection Area under EU law, is a protected area under Spanish national law, and has special legislation enacted by the Murcian Autonomous Region to improve its ecological health. Redundant river rights run the risk of becoming hollow legislation if characterised by public powers as mere political concessions or 'feel-good' measures in place of effective enforcement of existing protections. It is possible that the Mar Menor person may be a more effective advocate for enforcement, but, in 2025, it is too early to tell.

In the case of the Atrato River, the court ordered all stakeholders to devise a decontamination and recovery plan for the river, which was profoundly impacted by illegal mining that also breached collective and individual rights to water. Yet there was no mention of how the river's rights to conservation and restoration would interact with the state's existing duty to maintain and improve ecological conditions and protect water ecosystems and hydrographic basins. As a result, the implementation of the Atrato River judgement has been undermined not only by the complex overlapping of rights and responsibilities but also by inadequate institutional coordination and accountability mechanisms (McNeish & Moreno Socha, 2024). Drafting issues in river-rights norms often conceal significant structural and contextual factors that the creation of the rights cannot address. This leaves the institutional difficulty of coherently embedding the rights within existing legal frameworks.

21.2.3 Hiding the Real Problem

River rights fall within broader, rights-based environmental protections and share many features with human-rights principles, standards, and accountability. This allows them to contribute to socio-environmental justice in river management and protection in a way that existing water law would not be able to. However, they also share many of the problems frequently associated with human-rights discourses, which tend only to talk about the 'root causes' of problems, without really addressing the underlying structural socioeconomic inequality (see, for example, Petel, 2024). They foreshorten the investigation of real causes, tending instead to treat their effects as causes, or they identify the causes only to put them aside (Marks, 2011). An example of this is the Colombian Constitutional Court having ordered an ambitious management plan to protect the Atrato River's rights. The approach implied that the water's contamination was due not to the historic marginalisation of communities and colonialist exploitation but to the failure of governments and lawbreakers to recognise the waterway's intrinsic value (González-Serrano, 2024). River rights and river persons change the relationship between people and rivers and demand broader public participation in riverine decision-making (Cribb et al., 2024). However, the initial absence of democracy and plurality is a function of the capture of legislative and executive powers by those who profit from waterway degradation. This has hindered wider legal reform that would allow the public to be involved in environmental matters or even to intervene in social production planning beyond the mere expression of concerns with no binding effect on decision-makers.

River rights can also operate as a distraction. Rights of nature can fail to frame environmental degradation as an issue of the commodification of nature. This allows profiteering at the expense of the environment and vulnerable groups to go largely unquestioned. The popular legislative initiative that led to Spain's Mar Menor Act 2022 sought to establish an environmental guardianship to protect the lagoon, yet did nothing to restrict its use as a run-of-the-river dumping site for agribusiness. Moreover, rights of nature do not directly address the devastating effects of ecological imperialism in the Global South, which are especially deleterious regarding rivers and waterways. While global advocacy for river rights is a powerful avenue for transformative change, there is a real risk of Global North advocacy perpetuating green colonialism and reinforcing the existing unequal power dynamics (Sultana, 2022). Finally, Indigenous scholars highlight a real risk that recognising rivers as legal persons in settler-colonial law can further undermine the rights of Indigenous Peoples and fracture their existing relationships with their more-than-human kin (Marshall, 2020).

21.2.4 Representation: Who Speaks for the Rivers?

Representation is one of the most crucial aspects of river personhood. Legal personhood requires representation by an entity that speaks for the interests of the river itself (which is not the same as those tasked with managing rivers for human uses). Who gets to speak for waterways, how their representatives engage in such activity, how they are held accountable, and how the power relations interwoven around waterways and their guardians should be navigated are all essential questions to answer (Cano Pecharromán & O'Donnell, 2024).

In some countries, like Ecuador, where nature as a whole is the subject of rights, all individuals, communities, and peoples can represent the rights of nature before the courts and public authorities. In other jurisdictions, where rights have been granted to particular waterways, they are represented by committees of guardians or stewards that may include a mix of public officials, community representatives, scientists, environmental defenders, business and trade spokespersons, Indigenous Peoples, and different ecosystem-service stakeholders.

For example, in Aotearoa, New Zealand, Te Pou Tupua is the human 'face' of the river person Te Awa Tupua (who is seen as a living ancestor). It comprises an appointee from each of the Crown and the Whanganui iwi (Māori tribe). In addition to acting to protect the river, Te Pou Tupua must engage and report to the community and relevant decision-makers. There is no clear boundary between Te Awa Tupua and Te Pou Tupua, as the two are treated as one legal person for the purposes of tax liability (s 25 Te Awa Tupua Act, 2017).

The Colombian Atrato River case established a river-guardianship committee to enact river-protection judgement orders with the assistance of an advisory team. The committee includes the Collegiate Body of Communitarian Guardians – which now includes two spokespeople from each of the seven organisations representing the plaintiff communities – and the Government Guardian, represented by the Minister for the Environment.

A significant and unresolved issue relates to guardians' potential liability for the actions of a river person, such as flooding or causing death by drowning. In India, the Uttarakhand High Court appointed the Director of the Namami Gange Programme, the Chief State Secretary, and the State Attorney-General to act in the rivers' interest, since rivers are treated as minors who cannot protect themselves. However, given the stringent personal liability associated with the role for those public servants acting as guardians *in loco parentis* (in the shoes of parents), the nominated guardians appealed the decision and the Supreme Court swiftly ordered that the execution of the judgement be stayed.

The implementation of representation arrangements is often a drawn-out process. The Mar Menor lagoon statute was enacted in 2022, but provisions on the functions and powers of the guardianship committees were pending until over two years later, when the executive passed Statutory Instrument 90/2025. In 2025, implementation of the stewardship and scientific committees is still at a very early stage, although NGOs have been active in this area for many years.

21.2.5 A New But Not Necessarily Better Relationship

River persons have enormous transformational potential, but can prove a double-edged sword. On one hand, they can level the playing field between people and rivers to ensure that the rights and interests of the latter are fully respected. But on the other hand, the legal personhood of rivers can generate a significant backlash (O'Donnell, 2018).

First, river persons with rights and powers of their own create a sense of complacency: the river can look after itself now, so local communities may feel it is no longer their responsibility. Second, and more insidiously, the river person can be seen as a competitor for resources and rights. The ability of the river person to sue people who infringe its rights can engender a rapid defensive response in which potential river polluters act to undermine the ability of the river to sue them. For example, in the case of Lake Erie, a farming business (Drewes Farms) filed suit the very day after the lake received legal rights. The suit, which challenged the validity of the lake's rights (ultimately held to be invalid), was informed by a fear that the farming business could be held accountable for any future impact on water quality in the lake. This is a consistent reaction to the rights of nature in settler-colonial law in the US (Sachs, 2023).

When recognised as living entities, river persons can bring people into a closer relationship with the river, as they learn to understand it, speak with it, and listen to it. However, the growing institutionalisation of river persons and their representatives can undermine this nascent connection by making the responsibility of care and connection with the river someone else's problem (Clark et al., 2018).

Creating river persons is therefore not a risk-free endeavour. River personhood is able to change relationships between people and rivers in both welcome and unwelcome directions.

21.2.6 Law for Show Rather Than On-Ground Outcomes

River persons and river rights attract attention, but there is a genuine risk that this leads to enacting environmental 'law-for-show'. That is, law with so many hurdles to deliver substantive action that its implementation becomes a performance (Lees & Pedersen, 2025). If river-person legislation relies heavily on reporting, if it endows the executive branch with elevated discretion for its implementation, lacks strong enforcement mechanisms, or fails to provide sufficient funding for institutional bodies, and if it has weak public scrutiny, it risks creating the impression of substantive action while producing little if any substantive change in river health.

Performativity in river rights is omnipresent. In the 2016 Atrato River case, the court used river rights to impose complex environmental management plans with ad hoc participatory governance arrangements. However, no funding was expressly provided in the ruling for the functioning of the guardianship committee, and budget funding started only in 2023.[2] Australia's Yarra River Protection (Wilip-gin Birrarung murron) Act 2017, while ostensibly focused on the living Birrarung River, is essentially a land-use (rather than water) governance statute with an emphasis on strategic planning and annual reporting. Yet even here, it was not until 2025 that the necessary land-use planning-scheme amendments were made that required decision-makers to take account of the living river when considering developments that may impact it.[3]

There is no denying that river rights and river persons have excelled in placing river-protection values at the centre of the public debate. As symbolic and aspirational legislation, river persons have reinvigorated support for more stringent environmental regulation and manifested the need for socio-environmental justice. But given the exponential rate of environmental decline, to the detriment of human dignity, there is a legitimate fear that river rights and river persons may not have the necessary tools to make a significant difference, and can potentially damage human-river relationships.

21.3 Conclusions

As waterway health continues to decline in many parts of the world, water policy makers and local communities continue to search for new, radical solutions to combat it. Recognising rivers and waterways as persons, in some cases with legal rights, is a big idea that harnesses the creative power of communities. However, river rights must be carefully considered in the specific socioeconomic, environmental and legal context where they are to be enforced. This chapter has identified multiple problems (such as compliance, enforcement, values mismatch, timeframes and interaction with existing laws) that must be addressed at the law-making stage to avoid the high performativity and low substantive action of river-rights norms.

Community-led responses to waterway degradation must be constructed collectively and democratically, ensuring a plurality of voices, including that of the waterway. In colonial contexts, the rights and laws of Indigenous Peoples must be centred in any river-rights process, and Indigenous voices should lead the conversation.

Legal recognition of river rights and personhood is also merely the beginning. Policymakers and local communities should be aware that there will be a political struggle ahead to actively represent the voice of the river before public authorities and society at large, and that it will be time- and resource-consuming.

River persons are not a panacea for river health. However, identifying the problems also articulates the issues that need to be resolved in the creation and implementation of river persons for them to have maximum effect. Best practice for river persons and rights starts with articulating the relationships of interdependence between people and waterways and creating space for reciprocity. The important question is: what does it mean to be in good relations with this living river? This anchors future water governance work in relational processes that strive for abundance and wellbeing for everyone, including the river, and strengthen the collaborative governance of waterways (Cribb et al., 2024).

Optimism can come from seeing these challenges as surmountable. In Victoria, the legal recognition of Birrarung (Yarra River) as a living entity has delivered a ten-year strategic plan with a legally binding land-use framework and a set of agreed actions. These highlight both the leadership of the Traditional Owners – the Wurundjeri Woi-wurrung and Bunurong peoples – and a recognition by all members of the collaborative governance framework that everyone's fate is intertwined with that of the river. The plan, which commits to actions that go beyond existing commitments by individual agencies, is called *Burndap Birrarung burndap umarkoo* – a Woi-wurrung-language name that translates to 'what is good for Birrarung is good for us all'. Although some of these outcomes may have been achieved without recognising the river as a living entity, the novelty of the living river is helping to lift planning and policy processes out of settler-colonial business-as-usual and amplify Indigenous leadership.

Further Readings

Boelens, R., Escobar, A., Bakker, K., Hommes, L., Swyngedouw, E., … Wantzen, K. M. (2023). Riverhood: Political ecologies of socionature commoning and translocal struggles for water justice. *The Journal of Peasant Studies*, *50*(3), 1125–1156. https://doi.org/10.1080/03066150.2022.2120810

Macpherson, E. (2022). Can Western water law become more 'relational'? A survey of comparative laws affecting water across Australasia and the Americas. *Journal of the Royal Society of New Zealand, 53*(3), 395–424. https://doi.org/10.1080/03036758.2022.2143383

O'Donnell, E., Clark, C., & Killean, R. (2024). Rights and relationality: A review of the role of law in the human/water relationship. *Water Alternatives, 17*(2), 207–238. www.water-alternatives.org/index.php/alldoc/articles/volume-17/v17issue2/749-a17-2-7/file

Tănăsescu, M. (2022). *Understanding the rights of nature: A critical introduction.* Transcript.

Notes

1 O'Donnell is a member of the Birrarung Council, voice of the Birrarung River. This gives her the status of an embodied knower, but she also acknowledges that she is not speaking for the Birrarung Council in this chapter.

2 s 275(1) Colombian National Development Plan (2022–2026) Act 2294/2023.

3 Victorian Planning Scheme Amendment VC281, 14 April 2025.

Selected References

Agnew, C., & Jahan, I. (2021). A critical analysis of the development of the concept of giving rivers a personality: Does it in fact help to protect rivers? *Journal of Water Law, 27*(3), 77–88.

Bétaille, J. (2019). Rights of nature: Why it might not save the entire world. *Journal for European Environmental & Planning Law, 16*, 35–64.

Cano Pecharromán, L., & O'Donnell, E. (2024). Relational representation: Speaking with and not about nature. *PLOS Water, 3*(10), e0000236. https://doi.org/10.1371/journal.pwat.0000236

Clark, C., Emmanouil, N., Page, J., & Pelizzon, A. (2018). Can you hear the rivers sing? Legal personhood, ontology, and the nitty-gritty of governance. *Ecology Law Quarterly, 45*(4), 787–844.

Cribb, M., Macpherson, E., & Borchgrevink, A. (2024). Beyond legal personhood for the Whanganui river: Collaboration and pluralism in implementing the Te Awa Tupua Act. *The International Journal of Human Rights*, 1–24. https://doi.org/10.1080/13642987.2024.2314532

González-Serrano, M. X. (2024). Rights of nature, an ornamental framework: Water extractives and Backbone Rivers with rights in Colombia. *The Journal of Peasant Studies, 52*(2), 322–342. https://doi.org/10.1080/03066150.2024.2349228

Immovilli, M., Reitsma, S., Roncucci, R., Dueholm Rasch, E., & Roth, D. (2022). Exploring contestation in rights of river approaches: Comparing Colombia, India and New Zealand. *Water Alternatives, 15*(3), 574–591. www.water-alternatives.org/index.php/alldoc/articles/vol15/v15issue3/673-a15-3-2

Kauffman, C. M., & Martin, P. L. (2023). How Ecuador's courts are giving form and force to rights of nature norms. *Transnational Environmental Law, 12*(2), 366–395. https://doi.org/10.1017/S2047102523000080

Lees, E., & Pedersen, O. W. (2025). Performative environmental law. *Modern Law Review, 88*(1), 124–154. https://doi.org/10.1111/1468-2230.12918

Linton, J., & Budds, J. (2014). The hydrosocial cycle: Defining and mobilizing a relational-dialectical approach to water. *Geoforum, 57*, 170–180.

Marks, S. (2011). Human rights and root causes. *Modern Law Review, 74*(1), 57–78.

Marshall, V. (2020). Removing the veil from the 'Rights of Nature': The dichotomy between First Nations customary rights and environmental legal personhood. *Australian Feminist Law Journal, 45*(2), 233–248. https://doi.org/10.1080/13200968.2019.1802154

McNeish, J.-A., & Moreno Socha, J. (2024). Muddy waters: On the problematic political ecology of the Atrato ruling, Colombia. *International Journal of Human Rights*, 1–21. https://doi.org/10.1080/13642987.2024.2324121

Petel, M. (2024). The illusion of harmony: Power, politics, and distributive implications of rights of nature. *Transnational Environmental Law, 13*(1), 12–34. https://doi.org/10.1017/S2047102523000262

Putzer, A., Cook, J., & Pollock, B. (2025). Putting the rights of nature on the map. A quantitative analysis of rights of nature initiatives across the world – Second edition. *Journal of Maps, 21*(1), 1–3. https://doi.org/10.1080/17445647.2024.2440376

O'Donnell, E. (2018). *Legal rights for rivers: Competition, collaboration and water governance*. Routledge.

O'Donnell, E. (2020). Rivers as living beings: Rights in law, but no rights to water? *Griffith Law Review, 29*(4), 643–668. https://doi.org/10.1080/10383441.2020.1881304

Sachs, N. (2023). A wrong turn with the rights of nature movement. *Georgetown Environmental Law Review, 36*(1), 39–93.

Sultana, F. (2022). The unbearable heaviness of climate coloniality. *Political Geography, 99*, 102638. https://doi.org/10.1016/j.polgeo.2022.102638

Stone, C. D. (1972). Should trees have standing? Towards legal rights for natural objects. *Southern California Law Review, 45*, 450–501.

Suárez, J. (2023). Could rights of nature be overlapping, redundant and conflicting regarding existing environmental protection? An overview of four selected European domestic law frameworks. *Environmental Liability – Law, Policy and Practice, 28*(2), 90–107.

Suárez, J. (2024). Rights of nature as an alternative or a complement to existing environmental protection. *Opolskie Studia Administracyjno-Prawn, 22*(1), 87–117. https://doi.org/10.25167/osap.5340

Tănăsescu, M., Macpherson, E., Jefferson, D., & Torres Ventura, J. (2024). Rights of nature and rivers in Ecuador's Constitutional Court. *The International Journal of Human Rights*, 1–23. https://doi.org/10.1080/13642987.2024.2314536

22 CONCLUSION: Revisiting Global Water Mantras

François Molle and Sylvain Barone

22.1 Introduction

In every field of policy and practice, certain ideas emerge, gain traction, and are ultimately institutionalised and packaged as sanctioned 'solutions' to perennial problems. Supported by compelling narratives and select success stories, and endorsed by expansive networks of actors, these ideas may rapidly achieve international recognition as best practices. To those working in the water sector, examples of the phenomenon are all too familiar. This book examines 20 ubiquitous solutions that have been advanced to address the water challenges of insecurity, scarcity, conflict, flooding, and environmental degradation. These solutions are frequently framed as storylines – reflected in the chapter titles – that, through constant repetition, take on the status of mantras. They are so entangled with normative beliefs that they become almost beyond question (Sneddon, 2015).[1]

There is nothing wrong with finding solutions to problems, of course. Don't we do this all the time in our daily lives? However, everyday problems are most of the time what Rittel and Webber (1973) call 'tame' or 'benign'. They have well-defined boundaries, a straightforward causal link between the problem and its resolution, and a clear solution. The need to remove a cork from a bottle, for example, has inspired remarkable ingenuity, producing a wide array of devices designed just for this purpose. There are many problems in the realm of water too, such as lifting or conveying water, treating wastewater, or irrigating farm plots.[2]

But finding adequate responses to broader societal issues is not that straightforward. 'Solutions' in the domain of public policy are often simplistic and their achievements unclear. There is enormous complexity involved in assessing outcomes when the causes of the problems are multiple and convoluted at the outset. Societies are built on a huge variety of values, worldviews, interests, agency, and power dynamics; people form networks and nested systems that are hard to govern with 'problem-solution' narratives and panaceas (Meinzen-Dick, 2007). It gets even more complex when it comes to environmental policy. Interventions that aim to reorder nature tend to end up also reordering society (Scott, 1998). In other words, most problems are

DOI: 10.4324/9781003745532-22

'wicked': a class of complex, interconnected and persistent social or environmental issues that are difficult or impossible to solve definitively due to their intricate nature (Rittel & Webber, 1973; Fallon et al., 2021). As Freeman (2000) explains, how these are framed 'depends on the stakeholder's perspective, leading to conflicting narratives and solutions. The "solution" to one interested party is often a "problem" for others, and there are no obvious stopping rules that define when enough has been accomplished'.

Problems at the intersection of hydrological and social systems are complicated by the unique challenge of water's unpredictable circulation across space, time, and scales. Interventions in waterscapes, whether large in scale, such as dams, or small but widespread, such as the rollout of drip irrigation or wastewater reuse, tend to have unintended consequences. The same goes for institutional reform, be it water pricing, water-user associations, the legal personhood of rivers, or 'smart water management', which inevitably alters behaviour such that it ripples back into the waterscape in unexpected ways.

While most best practices address so-called 'wicked' problems, they are actually often framed as the solution to a 'tame' problem with an uncontroversial causal explanation. As such, they are akin to solutionism (Morozov, 2013) – an approach whereby

> the problem to be solved is presumed or created, rather than investigated thoroughly; the answer is sought before the right questions are asked. […] The solution does not solve broader social or political problems, and may lead to unintended consequences.
>
> (Fallon et al., 2021)

This echoes the 'organised anarchy' long identified by organisational theory, whereby purveyors of ready-made solutions look for problems to arise so as to proceed with implementation (Cohen et al., 1972). Paraphrasing Morozov (2013), water policy or strategies like France's *Plan Eau,* with its 53 measures,[3] could well be announced with the prompt: *To solve all water problems, click here*.

We take issue with two specific features: Firstly, best practices are adopted without due process. That is to say, there is a host of reasons why their appeal is so great and overrides deeper, case-by-case analysis. These include a lack of time or other resources, plain incompetence, a sense of belonging to an expert epistemic community where solutions come pre-verified, a tendency to adhere to an accepted narrative, or the pull of vested interests, but also the security of 'being on the safe side' in respect of perceived global validation and legitimacy. If a given practice is sanctioned as being among the "best", then why bother looking for options or even ponder about alternatives? Which ones could ever be better than best? From a practitioner's point of view, the best practices and trendy concepts of the day are also those that are more likely to attract funding, while providing some kind of immunity. The more hegemonic best practices become, the more likely they are to be turned to unreflexively.

Secondly, and relatedly, any undesirable results or externalities tend to be either ignored or treated as a transient nuisance to be mitigated. While it is true that best practices are rarely explicitly proposed as panaceas, and the relevance of context is often acknowledged, perhaps with wishful caveats and standard warnings that there is no one-size-fits-all solution,[4] the combination of the incentives listed above and structural and conjunctural constraints often favours copycat implementation. With hindsight, the tools developed to anticipate and evaluate the economic, social, and environmental consequences of interventions, however sophisticated, remain inadequate and are often merely a device to rubberstamp decisions that have already been taken.

The best practices analysed here, and the many others that could have been included,[5] are not necessarily misguided and surely form part of a response to the world's water problems. Indeed, there are heartening illustrations of what they can achieve in specific contexts. Yet, as underlined in the introduction, success stories often reflect 'survivorship bias' by obfuscating the cemetery of failed reforms and tools that, by reordering nature and society, had unexpected negative externalities.

The aim here, therefore, is to invert the burden of proof. Rather than uncritically adopting best practices based on their perceived authority – and subsequently addressing (or ignoring) their unintended consequences – we advocate for a thorough and critical examination of contexts and implementation processes.

This examination should account for the diverse reasons why interventions may go awry, including flawed science, narrow planning, or poor implementation, as documented in this book. Yet the chapters have also highlighted issues related to political economy contexts and cognitive rigidities, suggesting that the matter is far more complex than a purely rational assessment of arguments. A comprehensive analysis of best practices would therefore combine insights obtained through several lenses:

Substance: This examines the intrinsic content and validity of best practices – their scientific consistency and empirical robustness. It assesses their impacts, positive and negative, in economic, social, and environmental terms. Substantive analysis involves questioning whether a practice is genuinely evidence-based, whether its impact has been contextually verified, and how its outcomes perform over time and across scales. It also interrogates the underlying model of causality: does it account for system complexity, feedback, and uncertainty, or rely on oversimplified, linear assumptions? This is to judge best practices not merely as technical solutions but as socio-ecological interventions with measurable consequences.

Political economy: This perspective sees best practices as products of power relations and material interests. It looks at whose priorities shape their formulation, promotion, and adoption – from international donors and national bureaucracies to private contractors, consultants, communities, and local water users. It examines how resources, influence, and legitimacy are distributed between these actors, whether in policy design or project implementation. By mapping who reaps the benefit and who bears the cost, this approach reveals the often-uneven terrain beneath the ostensibly neutral label of 'best practice'. It questions whether these practices reinforce existing hierarchies, open new arenas for rent-seeking, or genuinely democratise water governance.

Discourse: The discursive lens focuses on how problems and solutions are framed, justified, and communicated in policy and practice. It asks how language, metaphor, and narrative frame certain interventions as self-evidently 'best' while rendering alternatives invisible or illegitimate. It scrutinises the rhetorical work done by terms like 'efficiency', 'integration', 'participation', or 'restoration' and how externalities or failures are cast as exceptions rather than structural flaws. It also considers the role of epistemic authority – how experts, consultants, and international organisations define what counts as 'knowledge' and 'success'. By highlighting the politics of meaning, the discursive analysis exposes the persuasive and performative dimensions of best practices.

Ideas: This lens examines the deep-seated worldviews, values and normative assumptions that shape how water problems are conceived. It explores how ideologies – such as neoliberalism, managerialism, modernism, environmentalism, solutionism or techno-optimism – influence what is considered desirable, rational or inevitable. Thus, best practices appear not as neutral or purely technical recipes but as manifestations of a particular vision of progress, order, and humans' relationship with nature. The ideational perspective brings into view the cognitive, normative, and cultural underpinnings of governance: the belief in growth, efficiency[6] or wilderness as social goods and the frequent marginalisation of alternative ethics grounded in care, sufficiency or ecological health.

By addressing substantive issues and exposing the many simplifications, inconsistencies, and devils in the details that so often derail the rollout of a promising best practice, this book provides a foundation for the healthy debate that should underlie new policies and projects. Although arguably not new, these inconvenient truths, so often glossed over, warrant renewed acknowledgement and examination.

While the political economy background tends to be case-specific and hence less amenable to generalisation, these chapters also addressed, or alluded to, the dominance of certain policy ideas or discourses, the ideological biases or rigidity within epistemic communities, the vocabulary and grammar of narratives around particular best practices or the multi-actor 'industry' that grows out of, and sustains, them.

As the introduction pointed out, the best practices examined here can be grouped into categories, the first coming under the 'zombie idea' moniker. These are ideas that have been largely refuted by analysis and evidence, and should be dead, yet live on. Several reasons for their persistence have been identified: beliefs, path dependency, incentives, politics and power, information gaps and filtering, and the absence of alternative ideas (Pérez-Blanco et al., 2021; Peters & Nagel, 2020). Examples in this category are the promotion of drip irrigation as a water-saving technique,[7] taking aquifer recharge as a sustainable target for pumping, or planting trees to enhance springs and streamflow.

Another category includes technofixes that are debatable in terms of their social and environmental impact and how the costs, benefits and risks are distributed. These include storage, desalination, treated wastewater reuse (TWWR), drip irrigation, off-grid access to water, and the private financing of water supply and sanitation services, as well as a sub-category that seeks to modify nature (river restoration, weir removal, E-flows). Institutional mechanisms have also been examined, such as privatisation, river basin organisations (RBOs), water user associations (WUAs) and legal rights for rivers, as well as the dominant meta-narratives of integration and ecological modernisation.

The following section highlights the most significant commonalities between these categories.

22.2 Understanding Best Practices

22.2.1 The 'Industry' and Sociotechnical Networks Behind Global Best Practices

A good idea needs friends. The literature on discourse coalitions (Hajer, 1995), sociotechnical networks (Moss, 2009; Venot & Veldwisch, 2017), development policy (Mosse, 2004) and actor-network theory (Akrich et al., 1988a, b; Latour, 1987) all converge to show how a particular idea, truth, innovation, model or narrative is socially constructed by a network of actors with common interests or ideologies. A best practice is generated and institutionalised by, and will then fuel, an industry or 'ecosystem' of actors with a stake in its promotion.

Acquiring the status of a best practice involves epistemic communities (Haas, 1992) who may receive substantial sums to support their research. This represents welcome political support for the development of scientific evidence to operationalise the best practice but also means the actors involved (researchers, universities, private firms and later donors and commercial interests) have a vested interest in packaging their result such that it (1) maintains the state's belief in that particular solution and (2) justifies investing further public funds to extend the programme. This intersection of multiple interests was evident in many of the best practices investigated.

Networks of interests have promoted practices such as WUAs, E-flows, nature-based solutions and payments for environmental services (PES), often involving key intergovernmental organisations. WUAs, for example, acquired prominence when the World Bank's Institute for Economic Development launched the International Network on Participatory Irrigation Management (INPIM). This brought together development banks and organisations, research institutes, the UN's Food and Agriculture Organization (FAO), the International Water Management Institute (IWMI) and the International Food Policy Research Institute (IFPRI) (Chapter 17). In the case of PES, international environmental NGOs spearheaded on-the-ground implementation through partnerships with governments, corporations and local communities, the Global Environmental Facility and development banks having provided funds. Research centres contributed analyses and reports of successful cases, framing it as a promising solution. 'Collectively, these actors form a robust, transnational epistemic community positioning PES as a mainstream conservation strategy, particularly for water-resource management' (Chapter 10; see also Rodríguez-de-Francisco & Boelens, 2015). The way these networks are formed, and the backing they receive, tends to value certain knowledge or disciplines while marginalising others. This creates a risk of 'epistemic capture', as seen with 'green infrastructure' approaches like 'Room for the River' or 'Sponge Cities', where engineers hold much more influence than social sciences, not to mention non-experts (Chapter 11).

With other best practices, the clout of industrial and commercial partners grows as they are institutionalised. River restoration, for example, became a flourishing market. In the US, no-net-biodiversity-loss policies have encouraged expensive mitigation schemes, and consultancy and public works companies have specialised in this area (Chapter 14). Water privatisation was promoted by actors such as the World Bank, multilateral development agencies and mainstream think tanks, notwithstanding concerned private business (Chapter 19; see also Goldman, 2007). The same actors push for private banks to help finance the water sector, having introduced the idea of PPP as a means of easing their participation through public subsidies (Chapter 18). Off-grid solutions are promoted not only by informal local entrepreneurs but also by NGOs and large corporations presenting marketing narratives at World Water Forums (Chapter 20). Some countries, including India, have seen an entire industry develop around water harvesting, with a mushrooming of companies offering consultancy services on its use in urban areas and making tall claims about the benefits (Chapter 6).[8]

Due to its capital intensity, water storage has long been the textbook example of powerful lobbies uniting around the promotion of a solution. The 'dam lobby' is commonly recognised in Iran, Turkey, Morocco and Brazil, among others (Atkins, 2017; McCully, 2001; Khagram, 2004). Described as 'iron networks' in the US, there is a convergence of interest between hydraulic bureaucracies, civil engineering contractors and politicians (see Molle et al., 2009; McCool, 1994). In recent years, the focus of capital-intensive investment has shifted towards wastewater treatment and desalination plants, backed by what Williams (Chapter 4) calls 'global assemblages'. In the case of desalination, these include government officials whose ties to global corporations, industrial lobbies and scientific and research networks encourage them to spread a narrative promising virtually unlimited water supply.[9]

The drip irrigation assemblage similarly unites industry actors, local implementers, epistemic communities, donors and NGOs (Chapter 2; Venot, 2016; Pérez-Blanco et al., 2021). The existence of economic actors in this industry surfaced, for example, in Morocco in 2023 and 2024, when Morocco's Sprinkler and Drip Irrigation Association (AMIAG) protested changes to the conditionalities for subsidies and loan interests for irrigation materials, as well as delays in government funding that they claimed made investment harder.[10]

Predictably, when challenged by new knowledge or contradictions, these webs of vested interests resist, while challengers find themselves self-censoring.[11] In the 2010s, an FAO report on micro-irrigation and water savings aroused a stiff reaction from the Australian agricultural business that felt threatened by its conclusions (which were eventually resolved when reference to Australia was dropped from the report). There was a similar challenge from the World Bank in India, which was in the process of rolling out subsidies for groundwater irrigation. As a practitioner from a Middle Eastern country explained:

> narratives [about modern irrigation] support those politicians and interest groups in favour of expanding pressurised irrigation. [...] Unsurprisingly as there are large sums of money involved, many of them refuse to allow a critical scientific approach to get at the core of policymaking.[12]

Finally, proponents of certain best practices may simply be moved by the opportunity these offer to distort the mass balance and obscure water pathways. When seeking to justify profitable projects, for example, it is generally beneficial to artificially inflate the level of water availability. Whether rooted in incompetence, a lack of data, or an intent to stand by (optimistic) official water balances, countless basin masterplans persist in such self-serving 'oversights': the double-accounting that often goes hand in hand with the principle that pumping as much as the recharge is safe (see Chapter 7), or the failure to consider what happens to return flows and their prior-appropriators when shifting to drip irrigation (Chapter 2), whether treated wastewater had already been used (before it came to be treated) (Chapter 3) or how water harvesting deprives downstream users of water (Chapter 6). They try to avoid the sobering fact that there is no 'new' water and often significantly less than initially assumed.

22.2.2 Ideological Underpinnings

The emergence and endurance of best practices also appeared to be shaped by globally dominant political ideas and the ideological biases of the epistemic communities and networks that promote them. Water policy is often based on 'reconciliation' – the idea that competing human uses of water can be harmonised with the preservation of the ecosystem through technological innovation, market mechanisms, and improved governance. This principle has its ideological roots in approaches that are compatible with capitalist economic systems, such as sustainable development and Integrated Water Resources Management (IWRM), or even aligned with neoliberal and neo-managerial doctrines, like ecological modernisation and green growth. As a result, water policy and management solutions – such as drip irrigation, smart urban water management, desalination, TWWR and water rights markets – are often presented and legitimised as fostering economic growth and efficiency or creating new markets (see Chapter 15).

To illustrate, PES was the brainchild of resource economists, but critics see it as commodifying or 'neoliberalising' in nature, where the logic of the market overrides the principle of moral/collective stewardship, reinforcing inequality and privatising environmental governance (Chapter 10; McAfee & Shapiro, 2010). The belief that market mechanisms and the private sector are more efficient, innovative, and reliable than state institutions is most visible in the wave of water service privatisation that emerged around the turn of the twenty-first century, with its ultimate realisation in the UK (Chapter 19). The myth that private finance would solve the world's water and sanitation crises, while driven by clear ideological commitment, obscures the fact that, for a variety of reasons, the majority of the investment in this sector has come from state actors (Chapter 18). Likewise, the neoliberal paradigm inspired the transfer of the management of irrigation systems and drinking water facilities from the state to various forms of water user organisations and the establishment of WUAs as a global best practice (Chapter 17).

The history of the idea of irrigation water pricing also reveals its ideological origins (Chapter 9; Molle & Berkoff, 2007). The World Bank's Water Resource Management Policy Paper of 1993 observed that 'waste and inefficiencies have resulted from the frequent failure to use prices and other instruments to manage demand and guide allocation' and began a decade of enthusiasm and missionary zeal for economic tools in general and water pricing in particular. Neoclassical resource economists fuelled a massive upsurge in academic literature, conferences, and even mainstream media articles on the subject. Water was presented as any other good to which the textbook supply-and-demand approach could be applied to optimise management.[13] The acme of the promotion of water pricing coincided with the European Water Framework Directive (2000) and its concept of 'full cost recovery', whose conspicuous elusiveness 25 years later should certainly be a matter of reflexion.

Several best practices are rooted in the ideology of renaturation or 'rewilding' (Barraud, 2025). River restoration, Room for the River, and weir removal, for example, share a view of nature without human intervention (Chapter 12). This is pitted against recent currents in critical physical geography that acknowledge the hybrid nature of managed river environments (Dufour et al., 2017) and the value of the long-term interaction between natural processes and human activity (Lespez & Dufour, 2021). Also bent on 'restoring' healthy aquatic ecosystems, E-flows tend to frame nature as distinct, contained, and manageable. Alexandra (Chapter 13) expresses the concern that scientifically and technically orientated E-flow methods reproduce their normative foundations by reinforcing particular 'assumptions, perspectives, relationships and norms about society, water and rivers and their interrelationships' (see also Alexandra et al., 2023). A similar critique can be levelled at PES, where science is enlisted to measure and commensurate ecological processes and determine values (Chapter 10), and at river restoration, so often based on modernist dreams of control that do not fit with the variability and interconnectedness of ecosystems. These paradigms 'mostly benefitted cities, capitalists and global trade, resulting in detrimental environmental impacts for marginalised populations and future generations, not to mention the rivers themselves' (Chapter 14).

Inspired by environmentalism, the ideology that 'more trees are always better' permeated international NGOs such as the IUCN and WWF until at least the mid-2000s. The European Environment Agency also endorsed unscientific beliefs about trees mitigating drought (Chapter 8). Yet, as Calder (2007) stressed, the water and ecosystem benefits from forests should not obfuscate the fact that trees deplete water.

Technoscience is a major imaginary in many of the best practices centred on technology and infrastructure. For Williams (Chapter 4), 'the utopianism associated with desalination was closely tied to notions of modernity, technological progress and domination of nature' (see also Chapter 3; O'Neill & Boyer, 2023). Such notions are also at the core of dam building as symbols of modern water control (Flaminio, 2021). The belief in technology as the ultimate problem-solver also lies behind TWWR and drip irrigation and still seduces water professionals globally.[14] In the case of E-flows,

> logics and paradigms [which] separate humans and nature also rely on increasingly professionalised agencies and their techno-scientific experts to define what is desirable in nature, what nature needs and what flows must be provided to satisfy these needs. There are risks that a reliance on experts leads to the establishment of technocratic elites, who determine the needs of the modern deity – the idealised and separate River-Nature.
>
> (Chapter 13)

Some best practices have been promoted by extolling traditional knowledge and local wisdom as opposed to imposing external blueprints – water harvesting being a case in point. Since it was embraced by the powerful Indian NGO Centre for Science and Environment, Agarwal and Narain's *Dying Wisdom: The Rise and Fall of Traditional Water Harvesting Systems in India* (1997) boosted the morale of civil society organisations seeking alternatives to large-scale, centralised solutions (Chapter 6). Scholars like Denison and Wotshela (2009) emphasised that 'Indigenous rainwater harvesting and conservation practices are the product of accumulated knowledge, practices and traditions which have evolved over many generations of experimentation and adaptation. These practices thus have *inherent sustainability*' (emphasis added). Their promotion as traditional, low-cost, environmentally sound, and socially acceptable may explain the degree of 'ideological blindness' demonstrated by overenthusiastic NGOs and donor agencies (Chapter 6).

22.2.3 Simplifying Complexity

Narratives, or shorter storylines, always spearhead the development and dissemination of best practices. On one hand, they foreground a simple causality that resonates with human experience and seems uncontroversial. On the other hand, they suggest that complexity can be dissolved within a best practice that has seemingly been validated by experts at the global level.

Policy narratives are causal stories that underwrite and stabilise the assumptions used in policymaking where unknowns persist (Roe, 1994). They organise the world by creating order within disorder (Hajer & Laws, 2006), condensing a vast quantity of information and connecting it to normative conceptions – sometimes directly connected to the ideologies mentioned in the previous section – to lend it meaning. Narratives need to garner support to win out over competing causal stories. Thus, what we hold as truths, like a good practice or a policy success, are first stories that have been consistently presented as plausible by 'influential interpreters' (Fischer, 2003; see also Rap, 2006, and Mosse, 2004). Once they have become dominant or even hegemonic, in the ways outlined in the first section of each chapter, their narratives and storylines empower certain actors, redirect collective attitudes, and shun any conflicting viewpoints as irrelevant or strange.

The political aspect of narratives stems from their selection of specific elements of reality – what they say and what they leave out. The best practices examined here are full of these 'unspoken facts' (Radaelli, 2000), and one of the main objectives of this book was to highlight and interpret these 'silences' – to make them speak. While essential to the coherence and very existence of the narratives underlying best practices, the silences raise fundamental questions. We were concerned with what is lost or rendered invisible when actors seek to simplify complexity. This simplification leads to solutions being reduced to their most positive or indisputable aspects.

PES is said to offer a win-win solution that, by definition, should please everyone. Agricultural water pricing is based on the familiar trope that 'what is cheap is bound to be wasted'. Drip irrigation is a compelling answer to the evidence that 'half of the water that is diverted never reaches the plants'. Treating and reusing wastewater rather than dumping polluted water into the sea or streams sounds like a self-evident 'no-regret action'. Likewise, removing obstacles to ease fish and sediment migration or infrastructure that straitjackets rivers, harvesting water rather than letting it 'go to waste in the sea', reconciling all water uses while meeting environmental needs, and protecting rivers by granting them legal status all sound unarguably sensible.

In that sense, best practices are akin to, or inspired by, 'nirvana concepts' (Molle, 2008), which promote and promise lofty goals such as sustainable development, integrated management, good governance, water security or green growth, among others (Chapter 15). These concepts or buzzwords are vague and aspirational and thus unlikely to arouse much opposition. (Who is against integrated management?) Best practices are seen as contributing to a superior, collective, if not moral, good (Venot & Veldwisch, 2017), whether through enhanced efficiency (drip irrigation, water pricing or privatisation), water security (storage, desalination), environmental health (PES, E-flows or legal rights for rivers), river wilding (restoration or weir removal), integration and optimisation (ecological modernisation), democratisation or equity (WUAs and RBOs), nature enhancement (reforestation to increase river flows), sustainability (pumping less than aquifer recharge), the circular economy (TWWR) or autonomy and flexibility (off-grid options).

As noted in the introduction, policy solutions are frequently backed by models or success stories. While there are certainly lessons to be learned from specific 'models', examples are used to instantiate possibility and magnify potential benefits. From PES in the Catskill watershed supplying water to New York City to nature-based flood-control solutions in the Netherlands, from high-tech desalination and urban water supply in Singapore to drip irrigation in Israel, from wastewater reuse in Namibia to legal rights for rivers in New Zealand and Ecuador, from water markets in Australia to irrigation management transfer in Mexico or RBOs in France and Spain, there are numerous models that seem to attest that – with the appropriate quantum of determination, funding and technology – 'it works'. So why can it not be replicated here?

Another strategy for evading complexity involves deferring governance specifics to vague references such as 'regulations' or policy 'frameworks' that are deemed necessary, yet without specifying how they will be implemented or whether they are even feasible. This is accompanied by 'wishful disclaimers'. For example, to mitigate the rebound effects of a transition to drip irrigation, the standard recommendation is to introduce regulations limiting the expansion of irrigation (Chapter 2). Similarly, solar irrigation projects are supposed to be accompanied by regulations capping groundwater withdrawals – yet such measures are rarely detailed or enforced. Another illustration is the 'multiple barrier' concept applied to TWWR (Chapter 3). To increase the likelihood that the quality of the water permits its reuse, the World Health Organization came up with the 'multi-barrier' approach, whereby a series of successive and independent layers of protection can be designed to improve the quality of water before it is used. However, this complex and ad hoc concept is clearly incompatible with the monitoring and enforcement capacities of most countries.

22.2.4 Obscuring Political Issues and Trade-Offs

Resting on radical simplifications of complexity and propelled by simple storylines, best practices are often promoted as uncontroversial 'no-regret solutions'. Yet, by addressing 'wicked' problems, they belong in the political realm of power dynamics and diverging worldviews. Nevertheless, these trade-offs are often obfuscated by (1) rendering problems technical and thus seemingly objective, and (2) painting them as conducive to win-win agreements where there is no apparent losing side.

The framing of a decidedly political problem (i.e., with ramifications for society as a whole) as a purely technical or economic issue is a well-known strategy of project planners (Li, 2007) and policymakers (Hay, 2014). This 'rendering technical' (Li, 2007) is manifest in technofixes such as TWWR. Conceived on two levels, it takes an industrial process to treat wastewater to a certain quality and health norms to ensure the suitability of its reuse, leaving little room for the politics of water reallocation or considerations of costs and risk. Drip irrigation is also a technological mirage, where an alchemy of pressure, pipes, filters, drippers, and sensors is supposed to deliver fertilised water for 'precision agriculture'. There is scant attention to the biofilm, algae, sediment, sun-wearing, clogging, leaks, uneven pressure, etc. that will likely surface, but also limited acknowledgement of what it entails politically in terms of increased consumption, water reallocation, or subsidies. Likewise, water harvesting, storage, and desalination are discussed primarily in technical terms, focusing on the bounty of augmented supply.

This technical and apparently objective approach to best practices extends to what often amounts to social engineering. A renewed faith in collective action arose from neo-institutional theory and, in line with Ostrom (1990), the idea of 'crafting institutions for self-governance'. This inspired Community-Based Management, Participatory Irrigation Management, and the idea that WUAs would harness people's rational choice to act collectively for the common good. Design principles and their functionalist approach to collective action fuelled donors' 'belief that "training" can constitute user groups, independent from politics, community divisions, power imbalances, elite capture and economic disparities, [which] is rather naïve' (Chapter 17).

A similar functionalist approach is at work with RBOs which are usually seen as an exercise in distributing roles and resources rather than managing or avoiding bureaucratic conflict with and among existing actors (Chapter 16). The privatisation of water services is also supposed to be optimised by the regulator to ensure the just distribution of costs and benefits, as if a regulatory function were not highly political[15] (Chapter 19). Even the

> legal recognition of river rights and personhood is also merely the beginning. Policy makers and local communities should be aware there will be a political struggle ahead to actively represent the voice of the river before public authorities and society at large, which will be time and resource-consuming.
>
> (Chapter 21)

Best practices are also the domain of win-win solutions, perhaps the epitome of a neutral, benevolent, and optimised approach to what is in reality a thorny collective/political intervention. This approach is at the very heart of the ubiquitous global meta-discourse on ecological modernisation (Chapter 15). In the Netherlands the 'Room for the River' policy fostered narratives of consensus and participation, where politics were dissolved into well-choregraphed stakeholder meetings ('no room for conflict', according to Roth et al., 2021). Yet, these projects 'mean water literally enters the lifeworlds of citizens and economic activities' and 'almost inevitably require sacrifice in terms of local interests, potentially involving high economic and other costs' (Chapter 11). Seawater desalination promises to be no less than 'a win-win fix to our water cycle' (Pistocchi et al., 2020; Chapter 4). Water pricing could 'realize a win-win strategy [...] if several issues are properly resolved' (Wang et al., 2016). TWWR cleans the environment and makes more water available for human use (Chapter 3); drip irrigation increases yield and revenue while preventing 'losses' or 'wastage'; even the privatisation of water services may qualify as a win-win[16] (Goldman, 2007). Likewise, the attractiveness of PES lies in its appearance as an unconfrontational approach to conservation, where values are commensurated through market mechanisms.

Reducing water policy to a question of 'common sense', a functional or a win-win issue, places it outside politics. Politics presuppose a confrontation of visions that best practices tend to conceal, reducing the expression of alternatives and always demanding 'more of the same' (Remling, 2018). River-restoration projects tend to address the symptoms of degradation rather than its causes, as it is difficult to tackle broader systemic issues related to industrial and agricultural activities, urbanisation, and deforestation. 'But without addressing these root causes even the best restoration efforts are likely to be superficial and temporary' (Chapter 14). More broadly, addressing root causes and considering alternatives in water policy requires in-depth political debate and acknowledging trade-offs. Hence, the crucial importance of 'a re-politicisation of water and its management, while ensuring that processes exist to channel disputes, allowing conflict to be managed effectively and collective agreement to be reached' (Chapter 15).

Each chapter in this book focuses on those complexities identified in the literature and in practice that are not always visible at first sight. The point here is the contrast between the narrow promised 'delivery point' and the complexity of the trade-offs that lurk behind, only to appear when the focus is broadened. TWWR, for example, is 'a good idea' on the face of it, but when zooming out from the processing plant itself, the water it produces recombines, and its quality becomes unstable. It defies quality standards, requires costly and continuous monitoring, its price and geographical location may limit its market, and its allocation impacts prior-appropriators or river low-flows (Chapter 3). Likewise, a transition to drip irrigation may seem promising from the point of view of the adopter, but when the behavioural responses of farmers and the pathways of water at larger scales are considered, expectations can be reversed (Chapter 2). Water harvesting, too, emphasises direct local benefits, but zooming out reveals the trade-offs and impacts at the macro level (Chapter 6). Finally, pumping to the point of aquifer recharge is a clear win for abstractors, but zooming out in space and time reveals a host of induced negative externalities that often dwarf those benefits (Chapter 7).

The best practices based on the transformation of nature (river restoration, weir removal, Room for the River, planting trees, or PES) rarely escape the 'surprises' that ripple across ecological systems. Here, too, the primary objective overshadows the undesirable results that become clear only when considering larger time and spatial scales. Hydrological and biochemical processes mediated by heterogeneous precipitation patterns, soil types, vegetation, topography, geology, and water-body interconnections, as well as farming and land-management practices and strategies, make it hard to anticipate, measure, and commensurate the services that are at the heart of the PES win-win promise (see Chapter 10). A weir's removal may jeopardise groundwater recharge with a dramatic local impact (Chapter 12). Water, sediment, and biota are dynamic and in continuous re-composition in space and time, in terms of quantity but also quality. They redistribute the costs and benefits associated with them (Molle, 2009b). This ultimately political complexity contrasts with the conviction that comes with the rollout of best practices.

22.2.5 Success Is Hard to Define or Evidence

While this book has looked at the many reasons best practices often fail to fulfil their promises, the meaning of this should be qualified. First, the very nature of wicked problems makes it hard to trace what exactly has changed, whether it can be attributed to the intervention, and whether it constitutes success. Second, as alluded to in the introduction, declaring success is very much a social performance in furthering a plausible story. Consequently, what constitutes success or failure partly depends on the subjectivity of those involved in the performance and also how the stakeholders and public see it ('the eye of the beholder') (Bovens, 2018; McConnell, 2011; Dunlop, 2017). Third, the rationale behind best practices will often give way to intense institutional bricolage: the context – particularly the agency of local and other actors – will shape the implementation process in ways that may diverge from the initial purpose.

Wicked problems are never really solved (Rittel & Webber, 1973). Although in some cases there is a tangible sign of implementation and transformation – the desalination plant is working, the reservoir has been constructed, the weir removed or trees planted – many other, slower or faster, changes are underway concurrently, whether in the social or the ecological system. Abstracting as much groundwater as the aquifer recharge may be acceptable in the medium term, as the stock is gradually depleted, but not longer, when drainage flows are impacted (Chapter 7).

As Bouleau and Lave note, 'Relatively few projects are evaluated by comparing pre- and post-restoration states (though the number is increasing). Of the projects that are evaluated, many show little to no improvement'. Besides, 'It is surprisingly difficult to define whether a restoration project is in fact successful, even if it is carefully monitored', either because there is no single government standard (as in the United States) or because any standard contains ambiguities that can be interpreted in different ways (as with the European Water Framework Directive) (Chapter 14). Generally speaking, ex post evaluations long enough after a project to give a clear assessment of the results are scarce. Success is often declared while resources (money, staff, political support) are still being poured into the implementation. Assessments tend to be superficial, while quality and social and ecological processes are harder to measure, notwithstanding the fact that there is often no willingness to do so.

WUAs, for example, often exist only on paper (Chapter 17). RBOs may have been established, and offices and staff assigned, but these often lack teeth. How much 'integration', as opposed to internal turf battles, they have produced is unclear (Chapter 16). Likewise, 'River persons and river rights attract attention, but there is a genuine risk that this leads to enacting environmental "law-for-show". That is, law with so many hurdles to deliver substantive action that its implementation becomes a performance' (Chapter 21). Meanwhile, some best practices are perceived as inseparable from a model of good governance, and their mere existence, regardless of their actual impact, is seen as a mark of success, as in the case of WUAs and RBOs. Hence, success stories can have a performative effect, becoming self-fulfilling prophecies by attracting political backing and funding (van Assche et al., 2012; Barone, 2018).

Having become hegemonic, an idea cannot fail. If it does, its supporters will deny it was fundamentally flawed or unsuitable and instead blame some unexpected nuisance. They will say 'politics', a 'lack of political will' or corruption got in the way, or that the implementation was inadequate, i.e. the real world refused to conform to theory (Akrich et al., 1988a; Long & van der Ploeg, 1989). This 'failing forward' syndrome means failure is taken as an invitation to double down efforts to roll the policy out, as with neoliberal reforms (Crouch, 2011), carbon and biodiversity markets (Fletcher, 2023), or agricultural water pricing (Molle & Berkoff, 2007). Likewise, private investment in water supply and sanitation remains minimal, yet 'the majority of policymaking related to financing WSS continues to focus on the private sector' (Chapter 18), the World Bank insisting it is 'essential to mobilise up-front financing from private or commercial sources' (Leigland et al., 2016, 1).

While their success is often measured against the original objective, the implementation of best practices is mediated by the context, the diverse interest groups mentioned earlier, and the response of the natural environment itself. It is also shaped by the capacity of local actors to reformulate, adapt, or circumvent the interventions through technical, social, or institutional 'bricolage' (Cleaver, 2017; Mayaux et al., 2023). This is particularly apparent with best practices such as WUAs, RBOs[17] or off-grid services.

Finally, when actively promoted in global water policy, best practices can generate opposition, as illustrated by privatisation (Chapter 19) and storage (Chapter 5; Biswas, 2004; Yong & Grundy-Warr, 2012). Indeed, some countries are now reversing course, removing their dams or remunicipalising their water services.[18] By transitioning from professional circles to social movements, certain best practices may find themselves becoming 'worst practices' or counter-models.

22.3 A Constructivist Approach to Best Practices: Avenues for Further Research

Best practices have been shown to result from processes aiming to stabilise a particular interpretation of a (water-related) problem and promote a specific policy solution or governance model. Along with the narratives that sustain them, they thus tell us a lot about how we handle complexity and reveal the beliefs and interests that produce them. They also demonstrate how knowledge is generated, legitimised, and made mainstream. While this is all squarely political, it was important to also acknowledge materiality by examining the concrete, sometimes unexpected, effects best practices have on water resources and ecosystems, as well as the materiality of water itself – as substance and flow – capable of resisting or disrupting social arrangements.

While the ambition here was not to arrive at an agenda for research on best practices, some of the issues that were found relevant warrant further exploration. The following selection is not suggested to be exhaustive.

22.3.1 Getting It Right with Water

The best practices reviewed in this book have all generated countless studies on their application 'in the field'. Yet, while debunking misconceptions or 'models' is not solely a question of revealing their shortcomings or the lack of communication (cf. the 'zombie ideas' mentioned above), there remains scope for more elaborate substantive studies. These should span social and spatial scales, foreground the mercilessness of the mass balance, and emphasise political trade-offs rather than simplistic and elusive win-wins.

Authoritative literature on the pitfalls or limitations of best practices certainly has a role in shifting or undermining certain faulty arguments. Although it is often unclear why one idea fades away, or a counter-narrative comes to prevail (see below), a tireless focus on context and complexity in the analysis cannot hurt.

22.3.2 The Social Life of Global Ideas

The formation and circulation of ideas and policy are the subject of much scholarship in the political sciences. The literature on policy transfer (e.g. Dolowitz & Marsh, 1996; Stone 2012; Evans, 2013), in particular, has inspired work in the water sector (Mukhtarov, 2014, 2022; Rap, 2006) on topics such as the Europeanisation of policy (Thiel, 2009; Demirbilek et al., 2020). While several authors have addressed the genealogy of concepts like IWRM (Kramer & Pahl-Wostl, 2014; Allouche, 2016), river basin management (Molle, 2009a), water security (Cook & Bakker, 2012) and agricultural water pricing (Molle & Berkoff, 2007), there is room for further interrogation into the life of the concepts[19] and ideas dominating the water sector. The first sections of the chapters gave a flavour of what those stories could include.

It is also apparent that key players in the formation and dissemination of best practices globally have largely escaped scrutiny. While the role of the World Bank in knowledge creation has been extensively examined (see e.g., Goldman, 2001, 2007), and while there are examples of studies on OECD (Mukhtarov et al., 2026), global water charities (Menga, 2025), World Water Forums (Ulibarri, 2011; Molle & Blanchard, 2022; Silva et al., 2024), or UN-Water (Baumgartner & Pahl-Wostl, 2013), there is a dearth of studies on corporations, international NGOs, think tanks, or international consultants in the water sector (see Stone et al., 2021 on this blind spot).

Greater focus in general is needed on the role of transnational networks in the production and propagation of ideas, concepts, and knowledge in the policy process that they seek to influence.[20] The literature offers a variety of network concepts, including 'epistemic community', 'policy community', 'transnational advocacy network' (Stone, 2002), transnational 'discourse coalitions' (Hajer, 1995) and 'global public policy network', that tend to come under the umbrella of 'global knowledge networks' (GKNs) (Stone, 2004). Knowledge actors typically include development banks or agencies, foundations, think tanks, experts and research institutions, professional associations, UN bodies, NGOs, and even corporations, consultancy firms, and pressure groups/lobbies. The boundaries of a particular GKN, and between the different types, are often fuzzy, and GKNs may overlap or intersect (Molle & Blanchard, 2022). Attention is also needed on the proliferation of assemblages dubbed 'alliances', 'partnerships', 'centres', 'consortia' and 'initiatives' with hybrid membership: The Resilience Shift, the Global Center on Adaptation (GCA), Sanitation and Water for All (SWA), the Alliance for Water Stewardship, etc.

There are also cases of the state as a key actor in global dynamics, following a period of multilateralism from the 1980s to the 1990s. Exemplified by Singapore and the Netherlands, some states 'brand themselves as centres of excellence in water engineering, management and governance and, by implication, as the natural choice of future clients with water-related problems' (Mukhtarov, 2022). The emergence of 'water hubs' (Mukhtarov, 2022) and the marketing of specific expertise (e.g. Hasan et al., 2022) warrants deeper analysis.

'Idea travel' and policy formation could also be studied, for example, through the key actors' biographical trajectories (Poupeau et al., 2020), text and discourse analysis (e.g. Mukhtarov et al., 2026), as well as the venues and megaconferences where knowledge is exchanged (Silva et al., 2024).

22.3.3 Bricolage and the 'Practice of Best Practice'

As alluded to earlier, the territorialisation that operates around the (often top-down) implementation of a best practice suggests that local agency should not be overlooked, nor social and policy processes oversimplified. In other words, while best practices are more ubiquitous and performative than their proponents tend to acknowledge, their hegemony is not as absolute as water activists may have it.[21] This issue has common ground with the diffusion and translation of innovations or policy (Mukhtarov, 2022), micro-resistance (Scott, 1985; Boelens, 2015; Davidson-Harden et al., 2007), and bricolage (Cleaver, 2017). These bodies of literature take us beyond basic, functionalist conceptions or social-engineering approaches.

Normative concepts and policy models are usually simplified and then recast and reshaped locally (Allouche, 2016; Mukhtarov, 2014; Molle, 2008). The bricolage observed around the implementation of projects and policies is often both technical and institutional (Mayaux et al., 2023). Studies may focus on technological adaptation, such as drip irrigation in Morocco (Benouniche et al., 2014), PES (Chai et al., 2021), communal irrigation (Mirhanoğlu et al., 2023; Cleaver, 2017) or on institutional adaptation, such as aquifer contracts in Morocco (del Vecchio, 2023), or WUAs in Nepal (Haapala et al., 2016).

While substantial scholarship exists on these issues, there is an ongoing need to document and understand how local actors interact with development agents, NGOs, donors or 'street-level bureaucrats' (Lundmark, 2020; Sevä & Jagers, 2013; Ricks, 2017) to reshape the ideas and best practices that are introduced on their behalf, or at least as a solution to their problems (Long & van der Ploeg, 1989). The cumulative body of research on this topic can shed light on the extent to which norms are bypassed, funding is instrumentalised, and the very content of best practices is hollowed out in the field. The classic *traduttore-traditore* (translator-traitor) syndrome tends to reinforce the consensual character of best practices: their supporters continue to benefit from their hegemony, while local actors and stakeholders steer them in whatever direction suits them best. Beyond this, analysing hybridisation, re-creation, and local re-problematisation is key in the study of 'the practice of best practices' (Papanastasiou, 2024).

22.3.4 Examining 'Failure Stories' (and How They Are Avoided)

Best practices are generated from policies that are said to have proven successful in particular cases. This is one of their main sources of credibility (Pal & Clark, 2013; Papanastasiou, 2024) and why the existence of exemplars, success stories, and models is so crucial – and salient in the literature.

Notwithstanding examples of explorations of systemic policy failure in certain domains, such as water, sanitation and hygiene (WASH) programmes (Barrington et al., 2025) or the privatisation of water services (Hall & Lobina, 2006; Chapter 19), research into water policy has tended to focus on specific, visible – or rather, perceived – instances of failure. Indeed, success and failure are not inherent properties of a policy but, rather, discursive labels applied by stakeholders and observers and negotiated through actor networks (Latour, 1987). Success is never guaranteed; it requires objectives and interests to be aligned. If network coordination fails, the project is likely to do the same, regardless of the technical quality or resources available (Latour, 1987, 1996; Mosse, 2004). And, like success, failure is socially constructed (Bovens, 2018). It can be minimised or amplified, if desired, by those politically opposed to the intervention (Dunlop, 2017). In the water sector, development agencies produce official narratives that attribute success to their model or best practice and failures to local actors or the population (Mosse, 2003). If the best practice was unsuccessful, it cannot be because it was a bad choice in the first place; it must have been badly carried out, or people did not understand or comply with it.

The concept of 'uncomfortable knowledge' (Rayner, 2012) explains how failure is systematically made invisible. Similarly, ignorance studies suggest there are various mechanisms that can be deployed to minimise, conceal, or dilute evidence of policies having proved ineffective. Organisations learn not to learn and actively manage inconvenient knowledge (Rayner, 2012; McGoey, 2012; see also Proctor & Schiebinger, 2008).[22] Decisionmakers, for their part, must navigate the persistent tension between policy learning from failure and blame-avoidance (Howlett, 2012).

Compared with success stories, failures are rarely shared (or studied). The reluctance of the actors or organisations involved to admit any incompetence is understandable. Their reputations legitimise their jobs and budgets. In the case of the wicked problems that lie at the intersection of the social and environmental realms, the causes of (what is labelled) failure are often multifactorial, emergent, and context-specific, making simple attribution difficult.

Case studies remain essential to understanding the mechanisms through which organisations 'succeed to fail', and failures are constructed or masked. Future research should prioritise more systematic documentation of water policy failure, using the four 'lenses' described in the introduction.

Comparative studies across different contexts could analyse how projects 'die quietly' and reveal patterns that define, hide or reframe failures as non-events or even successes.

22.3.5 Policy Feedback

Water policy, with its panoply of tools, or 'best practices', is not merely the result of political struggle; it also reshapes politics over time by generating *policy feedback* (Skocpol, 1992; Béland, 2010; Campbell, 2012). It shifts the political attitudes, behaviours and identities of administrative elites, interest groups, water users, and citizens, while also transforming resource distribution, institutional arrangements, and power relations. For example, the policies that saw the establishment of large-scale irrigation systems often reinforced social hierarchies and empowered influential agricultural coalitions capable of shaping subsequent political choices. These policies generated 'positive feedback' (Pierson, 1993) by reaffirming the dominance of certain coalitions and shaping long-term water management routines and expectations regarding state provision.

Policy feedback is not always stabilising. Negative feedback can arise if a policy generates a crisis of legitimacy or social conflict or is seen as ineffective. Urban water privatisation programmes in Latin America in the 1990s and 2000s saw policies designed to increase efficiency and investment produce widespread social opposition, protest and political mobilisation against the state or private operators (Bakker, 2010; Chapter 19). While the drinking water concession in La Paz-El Alto, Bolivia, was hailed, during its decade of operation (1997–2007), as a model of private management benefitting the poor, large-scale public protest broke out. In reality, the reform had inadvertently created its own opposition by unifying a previously disparate population of those with no connection to the water supply into a social group of 'poor', who were now entitled to be connected (Mayaux, 2019).

The decentralisation promoted by IWRM and international neoliberal programmes has generated more ambivalent policy feedback. On the one hand, it has enabled certain rural communities to gain institutional recognition and greater user participation. The introduction of local water management committees has not only changed institutional arrangements but also built new collective identities among users and a discourse on 'responsible communities'. These organisations have become political actors that claim rights, challenge reform (such as privatisation or water transfers) or, conversely, reinforce existing local elites. On the other hand, it has created new mechanisms of dependence at different levels and 'capture' by political power (Chapter 17). Until the 2000s in Ben Ali's Tunisia, for example, the Agricultural Development Groups (GDAs) were part of the move towards 'local governance', combining economic withdrawal by the state, public-private partnerships, local development, and user participation within a centralised political system. In reality, the GDAs increased the state's control in rural areas (Cannesse, 2009).

Further investigation into policy feedback from best practices in global water policy would lead to a deeper understanding not only of how these projects relate to politics, but also, far less common, how politics is affected by them.

22.4 Conclusion

This book has argued that best practices in the water sector are simultaneously part of the solution and part of the problem. Essentially, they emerge as simple, neutral and legible responses to complex challenges backed by compelling narratives, institutional authority and standardised technical packages.

An examination of the promises associated with the 20 examples analysed here revealed that they are frequently unfulfilled. This book has uncovered a confluence of conceptual and scientific shortcomings, oversimplifications of complexity, flawed causal assumptions, black-boxed uncertainties, perverse side effects, and implementation pitfalls – factors that, more often than not, undermine their enthusiastic adoption. Despite the universality inherent in best practices, the very nature of wicked problems suggests that the wheel must be pretty much reinvented each and every time. The exploration of the political-economic contexts and of the vested interests of the diverse actors involved in the promotion and dissemination of these solutions has also revealed the emergence of 'industries' spawned by the institutionalisation of best practices.

This investigation further highlighted the ideologies and cognitive biases that shape how best practices are produced, promoted, and framed as universal solutions. At its core, mainstream knowledge production is an act of boundary-drawing – delimiting what constitutes legitimate knowledge while marginalising or excluding alternative perspectives. This book will have fulfilled its purpose if it succeeds in disrupting the comfort and complacency that the 'global best practice' label often engenders.

Notes

1 As Sneddon (2015) also emphasises, 'Applied to water governance, hegemonic concepts are presented by their proponents so frequently and authoritatively that they exude an aura of unassailable objectivity in the quest for more effective water policies and management practices'.
2 It is debatable whether irrigating a field is that straightforward, given the diversity of techniques crafted and the *metis* displayed to adjust this practice to circumstances.
3 This plan was designed in response to the 2022–2023 drought in France: www.ecologie.gouv.fr/politiques-publiques/plan-daction-gestion-resiliente-concertee-leau

4 Pritchett et al. (2013) noted that while development discourse is now replete with aphorisms such as "one size doesn't fit all," "there are no silver bullets," "context matters", and while practitioners are acutely aware of context and cultural differences, this recognition has not dislodged conventional visions of 'accelerated modernisation through transplanted best practice'.

5 Other best practices or faddish solutions include smart water management, smart cities, sponge cities, solar pumping, nature-based solutions, water rights, water markets, block tariffs, full cost recovery, PPP, etc.

6 The European Commission (2025) recently extended recommendations on "guiding principles of water efficiency first".

7 Again, in terms of net water consumption, notwithstanding the other advantages of drip irrigation (see Chapter 2).

8 In Tunisia, it is supported by companies and decision-makers while the general public show limited interest (Le Goulven et al., 2009).

9 Hartley and Kuecker (2020) offer insight into how the industry around fashionable smart water management (SWM) produces a discourse that validates the solution it offers: 'the private sector produces and mainstreams science- and technology-backed initiatives around which the logic of naming and framing problems is designed to fit the solutions on offer. This process illustrates how capitalist logic induces the types of investments that perpetuate particular approaches to policy intervention'.

10 Created in 2004, AMIAG has around 400 members from an industry that includes over 1,700 firms. https://medias24.com/2023/01/22/des-subventions-a-lirrigation-bloquees-des-professionnels-demandent-leur-deblocage/

11 Economists or engineers from at least four countries known to the authors have admitted being persona non grata in certain circles or ministries due to their critical stance on the potential of drip irrigation to save water. A senior water expert in an agricultural ministry mentioned to one of the authors that if he were to express doubt at the mega 'irrigation modernisation package' pushed by the minister, he would be sacked – a textbook example of a 'hegemonic idea'.

12 Likewise, the summary of discussion (TOPIC 2) during the 16th World Water Congress 'Bridging Science and Policy' held in Cancun, Mexico, in May 2007 reported: 'Some debated on the fact that the knowledge provided by water accounting may not be in high demand by policy makers as [it is] potentially disturbing. For example, it often says there is no water precisely where the state (and others) want to make a project, or want to believe there is a technical solution (e.g. drip) to a social (sharing scarce resources) problem. A number of contributors to the discussion have noted the potential conflict between rational decisions about water use/allocation and the political economy in which those decisions occur'.

13 Hence the well-known aphorism, 'An economist might be described as someone who doesn't see anything special about water' (Tregarthen, 1983).

14 In an email to one of the authors, an official at an international organisation admitted: 'I can't move people away from hi-tech, primarily because it is so deeply in the collective water mind now as "the answer" […] but also because there is no easy or big-spending alternative I can provide. Depressing'.

15 The experience of the privatisation of water services in the UK is sobering. The promise that the regulator (Ofwat) would ensure fair pricing and transparency turned sour. The financial collapse of Thames Water resulted from information

asymmetries, delayed maintenance, catastrophic river water quality, a revolving door between the staff of the regulator and the water companies, and massive returns for shareholders (see Chapter 19).

16 As Goldman (2007) notes, regarding the World Business Council's 2001 'Water for the Poor' report, 'In short, everyone wins – the firms that join into partnerships with the global development community and governments to bring water to the poor, and those who are at the receiving end of the water pipeline. This is the same perspective the World Bank promotes in its professional training seminars and its policy work'.

17 Henriquez (2025), for example, shows how the best practices RBOs, participation, water rights and water pricing have been appropriated by water actors in the state of Ceará, Brazil.

18 See special issues of *Water Alternatives* on dam removal (www.water-alternatives.org/index.php/current-issue/1895-articles-toc/vol10/309-issue10-3) and remunicipalisation (www.water-alternatives.org/index.php/tp1-2/1895-vol10/309-issue10-3).

19 Following Voß and Simons (2014), the social life of policy instruments or tools 'consists of heterogeneous practices involved in the making of instruments: scientific theory building, data production and publishing, political issue framing, agenda setting, coalition building, business development, marketing and lobbying, management of innovation networks, professional organisation. Interdependences between these practices may become reflected in instrument constituencies and lead to the build-up of a collective interest in retaining and expanding the instrument'.

20 Goldman (2007) asserts that transnational policy networks 'have converged to create what they describe as a "global consensus" on water'.

21 Following Sneddon (2015): 'Our sense is that hegemonic concepts, no matter how potent they seem, are simultaneously more omnipresent and rigid than often presented by the architects of global water governance and more variegated *and* fungible than often presented by the water justice movement and the water activist community'.

22 As Orwell is often quoted, 'Within any important issue, there are always aspects no one wishes to discuss'.

References

Agarwal, A., & Narain, S. (1997). *Dying wisdom: Rise and fall of traditional water harvesting systems*. Centre for Science and Environment.

Akrich, M., Callon, M., & Latour, B. (1988a). À quoi tient le succès des innovations? 1: L'art de l'intéressement. *Gérer et Comprendre, Annales des Mines, 11*, 4–17.

Akrich, M., Callon, M., & Latour, B. (1988b). À quoi tient le succès des innovations? 2: Le choix des porte-parole. *Gérer et Comprendre, Annales des Mines, 12*, 14–29.

Alexandra, J., Rickards, L., & Pahl-Wostl, C. (2023). The logics and politics of environmental flows: A review. *Water Alternatives, 16*(2), 346–373.

Allouche, J. (2016). The birth and spread of IWRM – A case study of global policy diffusion and translation. *Water Alternatives, 9*(3), 412–433.

Atkins, E. (2017). Dammed and diversionary: The multi-dimensional framing of Brazil's Belo Monte dam. *Singapore Journal of Tropical Geography, 38*(3), 276–292.

Bakker, K. (2010). *Privatizing water: Governance failure and the world's urban water crisis*. Cornell University Press.

Barone, S. (2018). Building a narrative on environmental policy success: Reflections from a watershed management experience. *Critical Policy Studies, 12*(2), 135–148.

Barraud, R. (2025). The changing meaning of wild rivers: A review. *Water Alternatives, 18*(2), 181–213.

Barrington, D. J., Sindall, R. C., Chinyama, A., Morse, T., Sule, M. N., Beale, J., ..., & Evans, B. E. (2025). The persistence of failure in water, sanitation and hygiene programming: A qualitative study. *BMJ Global Health, 10*(2). https://doi:10.1136/bmjgh-2024-016354

Baumgartner, T., & Pahl-Wostl, C. (2013). UN–Water and its role in global water governance. *Ecology and Society, 18*(3). http://dx.doi.org/10.5751/ES-05564-180303

Beland, D. (2010). Reconsidering policy feedback: How policies affect politics. *Administration & Society, 42*(5), 568–590.

Benouniche, M., Kuper, M., Hammani, A., & Boesveld, H. (2014). Making the user visible: Analysing irrigation practices and farmers' logic to explain actual drip irrigation performance. *Irrigation Science, 32*(6), 405–420. https://doi.org/10.1007/s00271-014-0438-0

Biswas, A. K. (2004). Dams: Cornucopia or disaster? *International Journal of Water Resources Development, 20*(1), 3–14.

Boelens, R. (2015). *Water, power and identity: The cultural politics of water in the Andes*. Routledge.

Bovens, M. (2018). Revisiting the study of policy failures. In K. Oppermann & A. Spencer (Eds.), *Fiascos in public policy and foreign policy* (pp. 23–36). Routledge.

Calder, I. R. (2007). Forests and water—Ensuring forest benefits outweigh water costs. *Forest Ecology and Management, 251*, 110–120. https://doi.org/10.1016/j.foreco.2007.06.015

Campbell, A. L. (2012). Policy makes mass politics. *Annual Review of Political Science, 15*(1), 333–351.

Cannesse, A.-A. (2009). Gestion des ressources naturelles et système institutionnel de gouvernance en Tunisie. *Maghreb–Machrek, 202*, 49–64.

Chai, Y., Zhang, H., Luo, Y., Wang, Y., & Zeng, Y. (2021). Payments for ecosystem services programs, institutional bricolage, and common pool resource management: Evidence from village collective-managed irrigation systems in China. *Ecological Economics, 182*, 106906.

Cleaver, F. (2017). *Development through bricolage: Rethinking institutions for natural resource management*. Routledge.

Cohen, M., March, J., & Olsen, J. (1972). A garbage can model of organizational choice. *Administrative Science Quarterly, 17*(1), 1–25.

Cook, C., & Bakker, K. (2012). Water security: Debating an emerging paradigm. *Global Environmental Change, 22*(1), 94–102.

Crouch, C. (2011). *The strange non-death of neo-liberalism*. Polity.

Davidson-Harden, A., Naidoo, A., & Harden, A. (2007). The geopolitics of the water justice movement. *Peace Conflict and Development, 11*, 1–34.

Del Vecchio, K. (2023). Un nouvel instrument au service du statu quo: Comment des contrats de nappe renforce les équilibres nationaux de la politique marocaine de l'eau. *Critique Internationale, 98*, 73–93.

Demirbilek, B., Fitch-Roy, O., Benson, D., & Fairbrass, J. (2020). Going 'off script': The influence of instrument constituencies on the Europeanisation of Turkish water policy. *Water Alternatives, 13*(3), 691–708.

Denison, J., & Wotshela, L. (2009). *Indigenous water harvesting and conservation practices: Historical context, cases and implications* (Report No. TT 392/09). Water Research Commission.

Dolowitz, D., & Marsh, D. (1996). Who learns what from whom: A review of the policy transfer literature. *Political Studies, 44*(2), 343–357.

Dufour, S., Rollet, A. J., Chapuis, M., Provansal, M., & Capanni, R. (2017). On the political roles of freshwater science in studying dam and weir removal policies: A critical physical geography approach. *Water Alternatives, 10*(3), 853–869.

Dunlop, C. A. (2017). Policy learning and policy failure: Definitions, dimensions and intersections. *Policy & Politics, 45*(1), 3–18.

European Commission. (2025, June 4). *Commission recommendation of 4.6.2025 on guiding principles of water efficiency first*. European Commission.

Evans, M. (Ed.). (2013). *New directions in the study of policy transfer*. Routledge.

Fallon, A. L., Lankford, B. A., & Weston, D. (2021). Navigating wicked water governance in the "solutionscape" of science, policy, practice, and participation. *Ecology and Society, 26*(2), 37. https://doi.org/10.5751/ES-12504-260237

Fischer, F. (2003). *Reframing public policy: Discursive politics and deliberative practices*. Oxford University Press.

Flaminio, S. (2021). Modern and nonmodern waters: Sociotechnical controversies, successful anti-dam movements and water ontologies. *Water Alternatives, 14*(1), 204–227.

Fletcher, R. (2023). *Failing forward: The rise and fall of neoliberal conservation*. University of California Press.

Freeman, D. M. (2000). Wicked water problems: Sociology and local water organizations in addressing water resources policy. *Journal of the American Water Resources Association, 36*(3), 483–491.

Goldman, M. (2001). The birth of a discipline: Producing authoritative green knowledge, World Bank-style. *Ethnography, 2*(2), 191–217.

Goldman, M. (2007). How 'water for all!' policy became hegemonic: The power of the World Bank and its transnational policy networks. *Geoforum, 38*(5), 786–800.

Haapala, J., Rautanen, S. L., White, P., Keskinen, M., & Varis, O. (2016). Facilitating bricolage through more organic institutional designs? The case of water users' associations in rural Nepal. *International Journal of the Commons, 10*(2). https://doi.org/10.18352/ijc.688

Haas, P. M. (1992). Introduction: Epistemic communities and international policy coordination. *International Organization, 46*(1), 1–35.

Hajer, M. (1995). *The politics of environmental discourse: Ecological modernization and the policy process*. Oxford University Press.

Hajer, M. A., & Laws, D. (2006). Policy frame and discourse. In M. Moran, M. Rein, & R. E. Goodin (Eds.), *The Oxford handbook of public policy* (pp. 407–424). Oxford University Press.

Hall, D., & Lobina, E. (2006). Pipe dreams: The failure of the private sector to invest in water services in developing countries. *PSIRU Reports*.

Hartley, K., & Kuecker, G. (2020). The moral hazards of smart water management. *Water International, 45*(6), 693–701.

Hasan, S., Evers, J., & Zwarteveen, M. (2022). The work that goes into policy transfer: Making the Dutch delta approach travel. *Water Alternatives, 15*(1), 56–72.
Hay, C. (2014). Depoliticisation as process, governance as practice: What did the 'first wave' get wrong and do we need a 'second wave' to put it right? *Policy and Politics, 42*(2), 293–311.
Henriquez, D. M. (2025). *Water governance and the allocation of scarcity in Brazil: Revisiting Ceará's model water policy [PhD dissertation]*. University of Montpellier.
Howlett, M. (2012). The lessons of failure: Learning and blame avoidance in public policy-making. *International Political Science Review, 33*(5), 539–555.
Khagram, S. (2004). *Dams and development: Transnational struggles for water and power*. Cornell University Press.
Kramer, A., & Pahl-Wostl, C. (2014). The global policy network behind integrated water resources management: Is it an effective norm diffusor? *Ecology and Society, 19*(4). www.jstor.org/stable/26269658
Latour, B. (1987). *Science in action: How to follow scientists and engineers through society*. Harvard University Press.
Latour, B. (1996). *Aramis, or the love of technology*. Harvard University Press.
Le Goulven, P. L., Leduc, C., Batcha, M. S., & Poussin, J.-C. (2009). Sharing scarce resources in a Mediterranean river basin: Wadi Merguellil in Central Tunisia. In F. Molle & P. Webster (Eds.), *River basin trajectories: Societies, environments and development* (pp. 147–170). CABI Publishing.
Leigland, J., Trémolet, S., & Ikeda, J. (2016). *Achieving universal access to water and sanitation by 2030: The role of blended finance*. World Bank.
Lespez, L., & Dufour, S. (2021). Les hybrides, la géographie de la nature et de l'environnement. *Annales de Géographie, 737*(1), 58–85.
Li, T. M. (2007). *The will to improve: Governmentality, development, and the practice of politics*. Duke University Press.
Long, N., & van der Ploeg, J. D. (1989). Demythologizing planned intervention: An actor perspective. *Sociologia Ruralis, 29*(3/4), 226–249.
Lundmark, C. (2020). The connectivity dilemma in freshwater management: Exploring the role of street level bureaucrats in water governance. *Water Policy, 22*(6), 1067–1081.
Mayaux, P.-L. (2019). Quand l'action publique produit ses contestataires. Les effets politiques imprévus d'un partenariat public-privé en Bolivie. *Gouvernement et Action Publique, 8*(1), 87–111.
Mayaux, P.-L., Dajani, M., Cleaver, F., Naouri, M., Kuper, M., & Hartani, T. (2023). Explaining societal change through bricolage: Transformations in regimes of water governance. *Environment and Planning E: Nature and Space*, 6(4), 2654–2677.
McAfee, K., & Shapiro, E. N. (2010). Payments for ecosystem services in Mexico: Nature, neoliberalism, social movements, and the state. *Annals of the Association of American Geographers, 100*(3), 579–599.
McCool, D. (1994). *Command of the waters: Iron triangles, federal water development and Indian water*. The University of Arizona Press.
McConnell, A. (2011). Success? Failure? Something in-between? A framework for evaluating crisis management. *Policy and Society, 30*(2), 63–76.
McCully, P. (2001). *Silenced rivers: The ecology and politics of large dams*. Zed Books.
McGoey, L. (2012). Strategic unknowns: Towards a sociology of ignorance. *Economy and Society, 41*(1), 1–16. https://doi.org/10.1080/03085147.2011.637330

Meinzen-Dick, R. (2007). Beyond panaceas in water institutions. *Proceedings of the National Academy of Sciences, 104*(39), 15200–15205.
Menga, F. (2025). *Thirst: The global quest to solve the water crisis*. Verso Books.
Mirhanoğlu, A., Özerol, G., Hoogesteger, J., van den Broeck, P., & Loopmans, M. (2023). Socio-material bricolage: (Co)shaping of irrigation institutions and infrastructures. *International Journal of the Commons, 17*(1). https://doi.org/10.5334/ijc.1188
Molle, F. (2008). Nirvana concepts, narratives and policy models: Insights from the water sector. *Water Alternatives, 1*(1), 131–156.
Molle, F. (2009a). River-basin planning and management: The social life of a concept. *Geoforum, 40*(3), 484–494.
Molle, F. (2009b). Water, politics and river basin governance: Repoliticizing approaches to river basin management. *Water International, 34*(1), 62–70.
Molle, F., & Berkoff, J. (2007). Water pricing in irrigated agriculture: The lifetime of an idea. In F. Molle & J. Berkoff (Eds.), *Irrigation water pricing: The gap between theory and practice* (pp. 1–20). CABI.
Molle, F., & Blanchard, M. (2022). *World Water Forums, knowledge production and the imagined "global water community"* (G-EAU Working Paper/Rapport de Recherche No. 14). www.g-eau.fr/images/production/working-papers/WP14.pdf.
Molle, F., Mollinga, P. P., & Wester, P. (2009). Hydraulic bureaucracies and the hydraulic mission: Flows of water, flows of power. *Water Alternatives, 2*(3), 328–349.
Morozov, E. (2013). *To save everything, click here: Technology, solutionism and the urge to fix problems that don't exist*. Penguin.
Moss, T. (2009). Intermediaries and the governance of sociotechnical networks in transition. *Environment and Planning A, 41*(6), 1480–1495.
Mosse, D. (2003). *The rule of water: Statecraft, Ecology and collective action in South India*. Cambridge University Press.
Mosse, D. (2004). Is good policy unimplementable? Reflections on the ethnography of aid policy and practice. *Development and Change, 35*(4), 639–671.
Mukhtarov, F. (2014). Rethinking the travel of ideas: Policy translation in the water sector. *Policy & Politics, 42*(1), 71–88.
Mukhtarov, F. (2022). A review of water policies on the move: Diffusion, transfer, translation or branding? *Water Alternatives, 15*(2), 290–306.
Mukhtarov, F., Gasper, D., Farrelly, M., Lüken, M., & Moodley, K. (2026). OECD's methods of legitimation and self-authorisation in water governance. *Water Alternatives, 19*(1), 1–30.
O'Neill, B. F., & Boyer, A.-L. (2023). 'Locking in' desalination in the U.S.-Mexico borderlands: Path dependency, techno-optimism and climate adaptation. *Water Alternatives, 16*(2), 480–508.
Ostrom, E. (1990). *Governing the commons: The evolution of institutions for collective action*. Cambridge University Press.
Pal, L. A., & Clark, I. D. (2013). *Best practices in public management: A critical assessment* (Working Paper). Best Practices in Public Management Project.
Papanastasiou, N. (2024). The politics of generating best practice knowledge: Epistemic practice and rendering space technical in a European Commission working group on education policy. *Environment and Planning C: Politics and Space, 42*(2), 179–197.
Pérez-Blanco, C. D., Loch, A., Ward, F., Perry, C., & Adamson, D. (2021). Agricultural water saving through technologies: A zombie idea. *Environmental Research Letters, 11*(16), Article 114032. https://iopscience.iop.org/article/10.1088/1748-9326/ac2fe0

Peters, B. G., & Nagel, M. L. (2020). *Zombie ideas: Why failed policy ideas persist*. Cambridge University Press.
Pierson, P. (1993). When effect becomes cause: Policy feedback and political change. *World Politics, 45*(4), 595–628.
Pistocchi, A., Bleninger, T., Breyer, C., Caldera, U., Dorati, C., Ganora, D., ..., & Zaragoza, G. (2020). Can seawater desalination be a win-win fix to our water cycle? *Water Research, 182*, 115906.
Poupeau, F., O'Neill, B., Cortinas Muñoz, J., Coeurdray, M., & Benites-Gambirazio, E. (2020). *The field of water policy: Power and scarcity in the American Southwest*. Routledge.
Pritchett, L., Woolcock, M., & Andrews, M. (2013). Looking like a state: Techniques of persistent failure in state capability for implementation. *The Journal of Development Studies, 49*(1), 1–18.
Proctor, R., & Schiebinger, L. (Eds.). (2008). *Agnotology: The making and unmaking of ignorance*. Stanford University Press.
Radaelli, C. M. (2000). Logiques de pouvoir et récits dans les politiques publiques de l'Union européenne. *Revue Française de Science Politique, 50*(2), 255–275.
Rap, E. (2006). The success of a policy model: Irrigation management transfer in Mexico. *Journal of Development Studies, 42*(8), 1301–1324.
Rayner, S. (2012). Uncomfortable knowledge: The social construction of ignorance in environmental research and policy. *Economy and Society, 41*, 107–125.
Remling, E. (2018). Depoliticizing adaptation: A critical analysis of EU climate adaptation policy. *Environmental Politics, 27*(3), 477–497.
Ricks, J. I. (2017). Street-level bureaucrats and irrigation policy reform in Southeast Asia. *Asian Politics & Policy, 9*(2), 310–319.
Rittel, H. W., & Webber, M. M. (1973). Dilemmas in a general theory of planning. *Policy Sciences, 4*(2), 155–169.
Rodríguez-de-Francisco, J. C., & Boelens, R. (2015). Payment for environmental services: Mobilising an epistemic community to construct dominant policy. *Environmental Politics, 24*, 481–500.
Roe, E. (1994). *Narrative policy analysis: Theory and practice*. Duke University Press.
Roth, D., Warner, J., & Winnubst, M. (2021). Room for the river, no room for conflict: Narratives of participation, win-win, consensus, and co-creation in Dutch spatial flood risk management. In L. Cortesi & K. J. Joy (Eds.), *Split waters* (pp. 69–92). Routledge India.
Scott, J. C. (1985). *Weapons of the weak: Everyday forms of peasant resistance*. Yale University Press.
Scott, J. C. (1998). *Seeing like a state: How certain schemes to improve the human condition have failed*. Yale University Press.
Sevä, M., & Jagers, S. C. (2013). Inspecting environmental management from within: The role of street-level bureaucrats in environmental policy implementation. *Journal of Environmental Management, 128*, 1060–1070.
Silva, T. H. C., do Nascimento Silva, N. R. R., Jordão, L. R., & Chiziane, E. A. (2024). The lords of water: Discourses and actors of the World Water Forums between 1997 and 2024. *Revista de Gestão Social e Ambiental, 18*(12), 1–21.
Skocpol, T. (1992). *Protecting soldiers and mothers: The political origins of social policy in United States*. Harvard University Press.

Sneddon, C. (2015). Water, governance and hegemony. In L. M. Harris, J. A. Goldin, & C. Sneddon (Eds.), *Contemporary water governance in the Global South* (pp. 13–24). Routledge.
Stone, D. (2002). Introduction: Global knowledge and advocacy network. *Global Networks, 2*(1), 1–12.
Stone, D. (2004). Knowledge networks and global policy. In D. L. Stone & S. Maxwell (Eds.), *Global knowledge networks and international development* (pp. 109–125). Routledge.
Stone, D. (2012). Transfer and translation of policy. *Policy Studies, 33*(6), 483–499.
Stone, D., Pal, L., & de Oliveira, O. (2021). Private consultants and policy advisory organizations: A blind spot on policy transfer research. In O. de Oliveira (Ed.), *Handbook of policy transfer, diffusion and circulation (pp. 173–195)*. Edward Elgar.
Thiel, A. (2009). Europeanisation and the rescaling of water services: Agency and state spatial strategies in the Algarve, Portugal. *Water Alternatives, 2*(2), 225–244.
Tregarthen, T. D. (1983). Water in Colorado: Fear and loathing of the marketplace. *Denver Journal of International Law & Policy*, 6(3), 363–375.
Ulibarri, N. (2011). Bridging divides for water? Dialogue and access at the 5th World Water Forum. *Water Alternatives, 4*(3), 301–315.
van Assche, K., Beunen, R., & Duineveld, M. (2012). Performing success and failure in governance: Dutch planning experiences. *Public Administration, 90*(3), 567–581.
Venot, J.-P. (2016). A success of some sort: Social enterprises and drip irrigation in the developing world. *World Development, 79*, 69–81.
Venot, J.-P., & Veldwisch, G. J. (2017). Mythes sociotechniques et développement. *Anthropologie & Développement, 46–47*, 27–48. https://doi.org/10.4000/anthropodev.585
Voß, J. P., & Simons, A. (2014). Instrument constituencies and the supply side of policy innovation: The social life of emissions trading. *Environmental Politics, 23*(5), 735–754.
Wang, J., Zhang, L., & Huang, J. (2016). How could we realize a win–win strategy on irrigation price policy? Evaluation of a pilot reform project in Hebei Province, China. *Journal of Hydrology, 539*, 379–391.
Yong, M. L., & Grundy-Warr, C. (2012). Tangled nets of discourse and turbines of development: Lower Mekong mainstream dam debates. *Third World Quarterly, 33*(6), 1037–1058.

Index

For Product Safety Concerns and Information please contact our
EU representative GPSR@taylorandfrancis.com Taylor & Francis
Verlag GmbH, Kaufingerstraße 24, 80331 München, Germany